CITIES 2016 International Conference

Coastal Planning for Sustainable Maritime Development

IOP Conference Series: Earth and Environmental Science
Volume 79

Surabaya, Indonesia
18 October 2016

ISBN: 978-1-5108-4717-0
ISSN: 1755-1307

Printed from e-media with permission by:

Curran Associates, Inc.
57 Morehouse Lane
Red Hook, NY 12571

Some format issues inherent in the e-media version may also appear in this print version.

Printed by Curran Associates, Inc. (2017)

For permission requests, please contact the Institute of Physics
at the address below.

Institute of Physics
Dirac House, Temple Back
Bristol BS1 6BE UK

Phone: 44 1 17 929 7481
Fax: 44 1 17 920 0979

techtracking@iop.org

Additional copies of this publication are available from:

Curran Associates, Inc.
57 Morehouse Lane
Red Hook, NY 12571 USA
Phone: 845-758-0400
Fax: 845-758-2633
Email: curran@proceedings.com
Web: www.proceedings.com

Table of contents

Volume 79

CITIES 2016 International Conference: Coastal Planning for Sustainable Maritime Development

18 October 2016, Institut Teknologi Sepuluh Nopember, Indonesia

Accepted papers received: 20 July 2017
Published online: 17 August 2017

Preface

Papers

CITIES 2016 International Conference

PREFACE

CITIES conference is an annual conference event held by the Departement of Urban and Regional Planning. CITIES conference has a scale as same as a national and international conference. Since the first time this departement was build in 2002, CITIES conference had been held 7 (seven) times nationally and 2 (two) times Internationally in 2013 and 2015. The purpose of this conference is to provide all of the science and technology result based on research and planning practices, multidiciplinary science, and to push the development of research. There are some conference that we already had before, as follows :

Innovations In Spatial Planning Practices for Development and Decentralization (2005),
In Search of Integration: Between Planning and Spatial Programming (2008),
Toward Sustainable, Competitive, and Autonomous Spatial Planning (2009),
Facing the Future: Innovation In Planning Research and Practices (2010),
Spatial Planning Research Agenda for Sustainable and Just Urban and Regional Development (2011),
Facing Global Challenges In The Future Urban Sphere (2012),
Resilient Cities: Beyond mitigation, preparedness, response, dan recovery (2013),
Eco City, Utopia or Reality (2014),
Intelligent Planning Towards Smart Cities (2015)

In 2016, we will be holding an international conference with the topics " Coastal Planning for Sustainable Maritime Development". Hopefully, from this topics we can deliver many ideas in science and technology field, with coastal area planning implementation in maintaning the growth of maritime activities, especially for some coastal regions in Indonesia.

IOP Publishing
doi:10.1088/1755-1315/79/1/011002

Peer review statement

All papers published in this volume of *IOP Conference Series: Earth and Environmental Science* have been peer reviewed through processes administered by the proceedings Editors. Reviews were conducted by expert referees to the professional and scientific standards expected of a proceedings journal published by IOP Publishing.

Assessment of Groundwater Quality to Achieve Sustainable Development in Semarang Coastal Areas

T T Putranto[1], D A Widiarso[1] and N Susanto[2]
[1]Geological Engineering Department Diponegoro University, Jl. Prof. Soedharto SH Tembalang, Semarang, 52075, Indonesia
[2]Industrial Engineering Department Diponegoro University, Jl. Prof. Soedharto SH Tembalang, Semarang, 52075, Indonesia

Phone : +62 811 266 2013; fax: +62 24 7460087.
E-Mail : putranto@ft.undip.ac.id

Abstract. Semarang is growing fast particularly in the industrial and commercial sectors. Hydrogeological mapping and spatial analysis were used to assess groundwater quality in Semarang coastal areas. Results conduct that water table varies from 0.03 to 17.5 m depth and flows from the south to the north which is following topography i.e. higher to the south. pH values range from 6.05 to 8.17 while DO is up to 1.5 mg/L. EC values reach up to 6,370 µS/cm in the north as well as the maximum salinity value is around 3,600 mg/L related to the excessive groundwater used and sea water intrusion.

1. Introduction

Water quality analysis is one of the most fundamental aspects of groundwater studies. Knowledge of processes that control natural water composition is needed for groundwater management [15]. Assessment of groundwater quality is possible to understand the change in quality due to rock-water interaction or impacts of anthropogenic influence [11]. Groundwater often consists of major ion elements i.e. Ca^{+2}, Mg^{+2}, Na^{+1}, K^{+1}, Cl^{-1}, HCO_3^{-1}, and SO_4^{-2}. Theses ions can be plotted into some graphs such as Stiff and Piper diagrams. The Stiff system shows the differences or similarities in water and changes in water composition. While Piper diagram depicts the groundwater facies to understand as well as to identify the hydrochemistry composition in different classes. Indeed, pH, electrical conductivity/eC and Dissolved Oxygen/DO values can be employed to assess groundwater quality. The pH of water represents the relative amount of free hydrogen (H^+) and hydroxyl ions (OH^-). It is most affected by soils through which the water flows. Additionally, many factors influence the pH of water such as the source of water, precipitation, and input of contaminants. The fundamental use of pH in water analysis is an indicator of relative acidity or alkalinity of water. Values of 9.5 and above indicate high alkalinity whereas values of 3 and below indicates acidity. The normal pH range for drinking water is from 6.5 to 8.5. Meanwhile, the eC value of water estimates the total amount of solids dissolved in water. It is a valuable indicator of the total salinity, especially in the coastal area. Moreover, DO analysis measures the amount of Oxygen (O_2) dissolved in an aqueous solution. Adequate dissolved O_2 is fundamental for good water quality. It is an essential element to all form of life. An extremely high or low of the DO level can harm aquatic life and affects water quality. Thus, DO value can be applied to investigate the changing of water quality e.g. seawater intrusion in the coastal area. Saltwater contains less O_2 than freshwater [3].

Semarang, as one of the urban coastal cities in the central north of Java Island, is growing fast, in particular in the industrial and commercial sectors [10]. Semarang has two morphological features. Coastal plains and lowlands are located in the north while in the south is represented by highlands. The land use of coastal plains and lowlands is dominated by residence, ponds, and trade and service areas (Figure. 1). Meanwhile, the highland region is mainly for plantation, park, and residence in some part. The study area of this research is focused on coastal plains and lowlands of Semarang due to some environmental problems such as lowering the water table, subsidence, flooding, seawater intrusion, and degradation of groundwater quality occur in that areas as the latest happened in early 2014 [7]. The increasing population (0.65% per year, [13]) in the urban coastal city such as Semarang affects an enormous stress on the natural resources – groundwater, in particularly. Groundwater represents the most valuable drinking water resource. The purpose of this research is to assess of groundwater quality via some graphical interpretation as well as interpolation of water quality data on maps. The assessment will provide the actual condition of groundwater quality data such as the major concentration of cations and anions, pH, eC, and DO values to achieve sustainable development of groundwater management in Semarang coastal plains and lowlands.

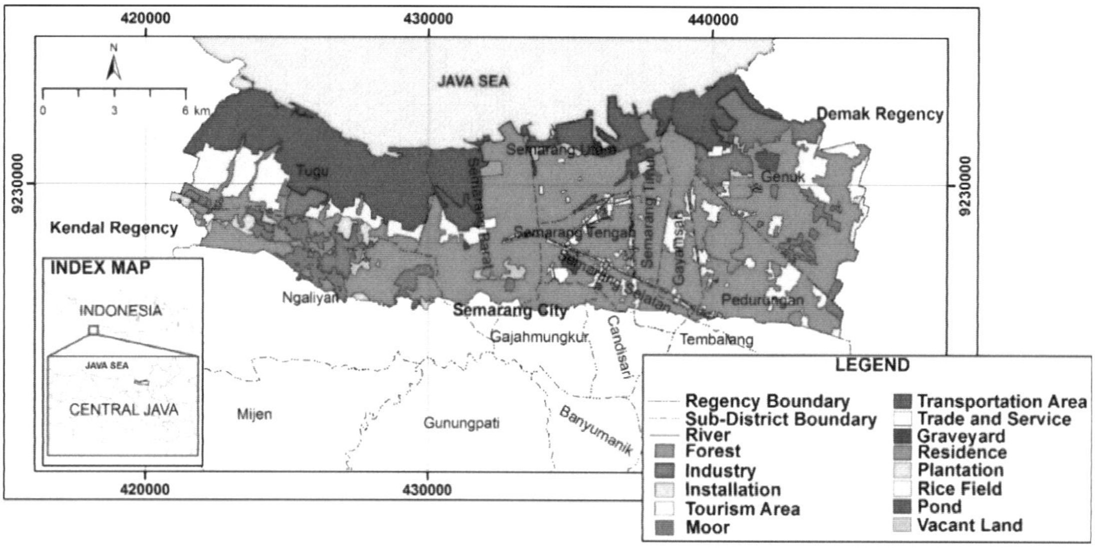

Figure 1. Study area and its land use. Coastal plains and lowlands of Semarang are located in the central north of Java Island (Indonesia) and a capital city of Central Java Province (Index map). They are dominated by residence in the north, in the city centre to the east while the rest is mainly occupied by ponds, trade and service areas.

2. Geologic and hydrogeologic setting

Coastal plains and lowlands of Semarang is located in the northern to the centre area of Semarang. Topography is around 0 to 15 meters above sea level (masl). Semarang has two seasons i.e. the dry and rainy seasons. The average of air temperature and rainfall from 2010 to 2014 is 28°C and 2,396 mm/yr respectively [13]. The peak of the rainy season usually occurs from December to January whereas August to September mostly the lowest precipitation takes place.

Regionally, stratigraphy of coastal plains and lowlands of Semarang from young to old consists of surficial deposits and sedimentary rocks (Figure. 2). Based on the regional geological map of Semarang [6], sedimentary rocks consist of Damar Formation which is composed of tuffaceous sandstone, conglomerate, and volcanic breccia. The latter occurs as lahar deposit in the centre to the south of lowlands. Surficial deposits as the youngest spread from lowlands in the centre to coastal plains in the north of Semarang. The coastal plains is composed by basin sediments [5]. The bulk of the basin sediments is dominated by alluvium. It is dominated by thick layers of calcareous and shell bearing clay with thin intercalations of sand and occasionally gravel to pebble or cemented gravel.

Some researchers describe that there are two aquifer systems in Semarang: i.e. confined and unconfined aquifers [12, 1, 10]. The aquifer which is separated by confined unit (thick of clay layer) is confined aquifer. The aquifer is composed of sedimentary marine and volcanic rocks in the lowlands and highlands of Semarang respectively. Groundwater in this aquifer is mainly abstracted for industrial purposes. Meanwhile, the unconfined aquifer consists of alluvial deposits. The unconfined aquifer (as the main focus of this research), groundwater flows from lowlands in the south to coastal plains in the north predominantly in an intergranular system. Fluctuation of water table appears depending on the season: high and low in the rainy and dry seasons, respectively. Groundwater is exploited by numerous dug wells, mainly for domestic water supply.

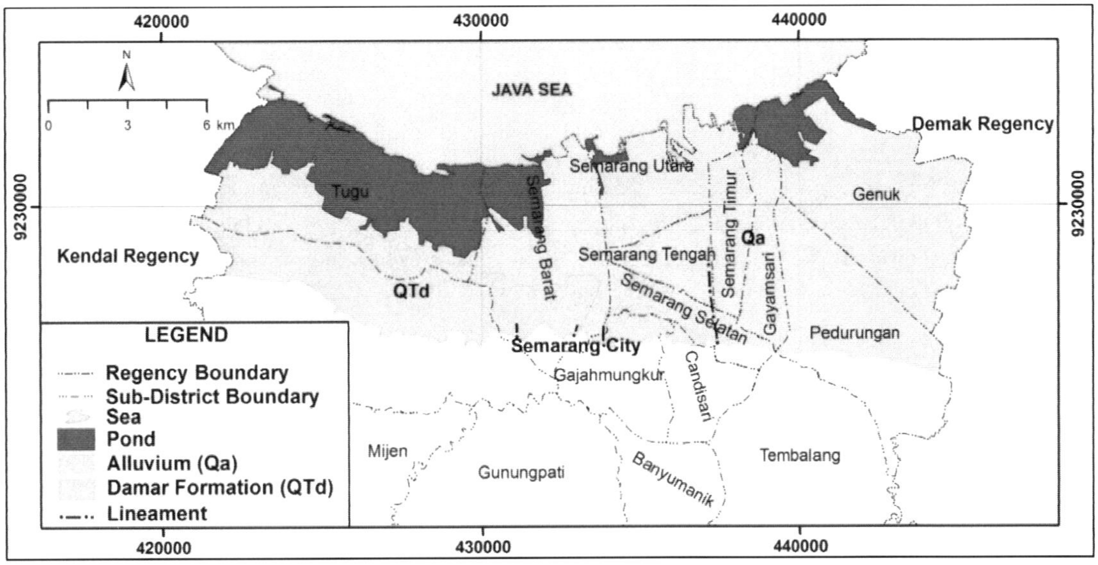

Figure 2. Geological map of Coastal plains and lowlands of Semarang (adapted from Thanden et al., 1996). Alluvium (Qa) spread in almost whole coastal plains and lowlands of Semarang while in the south of Semarang consists of Damar Formation (QTd).

3. Methods

To provide groundwater quality data, hydrogeological mapping was conducted in May 2016. Measuring water table and hydrochemistry properties such as pH, eC, salinity, and DO values were collected from one-hundred fifty (150) dug wells by using the water level meters and the WTW Handheld of pH (3210) meters, Conductivity (3110) meters, and Oxygen (3310) meters (Figure. 3). These hydrochemistry data were then interpolated by using spatial analysis and depicted into the hydrochemistry maps to analyse the actual hydrochemistry condition of the unconfined aquifer in coastal plains and lowlands of Semarang. Furthermore, to define the groundwater facies and its composition, six selected samples were analysed of major cations and anions in the Laboratory of Health at the Central Java Province. The results were plotted onto some graphs i.e. Piper and Stiff diagrams in the maps. The groundwater conservation zone was developed according to the hydrochemistry data.

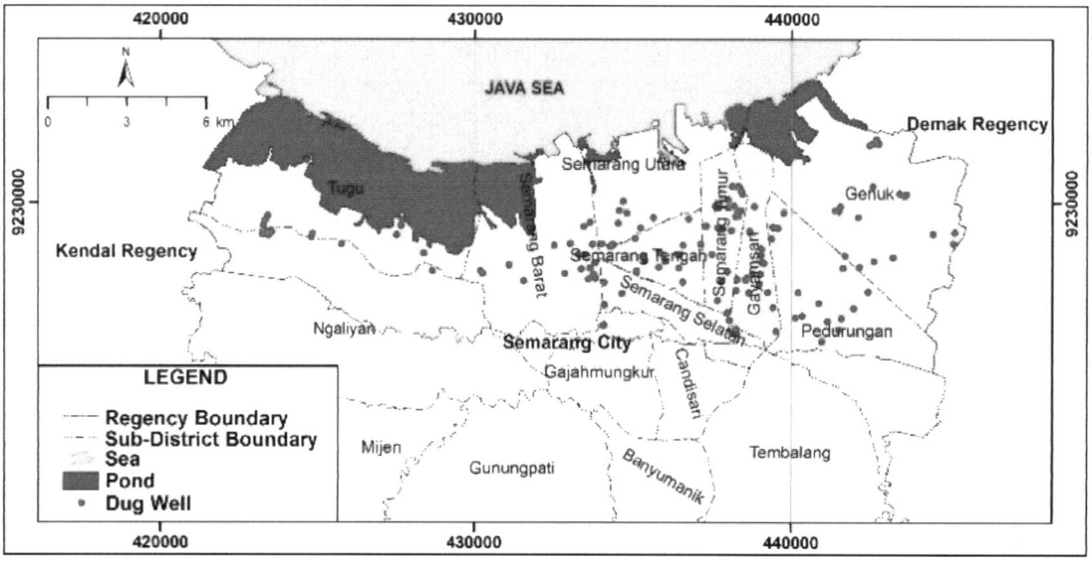

Figure 3. Location of groundwater samples from dug wells. The rapidness of dug wells represents the intensive of groundwater used in the centre and the east of Semarang which is corresponding to the population density.

4. Results and Discussions

4.1. Groundwater level

Dug wells in coastal plains and lowlands of Semarang were utilized for daily need. Groundwater was exploited from an unconfined aquifer. Based on the hydrogeological mapping, the minimum and maximum of groundwater depth are 0.03 m and 17.5 m respectively (Figure. 4a). Following the topography elevation, the minimum depth was found in the north to the east of Semarang i.e. Genuk Sub-District, Semarang Utara Sub-District, and Semarang Timur Sub-District. Water table depth was below 2 m. Meanwhile, the water table depth in the centre (Semarang Tengah Sub-District and Gayamsari Sub-District), the South (Semarang Selatan Sub-District and Pedurungan Sub-District), and the west of Semarang (Semarang Barat Sub-District and Tugu Sub-District) was around from 2 to 17.5 m. The maximum depth was found in Wonosari (Tugu Sub-District). Groundwater flowed from the south to the north of Semarang while groundwater levels (water table) were around 40 to below 1 meter sea level datum (Figure. 4b).

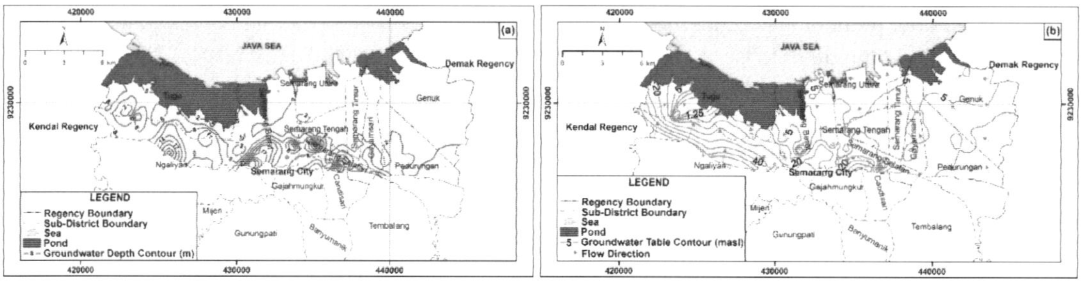

(a) (b)

Figure. 4. Groundwater depth (a), and groundwater level and its directions (b). Groundwater depth in unconfined aquifer is following topographic, higher in the south and groundwater flows from the south to the north of Semarang

4.2. Hydrochemistry

The hydrochemical composition of the dug well water samples was analysed, and the results were presented in average values as displayed in Table 1. The results showed that the maximum eC value (6,370 μS/cm) was located in Trimulyo Village (Genuk Sub-District). It was around 3 km from the Java Sea coastline in the northeast of Semarang. Groundwater in this region indicated unsuitable for drinking water. Water has a taste (salty) and indicates there was sea water intrusion coming in this region. Indeed, the values of Na^{+1} and Cl^{-1} were around 131 mg/L and 717 mg/L respectively. Obviously, those were higher than another region. Another indication of sea water intrusion in this region was the value of Salinity which was up to 3,600 mg/L. In the north of Semarang, eC values resulted above 2,000 μS/cm. While in the centre of Semarang, the range of eC values was 750-1,500 μS/cm (Figure. 5). In this region, groundwater from dug wells was still possible for consuming within deep processing such as filtering, boiling, etc. In the west of Semarang, the eC values were high in the regions which were located nearby the ponds, while in the south the eC values were lower than in the north around 750-1,000 μS/cm.

The pH values of the water in coastal plains and lowlands of Semarang were mostly neutral around 6.5-8.2. According to WHO (2011), they were in desirable ranges for pH in water-drinking quality [14]. Although pH usually was no direct impact on water consumers, it was one of the most fundamental operational water-quality parameters. The lower the pH caused, the higher of corrosion level.

Table 1. Summary statistics of major ions content in the groundwater of Semarang coastal plains and lowlands

Variable	eC (μS/cm)	pH	DO (mg/L)	Salinity (mg/L)	Na^{+1} (mg/L)	K^{+1} (mg/L)	Ca^{+2} (mg/L)	Mg^{+2} (mg/L)	Cl^{-1} (mg/L)	HCO_3^{-1} (mg/L)	SO_4^{-2} (mg/L)
Min	379	6.5	0.06	100	34.5	15.7	36.9	5.2	37	298	34.0
Ave	1,252	7.2	0.53	594	85.8	29.0	57.7	8.04	221	418	95.3
Max	6,370	8.2	1.50	3,600	131	50.0	85.2	11.8	717	534	310
SD	749	0.25	0.35	457	34.7	12.1	18.0	2.49	247	86.0	99.3

Note : Min: Minimum;
Ave: Average;
Max: Maximum,
SD: Standard Deviation

The highest and lowest of DO values were 1.5 and 0.06 mg/L respectively. The dissolved oxygen content of water was influenced by the source, temperature of water, treatment and chemical or biological processes taking place in the distribution system [14]. In the north and northeast of Semarang where the eC and Salinity values were high, the DO values were low 0.06-1.0 mg/L. Contrary, in the west of Semarang, the DO values were higher than those regions up to 1.5 mg/L.

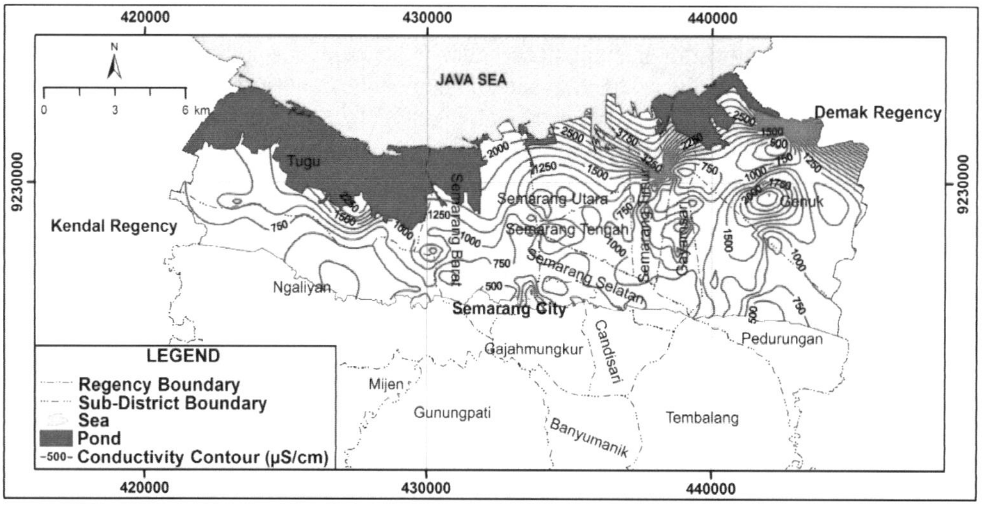

Figure 5. The electrical conductivity of coastal plains and lowlands of Semarang. In the north and north-east of Semarang show higher values of eC than another region.

To demonstrate the change of composition of water, a Stiff diagram (Figure. 6) showed the dominant cations and anions with concentration represented in electrical equivalent (meq / L). The major cations concentration in groundwater were in the decreasing order as $Na^{+1} > Ca^{+2} > K^{+1} > Mg^{+2}$. The major anions concentration were in the decreasing order as $HCO_3^{-1} > Cl^{-1} > SO_4^{-2}$. Sodium ($Na^{+1}$) was predominant cation in the east (SG-314) and north of Semarang (SG-142) while the chloride was the predominant anion in those areas. Chlorides were leached from various rocks into soil and water by weathering. The chloride ion concentration in Semarang was also influenced by seawater intrusion in coastal plains. It can be concluded that the most predominant water type in the east and the north of Semarang areas was the Na-Cl water type. While Ca^{2+} was distributed predominant cation of the groundwater samples in the south of Semarang (SG-139, SG-149, SG-198, SG-240). While HCO_3^{-1} was the predominant anion in the south of Semarang. It means that the most predominant water type in the south of Semarang was Ca-HCO_3 water type. Ca^{+2} ion was noted in Damar Formation which was contained by sedimentary rocks such as tuffaceous sandstone, conglomerate, and volcanic breccia. HCO_3^{-1} ion in Semarang was mainly derived from CO_2 present in the atmosphere.

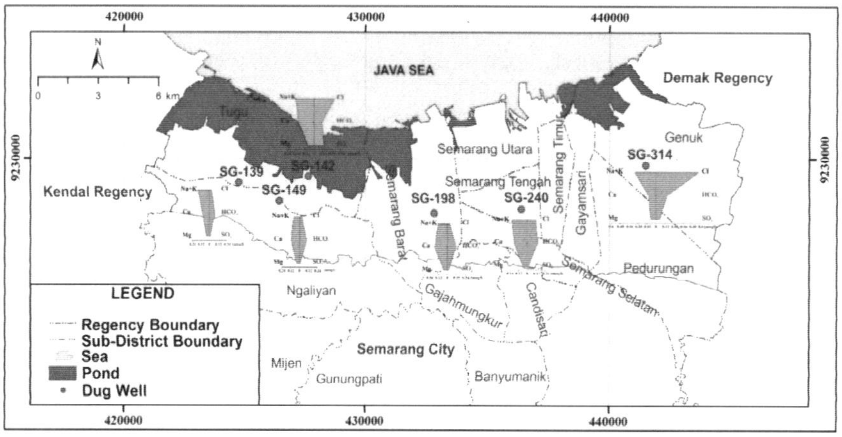

Figure 6. A plot of the Stiff diagram on the map in the coastal plains and lowlands of Semarang. Na-Cl water type is predominant in the north and east of Semarang (SG-142 and SG-314) while Ca-HCO_3 water type is predominant water type in the south (SG-139, SG-149, SG-198, SG-240).

To identify the groundwater facies, the Piper diagram [8] showed the electrical equivalent percentage (meq %) composition of different ions. Hydrochemical facies on the Piper diagram were conducted to determine the water type [4]. The Piper diagram provided in three parts: two trilinear diagrams along the bottom and one diamond-shaped diagram in the middle. The trilinear diagrams showed the percentage concentration of cations (Na^{+1} + K^{+1}, Ca^{+2}, Mg^{+2}) and anions (HCO_3^{-1}, Cl^{-1}, SO_4^{-2}) in each sample. The data points of the two trilinear diagrams were projected onto a diamond-shaped which lied perpendicular to the third axis in each triangle. According to the Piper diagram (Figure. 7), there were two groups: alkaline water predominantly chloride (SG-142, and SG-314), and alkaline earth water with higher alkaline content predominantly hydrogencarbonate (SG-139, SG-149, SG-198, and SG-240).

According to the actual assessment of groundwater quality data, as explained above, the groundwater conservation zones in the coastal plains and lowlands of Semarang are developed to achieve sustainable development in groundwater management of unconfined aquifer based on the eC value [2]. There are four conservation zones: secure, vulnerable, critical, and damage (Figure. 8). The characteristics of hydrochemistry for each zone are represented in Table 2. The secure area spreads mainly in the south of Semarang (Ngalian and Semarang Selatan), locally in the east (Pedurungan and Genuk) and the centre (Semarang Tengah). Land used in this zone is dominant forest, plantation, and residence with the dense of population low to moderate. Groundwater used in this zone has less impact on groundwater quality in general. The existence of industrial and residential zones which have the population density of the moderate to very dense is causing groundwater to be vulnerable to contamination both in the vulnerable and critical zones in the north and the east of Semarang. Moreover, groundwater is unsuitable for water-drinking quality due to seawater intrusion in the damage area.

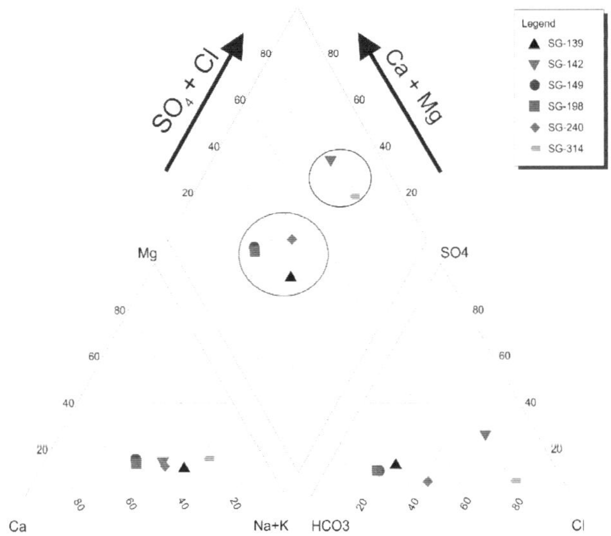

Figure 7. Piper diagram. Red circle shows the alkaline earth water with the higher alkaline content predominantly hydrogencarbonate group. Blue circle indicates alkaline water predominantly chloride group.

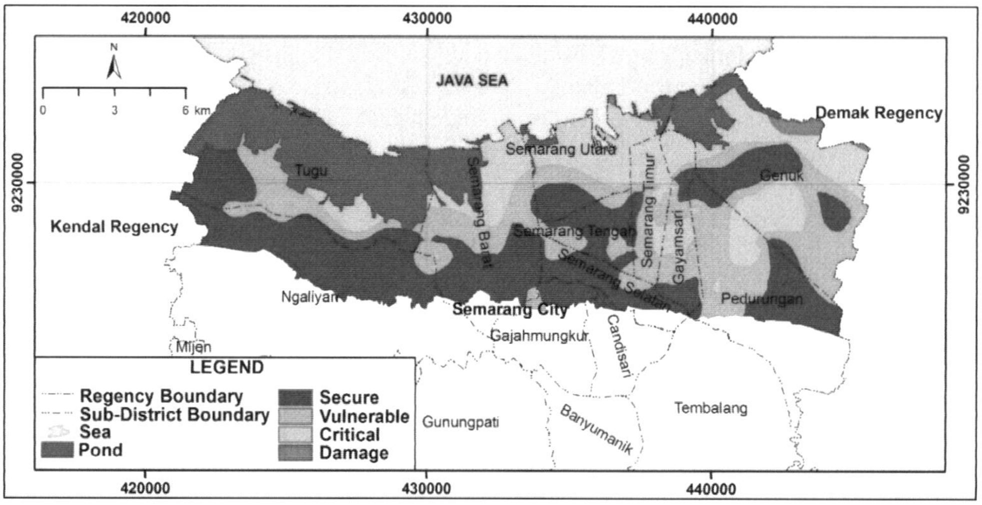

Figure 8. Groundwater conservation zone of unconfined aquifer in the coastal plains and lowlands of Semarang.

Tabel 2. Characteristics of groundwater conservation zone in the coastal plains and lowlands of Semarang.

Zone	eC (μS/cm)	Salinity (mg/L)	pH	DO (mg/L)	Groundwater depth (m)
Secure	<1,000	<300	6.5-8.2	0.06-1.50	0.03-17.5
Vulnerable	1,000-1,500	300-750	6.5-7.7	0.07-1.41	0.03-9.80
Critical	1,500-5,000	750-2,700	6.9-7.7	0.06-1.11	0.24-2.76
Damage	>5,000	2,700-3,600	6.9-7.2	0.26-0.80	0.21-1.04

5. Conclusions

Groundwater quality of unconfined aquifer that abstracts groundwater via dug well in Semarang coastal areas is investigated and assessed. Following the topography, the water table is lower in the north, and higher in the south and groundwater flows from the south to the north. The physical and chemical parameters of the samples that in the north and the east of the coastal plains and lowlands of Semarang show a higher value than in the centre and the south. EC and Salinity values have a range 1,500-6,370 μS/cm and 750-3,600 mg/L respectively. Based on the cations and anions concentration, there is two water type which is Na-Cl water type and Ca-HCO_3 water type. In the east and the north of Semarang areas are the Na-Cl water type while the most predominant water type in the south of Semarang is Ca-HCO_3 water type. The groundwater conservation zone is developed based on hydrochemistry value (eC) to achieve sustainable development in groundwater management of unconfined aquifer. There are four conservation zones: secure, vulnerable, critical, and damage.

Acknowledgements

The authors thank Diponegoro University who funded the project "Integrated Hydrogeological Mapping and Human Awareness Index for Establishing Groundwater Vulnerability Maps in Semarang Urban Area" within "International Scientific Publication (RPI)" scheme, Contract Number: SP DIPA-042.01.2.400909/2016, December 7th, 2015, with funding resource: DIPA PNBP of Diponegoro University 2016.

References

[1] Arifin MB and Wahyudin 2000 *Penyelidikan potensi cekungan air tanah Semarang, Jawa Tengah; data dan analisis (in Bahasa)* (Bandung : Direktorat Geologi Tata Lingkungan)

[2] Danaryanto, Titomiharjo H, Setiadi H and Siagian Y 2007 *Kumpulan Pedoman Teknis Pengelolaan Airtanah (in Bahasa)* (Bandung : Badan Geologi)
[3] Fondriest Environmental 2013 Dissolved oxygen. http://www.fondriest.com/environmental-measurements/parameters/water-quality/dissolved-oxygen/#1 [Online access 29.07.2016]
[4] Furtak H and Langguth HR 1967 Zur hydrochemischen kennzeichnung von grundwässern und grundwassertypen mittels kennzahlen (in Deutsch) *–Mem. IAH-Congress 1965* Hannover. **7** 86-96.
[5] Mathon, VSFMA 1975 *Analysis of groundwater condition on the basis of existing data. Conference of groundwater development for irrigation, Department of Public Works Republic of Indonesia* Surabaya
[6] Thanden RE, Sumadirdja H, Richards PW, Sutisna K and Amin TC 1996 *Geological map of The Magelang and Semarang sheet, Jawa* (Bandung : Geological research and development centre)
[7] Parwito 2014 Banjir, banyak kendaraan menuju Jakarta tertahan di Kendal. http://www.merdeka.com/peristiwa/banjir-banyak-kendaraan-menuju-jakarta-tertahan-di-kendal.html. [Online access 10.05.2015].
[8] Piper AM 1944 A graphic procedure in the geochemical interpretation of water-analyses Am. Geoph. Union Trans. **5**. 25 914-23.
[9] Putranto TT and Rüde TR Hydrogeological model of an urban city in a coastal area, case study: Semarang, Indonesia *Indonesian Journal on Geoscience* **3** 1 17-27
[10] Putranto TT, Rüde R, Cardona A and Wendland F 2014 *Research plan for managing groundwater reseources of the coastal cities in the central north of Java Island* 1st International Conference for Energy, Mineral, and Environment applieds. Bandung, Indonesia paper no. 12 95-102
[11] Sadashivaiah C, Ramakrishnaiah R and Ranganna, G. (2008). *International Journal of Environmental Research and Public Health*, 5 (3) 158-164.
[12] Spitz K and Moreno J 1996 *A practical guide to groundwater and solute transport modeling* (Canada : John willey & sons Inc.)
[13] Statistic Centre Bureau of Semarang City 2015 *Semarang city in figureures* Semarang
[14] WHO (World Health Organization) 2011 *Guidelines for drinking-water quality 4th Ed* Geneva, Switzerland.
[15] Zaporezec A 1972 Graphical interpretation of water-quality data. *Ground Water* **10** 2 3243.

IOP Conf. Series: Earth and Environmental Science **79** (2017) 012002 doi:10.1088/1755-1315/79/1/012002

Policy and Regulations to Support the Implementation of Coastal Conservation and ICZM: Comparative Analysis between Indonesia and France

Nandi[1]
Department of Geography Education
Universitas Pendidikan Indonesia, Jl. Dr. Setiabudi 229, Bandung 40154, Indonesia

E-mail: nandi@upi.edu

Abstract. Coastal is a very potential area for a variety of development options. The increase of population and its activities will be able to threaten the existence and sustainability of ecosystems and coastal resources, marine and small islands. Therefore, the policy on coastal conservation is required. This research analyses the comparison of the implementation of integrated coastal zone management (ICZM) between France and Indonesia focusing on policy and regulations. The objectives are to get an insight on the coastal conservation policies implemented in Indonesia and France and to learn experiences. This research also uses the literatures review and analytical descriptions as methodology. The findings demonstrate that both countries objectives for coastal management are to achieve sustainable development.

1. Introduction

Nowadays, the issue of conservation has become important for all of us, even in a local, regional, and international level. Organization for Economic Co-Operation and Development [1] states that coastal zone management is not purely a local/regional or purely national concern. There are many aspects of coastal zone management, which are of international nature. The damage of coastal resources and sea, the decrease of fisheries production, pollution and climate change has occurred and provide significant impact in many coastal areas throughout the world.

However, with the increasing population growth and rapid development activities in coastal areas, for various utilizations (residential, fishing, ports, tourism and others), the ecological pressure on the ecosystem and coastal and marine resources are increasing. The increasing pressure of will be able to threaten the existence and sustainability of ecosystems and coastal resources, marine and small islands that are around [2].

One thing that is more concerned is that the trend of environmental degradation and the coastal ocean due to the paradigm of development and practices which have not yet applied in accordance with the principles of sustainable development [3]. The tendency to be dominated extractives and economic interests take precedence over national economy of the local community (coastal). There should be more participatory, transparent, accountable, effective and efficient equity to support the rule of law.

While at the implementation level there is still found of low awareness of law and regulation enforcement, and the ecological damage to coastal in a serious condition. Therefore, a policy on coastal conservation is required. Issue on coastal conservation is a basic to realize the management of coastal area in an integrated way, as known as the integrated coastal zone management concept (ICZM) [4]. It more specialized in sustainable ecology of coastal areas [5], which states that coastal management should be integrated across sectors.

This study will discuss coastal conservation issue to support ICZM concept in Indonesia. However, it is worthwhile in this study to know the experience of the European countries especially in France, because they have better management policy on coastal conservation and better enforcement on regulation such as CELRL [meur] in implementing coastal conservation toward ICZM into their national policies in order to seize valuable lessons for coastal conservation practice in Indonesia.

The approach that will describe for a rational ICZM is based on the implementation of coastal conservation through comparative analysis between France and Indonesia. In order to practice ICZM, planners need to understand the ways in which the natural environment and human activities are inter-connected to form a system. ICZM and its aim are to develop sustainable development of the coastal zone and its resource [6]. So, it is important to get an insight coastal conservation and ICZM.

2. Methods

The method used in this research is descriptive comparative method. This research develops the theoretical development about coastal conservation and integrated coastal zone management (ICZM). This study is mainly based on literature review. Most of analysis in this research is descriptive analysis [7]. Collecting data and information about the implementation of coastal conservation and the relationship of ICZM in Indonesia and lessons learned from France. These data include constitution, policies, law and regulations. The collected data are derived from secondary data such as literatures, official documents articles, journals, internet, and other sources since there is limitation on primary data. Analysing Data is derived from the academic understanding and the case on coastal conservation in France and Indonesia context, the author will compare and analyse the elements in coastal conservation policies toward the implementation of ICZM. Assessing the status on the implementation of coastal conservation and ICZM is useful to propose strategic recommendation to enhance better coastal conservation in Indonesia.

3. Result and Discussions

3.1 Policy and Regulations to Support the Implementation Coastal Conservation and ICZM: Political Will of Government

Political will from both governments is necessary to realize a sustainable coastal management. This is because by the government have any power in determining the direction and government policy. With the political will, all the good rules and laws will be directed to achieve the goals and targets of sustainable development. So, all goals will be expected to form of a sustainable coastal management. With the French and Indonesian government, the political will of both government officials from the coastal management and its regular shows at the direction of sustainable coastal management. This can be seen from the fact that the government and the legislative create the rules and regulation to support the environmental issues, especially on coastal conservation and sustainable coastal zone management.

In France, there are several laws published by government as policies on coastal management. The foundation of the policy of coastal management is Law No 75-602 of 10 July 1975 about the formation of the *Conservatoire de l'espace littoral and Shoreline Lake* (CELRL). This is as an operator dedicated to land preservation and development of a third of natural coastline in partnership with local authorities concerned. It means that government has concern and good political will in handling environmental issues [8].

Furthermore, government and legislation re-issued several regulations related to the environment particularly in coastal management such as law about preservation and enhancement of Coastal, which arrange the legal and operational policy of the coastal with the various land contributed to its management. In addition, there are also rules which regulate land development in areas with high stakes, protect natural areas, classify and record reserves sites nature, national parks and nature parks.

In Indonesia, the government also has political will to improve and manage coastal area by publish several policies related to environmental issue. The basic foundation of regulation is Law No. 23 of 1997 about Environmental Management Act [9].

As a developing country, Indonesia continues to work to improve the rules related to environment issues. The reformation era in 1998 is a new way to establish basic law in coastal areas. The establishment of the Department of Fisheries and Marine exploration in The Government of President Abdurrahman Wahid era which is the birth of Law No.27 about Management of Coastal Region and Small-Island (UU-PWP3K) on 26 June 2007. In addition, the government of Indonesia with the House of Representatives (DPR) also makes rules that set the various issue related to environment such as conservation of natural resource, spatial act, and other regulations.

Table 1. Summarizing the main historical phase of conservation of natural coastal sites in France and Indonesia.

Year	France	Indonesia
1930	Law of May 2, concerning the protection of sites, instituting the '*Site Classé*' (classed site) and '*Site Inscrit*' (listed site). These terms are widely used to protect coastal sites; they impose numerous interdictions and limitations concerning utilization of coastal sites by their owners. (This kind of protection is close to certain British designations such as SSSI and AONB.)	
1945		Indonesia Constitution of 1945 (Art. 33 paragraph 3 and 4) the basis for the management and utilization of Indonesian natural resources, the following has been stipulated: Land and water, and the natural resources contained therein, shall be controlled by the state and utilized for the greatest feasible prosperity of the people.
1960	Institution of 'Parcs Nationaux' (National Parks).	
1968	Institution of the '*Redevance Départementale d'Espaces Verts*' (Departmental green tax), a tax levied on constructions. It financially and legally empowered *Départements* to intervene substantially in favor of the protection of coastal sites. It is the first step for a *Département* to play a prominent role in a voluntary policy that is close to those of both the Heritage Coasts and the *CELRL*.	
1973	Two important Ministerial circulars deal with the limitation of the use of the coastal zone for development operations (Anon. 1975).	
1974	Publication, at the initiative of the State authorities, of the Rapport Piquard,entitled '*Perspectives pour l'aménagement du littoral français*' (perspectives for the management of the	

	French littoral zone); it is the first document laying the foundations for coastal planning.	
1975	Foundation of the *Conservatoire de l'Espace Littoral et des Rivages Lacustres, CELRL*, a state body whose field of authority is specifically the coastal zone; it attempts to carry out the twin goals of protecting the beauty of landscapes and ecosystem diversity. This creation marks a significant turning point in the history of the conservation of French coastal sites.	
1976	Law changing the *Redevance Départementale d'Espaces Verts*into the Taxe *Départementale d'Espace Vert* (Green Tax); reinforcing the role of the *Département*	
1977	Decree instituting impact studies for new developments in order to preserve the ecological balance. (*Journal Officiel* 13/10/1977).	
1978		Presidential Decree No. 43 of 1978, about the endorsement of Convention International on Trade of Endangered Species of Wild Flora and Fauna
1979	Publication of the '*Directive de la Préservation et l'Aménagement du Zone Littoral*' (Directive on the preservation and management of the coastal zone) by the French Government (*Journal Officiel*, 26/08/1979).	
1985	Law changing the '*Taxe d'Espaces Verts*' (Green tax) into the *Taxe Départementale d'Espaces Naturels Sensibles* (Departmental tax on sensitive natural sites). It confirms again the competence of *Départements* in matters of coastal conservation.	Law No. 17 of 1985, about the endorsement of United Nations Convention on The Law of The Sea
1986	'Loi Littoral' (Coastal Law), relative to coastal management, protection and planning. It constitutes the basic text formalizing the doctrine of coastal management and marks a turning point (Becet & Le Morvan 1990). It aims not only at nature conservation but also at the reconciliation of the various coastal interests. It is the first legal text concerning the protection of natural coastal sites.	
1989	Application Decree of the Coastal Law of 1986, specifying the '*espaces remarquables*' (special sites) to be protected because of their ecological value. These sites must be delimited in the '*Plan d'Occupation des Sols*' (Land Use Plan) of the *Communes* and should be under strict legal protection. This is presently under way. In certain *Communes*, where the pressure is strong and the interests are varied, negotiations	

	between State agencies and local councellors may be very difficult. Since the law does not make a provision for compensation, *Communes* with large areas of 'protectionprone' natural sites, often consider themselves as disadvantaged compared to neighbouring *Communes* more free to encourage development.	
1990		Law No. 5 of 1990, Conservation of Natural Resources and Ecosystem Act, Presidential Decree No. 32 of 1990, the Management of Protected Area.
1992		Law No. 24 of 1992, Spatial Planning Act.
1994		Presidential Decree No. 5 of 1994, about the endorsement of United Nations Convention on Biological Diversity
1997		Law No. 23 of 1997, Environmental Management Act.
1998		Ministerial of Home Affairs Decree No. 8 of 1998, the Implementation of Local Spatial Planning Act.
1999		Government Regulation No.19 of 1999, about pollution control and / or destruction Sea, Law 22 of 1999, on Local Government Act.
2000		Government Regulation No. 25 of 2000, about Government and Local Authority as a local autonomy.
2001		Government Regulation No.82 of 2001 about management of water quality and water pollution control.
2004		Law No. 7 of 2004, about Water resources, Law No. 31 of 2004 about Fisheries, Law No. 32 of 2004 about Local autonomy, and Law No. 33 of 2004, about central and local financial balancing.
2007		Law No.27 of 2007, about Management of Coastal Region and Small-Island (UU-PWP3K).
2009		Law No. 32 of 2009, Environmental Management Act.

3.2 Regulation on Boundaries, Pollution, and Fisheries to support Coastal Conservation

In the implementation level, the need for regulations that manage the implementation of a rule is necessary. Implementation of clear regulations in the field will impact on the achievement of the purpose of sustainable coastal management, even with the implementation of some regulations related to coastal conservation in France and Indonesia.

3.2.1 Boundaries Regulations

In France, the boundaries regulation can be seen from the policy made by the states to manage the "public domain maritime". This is also adjusted to the physical character of its coastal and marine to make permits organizing economic and social uses of the area. There is interference from the government of France through the Department of Defence in collaboration with the CELRL to protect and develop coastal heritage at the same time. The balancing resource use has traditionally been a part of spatial planning (territorial planning and urbanism). Although this approach has generally proved quite capable of organizing land-based activities, it is clearly inappropriate for coastal/maritime zones, if only because, in the latter, a third dimension- the depth of the water column - must be taken into account. On the sea, exclusive zoning leads to de facto privatization of the public space; such de facto situation is incompatible with the status of the sea, which a priori should remain open to all as long as public interest/safety is not at risk [10].

In addition, there is integration with another institution such as the department of ecology and sustainable development to create collaboration for guidelines of the interdepartmental committee of organization and development of the territory as ways of Partnerships. The collaboration among them, made formally in the form of agreement which signed during the years 2006. Other government policies that are applied together are the determination of spaces for private land and public land, especially for social and housing.

Different from France, in Indonesia, the coastal zone management and coastal boundaries is set in the right of coastal (HP3) through legalization of Law No. 26 of 2007 regarding management of coastal areas and small island (Law-PWP3K) [11]. It's interesting that the Law of PWP3K will be the Rights of The Coastal (HP3) in all provinces, districts, and city in Indonesia that have the potential of sea resources, and can be given to individuals, agencies or business communities, which include undertaking HP3 on the sea and water surface column up to the seabed within the sustainability of ecosystem.

In addition, with the introduction of decentralization and regional autonomy the coastal management is at last given the authority to local government in full control of the central government. If we compare between the existing policies in France and in Indonesia, there can be see that, French policies are more stressing in the process of integration and collaboration between institutions based on the Partnerships (vertical views) which have been set clearly in territorial planning and urbanism. France as member of European Commissions (EU) has agreement on managing coastal resource particularly in European regional boundaries. The government must comply with the Recommendation voted by the council of Europe and the European Parliament, concerning the implementation of ICZM strategy in Europe. But it still lacks in implementation and in field of practice, because it does not consider the change of natural and social dynamic. So this can create horizontal conflicts between stakeholders in utilizing resources and claims to the coastal area as a private property. Meanwhile, Indonesia is more stressing on the involvement of all stakeholders in the utilization of coastal resources. The role of local government in decentralization era is more clearly seen in the frames of control by central government. However, this policy will cause egoism on local government level in managing their coastal resources and coastal conservancy strategies without considering the interests of other institutions, especially in the administrative border.

3.2.2 Pollution Regulation

As for pollution regulation, European Directives annex II action in its nature protection act of 1976 is basic constitutional to protect and manage the coastal and environment in France. This can be regulatory for handling pollution issues. But in implementation level, particularly in pollution regulation, France is still facing large oil spills problem and other environment issues. That issue has not yet found a complete satisfactory solution, despite the European Directive on equipment ports of the Union, adopted in September 2000 and implemented French Law on 16 January 2001. Besides that, the source of coastal pollution in France are comes from the river, industrial sites (estuaries), and agriculture this also happen to environment issues such as coastal erosion, marine flooding, and other impact of climate changes (Bouleau, 1993). To overcome these problems, the government issued several policies: the establishment of criminal courts Specialty (Le Havre, Brest and Marseille) and the establishment in the Mediterranean zone of protection ecological (ZPE) French rectifies the lack of EEZs, reduction of pollutants for upstream industry, better monitoring and control of several pollutants from urban and agriculture, and disaster mitigation management [12].

While in Indonesia, the basic law and regulation that government made for protecting the environment and management is written in Law No. 23 of 1997 about Environmental Management Act (EMA). This is to be umbrella type of law for all issues and aspect in environmental management, and to be a philosophy for sustainable development. In general, coastal pollution problems in Indonesia are caused by this human activity such as land clearing for agriculture, cities and industrial development, logging and mining in the catchment area. The other is physical habitat damage; it could be decreasing the quality of ecosystem such as, mangroves and coral reefs. Similar with France policy to overcome these problems, the government of Indonesia through Law of No. 23 of 1997 gives direction and guidance to protect and maintain the environment. The Law states that in certain activities pertaining to specific kinds of resources, strict liability rests on those causing the damage and/or pollution of the living environment at the time of the occurrence of the damage and/or pollution, which shall be stipulated in relevant legislation.

In general France and Indonesia try to give their best to protect the best ways to protect their environment, whether it's the influence of human activity or even from natural hazard. In legal practice France has specific sanctions such as fines and imprisonment that can be imposed as purely administrative sanctions, without the interference of a judge or court [13]. But in level of implementation of regulation, both countries still face problems. It is because many of the problems that occur and the level of compliance in the community to obey the rules, particularly in Indonesia still lack of law enforcement.

3.2.3 Fisheries Regulations

Fisheries regulations are intended as a guide in respect of the various legal controls which apply to fishing. France is a member of the European Union and therefore in Community waters, they implement the Common Fisheries Policy (CFP), which was reformed in 2002 and which is currently governed by Regulation (EC) No. 2371/2002 of 20 December 2002. In force since 1983, the CFP aims to reconcile resource conservation with the preservation of income and jobs in coastal zones that offer few alternatives in terms of production or employment. It therefore covers not just resources but also markets and structures. With regard to resource management, the CFP regulations comprise: A traditional management tool based on TACs (total Allowable Catches) and quotas; Technical measures relating to gear or catch; Effort-related management, based on vessel engine power and the number of days at sea. The CFP also provides the introduction of measures to rebuild, over a period of several years, stocks that are threatened in terms of sustainable harvesting, and for recourse to effort-related management rules to supplement TACs and quotas [12].

While in Indonesia, one of the main purposes of fisheries management is to enhance the living condition of small-scale fishermen and fish farmers (Law of the Republic of Indonesia No. 8/1985 strengthened by Law No. 31 in 2004 concerning Fisheries). Fisheries management to capture fisheries and fish-culture should take into account "Hukum Adat" (custom law) and indigenous knowledge, including community participation, so it could be a Knowledge-based fisheries co-management. Besides, that there is delegation of function/authority and supporting mandates. The delegation of part of fisheries functions from the Government to the Regional Governments and the Government may delegate its function to the Regional Government to carry out supporting mandates in the field of fisheries.

France and Indonesia have different policy direction on fisheries. France has implemented fisheries regulations with clearly, from the preparation, planning, implementation, requirements and processes in catching fish. In addition, France has also been applying the rules on fishery production, market, and the organizational structure so that management is clearly organized. While Indonesia still lacks of management although there have been mandates given to local government to manage the local authorities in fishery production with co-management. There are still many illegal practices, unregulated and unreported fishing due to the weak law enforcement in the sea, the level of discrepancy between the stock of fish one area with other areas of the sea, where on the one hand there are areas that the fish stock is done over fishing, and on the other hand there areas is still quite a lot of sea-level resource fish utilization is not optimal or even not at all touched.

Table 2. Comparison Aspect on Implementation of Coastal Conservation Policy in France and Indonesia.

Policy	Instruments	France	Indonesia
Political will of government: General policies for protection of the coastal environment	Yes, if it specifically is mentioned on constitution, law, and regulation No, if it is not mentioned on constitution, law, and regulation	√	√
Regulations:	Current policy implementation		
- Boundaries	Planning zone and land use	• Based on Roman law, European regional boundaries • Public domain maritime • Department of Defence in collaboration with the CELRL to protect and develop coastal heritage	• The right of coastal (HP3) through legalization of Law No. 26 of 2007 regarding management of coastal areas and small island (Law-PWP3K) • Decentralization and regional autonomy
- Pollution	Regulation of pollution	• European Directives annex II action in its nature protection act of 1976 • The European Directive on equipment ports of the Union • The establishment of criminal courts Specialty (Le Havre, Brest and Marseille) • The establishment in the Mediterranean a zone of protection ecological (ZPE)	Law No. 32 of 2009 about Environmental Management Act (EMA)
- Fisheries	Regulation of fisheries	• European Union about Community waters implements the Common Fisheries Policy (CFP) • Regulation (EC) No.	Law of the Republic of Indonesia No. 8/1985 strengthened by Law No. 31 in 2004 concerning

	2371/2002 of 20 December 2002	Fisheries

3.3 Lesson Learnt

From the description about coastal conservation and ICZM for both France and Indonesia in the previous part, there are some positive elements that can be drawn from the original policy issue, political will, and regulation enforcement. Indonesia has different implementation of coastal conservation policy and ICZM compared to France. This is because, coastal conservation and ICZM in Indonesia is generally still not being fully implemented. Existing condition in Indonesia has to be adjusted to implement coastal conservation as in France. Nevertheless, Indonesia has the possibility to implement it because there are some aspects that can support the implementation of coastal conservation and ICZM.

Several important lessons that can be taken from France including whom and how to implement the coastal conservation and ICZM concept in Indonesia can be considered as follow:

1. A strong will from government as the starting point for coastal conservation and ICZM implementation to all government institutions.
   With regard to this, the government of Indonesia has to have strong will to create a clear policy in terms of contents and roles of all stakeholders involved. This can be achieved either by establishing a new policy such as Indonesian Policy on Coastal Zone Management or improving the existing the policies to support the coastal conservation and ICZM concept.
2. A continuous evaluation and improvement both in policy/regulations and practices in order to succeed the coastal conservation and ICZM implementation
   The government, who has a power to implement the regulations, also should involve law practitioners, and other institutions such as NGOs which concern to these issues. Continuous evaluation and improvement can be achieved by encouraging such as: training and education on coastal management, exchanges of experts, or comparative study.

Besides the above important elements, France faced some challenges or crucial condition in the implementation of coastal conservation and ICZM. Crucial factor from France's experience in the ICZM practices should become an attention prior to be implemented in Indonesia. Some crucial factors in France's experiences are related to constraints in the regulation pollution (large oil spills) and authority's division for broader autonomy. These aspects can become constraints in developing ICZM and influence to the success of coastal sustainable implementation. All of these aspects should be considered as lessons for the possibility of coastal conservation and ICZM implementation in Indonesia. The implementation ICZM in Indonesia can be realized by considering the aspects that can become positive points to encourage its implementation and also some weaknesses that have to be improved by learning from the experiences of an advanced country (France).

A positive point in Indonesia's coastal conservation policy and ICZM is the new Law on Law No. 27 of 2007, about Management of Coastal Region and Small-Island (UU-PWP3K). UU-PWP3K is expected to be the best solution and at the same time can overcome various problems associated with the utilization and management of coastal, marine and small islands in Indonesia, which is less coordinated, so there has been degradation, pollution, sedimentation, over exploitation, conflict seizure of natural resources and other problems that reduce the quality of coastal and marine environment and its small islands. Another positive point is the acknowledgement of traditional practices in coastal management as the uniqueness of cultural diversity. It is this point that is not provided in the concept of CELRL in France. Therefore, to make an effective implementation of ICZM, the government as the enabler and facilitator should accommodate the cultural value in the society.

Conclusions

France and Indonesia have the same goal for coastal zone management especially on coastal conservation policy but a very different policy direction. Both countries consider coastal conservation policies implementation as a tool to achieve sustainability and protect the coastal. However, the policy direction of Indonesia is to support the economic development by exploring the coastal zone while France is toward protection of the resources. On implementation of coastal conservation policy, Indonesia needs to learn from other countries which have more experiences. From the experience of implementation of coastal conservation and ICZM in France, there are some elements to be learnt, strong political will from government, clear law and regulation to support coastal conservation and ICZM. Although not fully matched with the condition of Indonesia, there are several possibilities for the ICZM to be adopted as an alternative for government in making a policy and regulation. In the end, all these policies are implemented to achieve a sustainable coastal management.

References

[1] OECD 1993 *Coastal Zone Management: Integrated Policies (*Paris: OECD)
[2] IPCC 1990 *Impacts Assessment of Climate Change. Summary for Policymakers Report of Working Group II.* IPCC/WMO/UNEP
[3] Nandi et al 2016 IOP Conf. Ser.: Earth Environ. Sci. 47 012017
[4] Nandi 2014 Coastal Conservation Policies and Integrated Coastal Zone Management (ICZM) in Indonesia. *International Journal of Conservation Science, 5*(3), 387-396
[5] Meur-Ferec C 1997 A comparative review of coastal conservation policies in France, and England and Wales. *Jurnal of Coastal Conservation* 3 pp 121-132.
[6] Pergent-Martini C P 2002 Ecologycal research for Integrated Coastal Zone Management: introduction *Jurnal of Coastal Conservation Policy 8* pp 107-108.
[7] Babbie E 1992 *The Practice of Social Research* (Belmont, California: Wadsworth Publishing Company)
[8] Stamatiou E L R 2005 Coastal land acquisition and management towards the protection and restoration of natural spaces and landscapes *The 40 years experience of the CELRL in France, 3rd Conference, Management and Improvement of Coastal Zones* (Athena: NTUA-Laboratory of Harbour Works) pp. 22-25/1
[9] Kusuma-Atmadja M. 1996 Legal and institutional aspects of coastal zone management in Indonesia. *Marine Policy* 20(1) pp 63-86
[10] Philippe Deboudt J-C D 2008 Recent developments in coastal zone management zone management (1973 - 2007) *Ocean & Coastal Management* pp 212-228
[11] Latuconsina H 2008 *Hak Pengusahaan Perairan Pesisir Di Indonesia* (Jakarta)
[12] Goulletquer P and O Le Moine 2002 Shellfish farming and Coastal Zone Management (CZM) development in the Marennes-Oléron Bay and Charentais Sounds *A Review Of Recent Developments* (Charente Maritime, France: Aquaculture International*)* 10(6) 507
[13] Michael F and Gunter H 2007 *Criminal Enforcement of Environmental Law in European Union*

Acknowledgments

This article is part of Master thesis materials by Nandi on finding a framework of A Review on Coastal Conservation Policies and Integrated Coastal Zone Management (ICZM): A Lesson from France for Indonesia, conducted at University of Groningen and ITB. The author expresses his gratitude to Dr. Johan Woltjer and Dr. Myra P. Gunawan and also Department of Geography Education Universitas Pendidikan Indonesia, Bandung.

Risk Analysis of Return Support Material on Gas Compressor Platform Project

Silvianita[1], B U Aulia[2], M L N Khakim[1], and Daniel M Rosyid[1]
[1]Department of Ocean Engineering, Faculty of Marine Technology, Institut Teknologi Sepuluh Nopember, Surabaya, Indonesia
[2]Department of Urban and Regional Planning, Faculty of Civil Engineering and Planning, Institut Teknologi Sepuluh Nopember, Surabaya, Indonesia

Abstract. On a fixed platforms project are not only carried out by a contractor, but two or more contractors. Cooperation in the construction of fixed platforms is often not according to plan, it is caused by several factors. It takes a good synergy between the contractor to avoid miss communication may cause problems on the project. For the example is about support material (sea fastening, skid shoe and shipping support) used in the process of sending a jacket structure to operation place often does not return to the contractor. It needs a systematic method to overcome the problem of support material. This paper analyses the causes and effects of GAS Compressor Platform that support material is not return, using Fault Tree Analysis (FTA) and Event Tree Analysis (ETA). From fault tree analysis, the probability of top event is 0.7783. From event tree analysis diagram, the contractors lose Rp.350.000.000, - to Rp.10.000.000.000, -.

1. Introduction

Offshore platform is a structure used for the exploration and exploitation of oil and gas. Fixed platforms which are jacket structure are commonly used in Indonesia [1]. Fixed platform project is dynamic, if the contractors want optimum benefit, it must be responsive to conditions that occur on the project. The contractor makes a careful planning to anticipate the problems that may occur so that the project can still be done without having problems [2]. Management of a large-scale investment activities and the level of complexity is very difficult requiring a proven method, quality resources, and the application of up to date science [3]. On a fixed platforms project are not only carried out by a contractor, but two or more contractors. It takes a good synergy between the contracting parties to avoid any miss communication that could result in the emergence of problems and even loss on the project. One example of the problems that often occur is the support material used in the process of transporting a jacket structure into operation often does not return to the contracting company that has ownership rights to the material. Problems of this material are often overlooked, whereas this issue has the potential risks and substantial losses after the project is completed [4-5].

GAS Compressor Platform is one of PT. ABC Indonesia projects in the oil and gas sector. This platform is an order from the KLM company. This project has been completed. but there are problems related to support material. The material has not been returned by PT. XYZ while the transporting process has been completed. These materials are owned by PT. ABC. Total value of these materials is estimated to USD. 700,000. The approaches of risk based decision making used in this paper consist of FTA (Fault Tree Analysis) and ETA (Event Tree Analysis) which called bow tie analysis [6-7].

This paper analyses the root causes, consequences, preventive and mitigation measures did not return support material. The objectives of this paper are:

1. Investigating the root causes of support material is not returned using FTA
2. Determining the consequences of support material is not returned using ETA
3. Determining risk index of the consequences using risk matrix

2. Bow Tie Application

A bow tie is the combination method of fault tree analysis on the left and event tree analysis on the right. These bow ties are the basis for the application of the methodology. The bow tie diagrams help to understand clearly through graphical visualization the relationship between the potential causes and their possible consequences. This paper will only discuss the factors contributing to the support material is not returned.

Step 1: Building the complete fault tree

Fault tree analysis is a deductive approach to resolve the top events into all possible initiating events. It is useful to test the most probable sequence of events which lead to the undesirable top event. Probabilities of undesirable outcomes can be calculated with most probable outcome identified. FTA is a graphic "model" of the pathways within a system that can lead to a foreseeable, undesirable loss event. The pathways interconnect contributory events and conditions, using standard logic symbols. Numerical probabilities of occurrence can be entered and propagated through the model to evaluate frequency of the foreseeable, undesirable event. This paper used support material is not returned as a case study and involves twenty experts to elaborate the fault tree diagram. The detail of fault tree diagram can be seen from Figure 1 to Figure 5

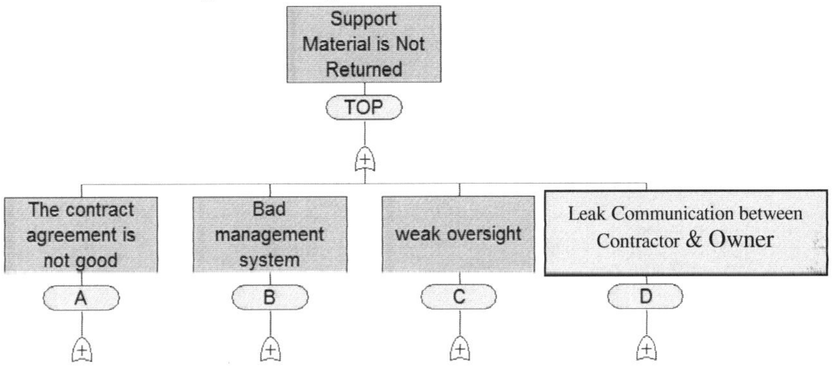

Figure 1. FT Model Support Material is Not Returned

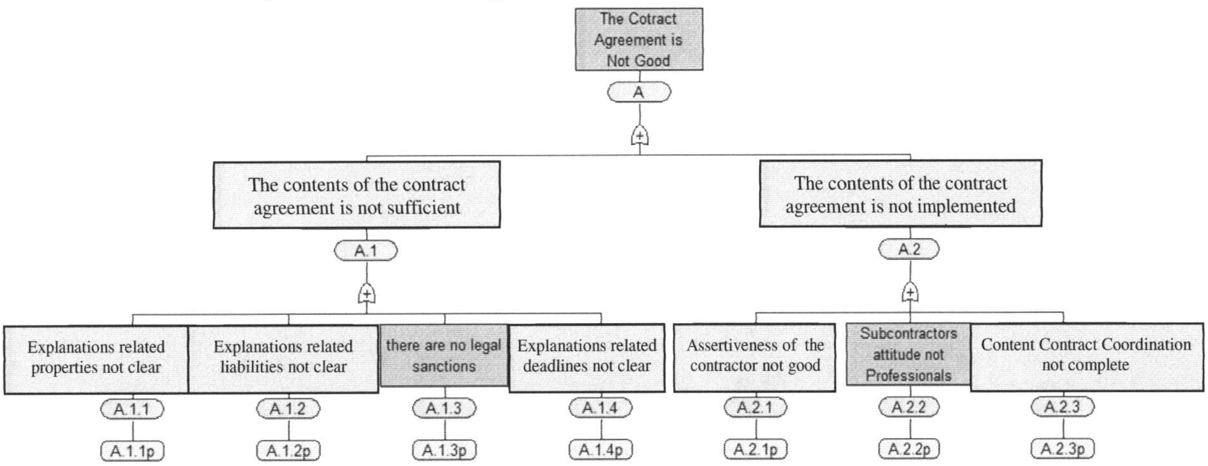

Figure 2. FT Model the Contract Agreement is Not Good.

IOP Conf. Series: Earth and Environmental Science **79** (2017) 012003 doi:10.1088/1755-1315/79/1/012003

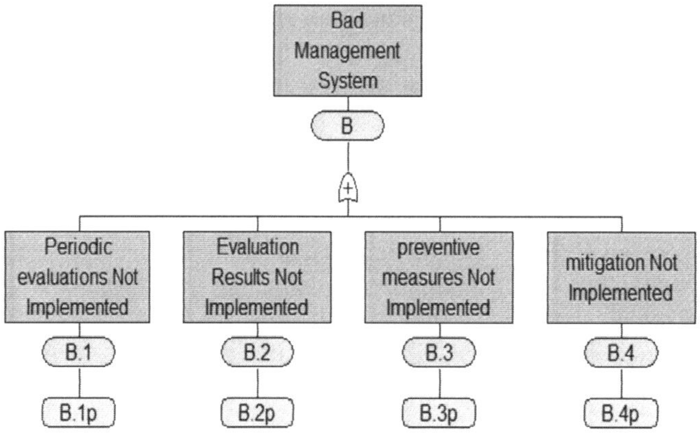

Figure 3. FT Model Bad Management System.

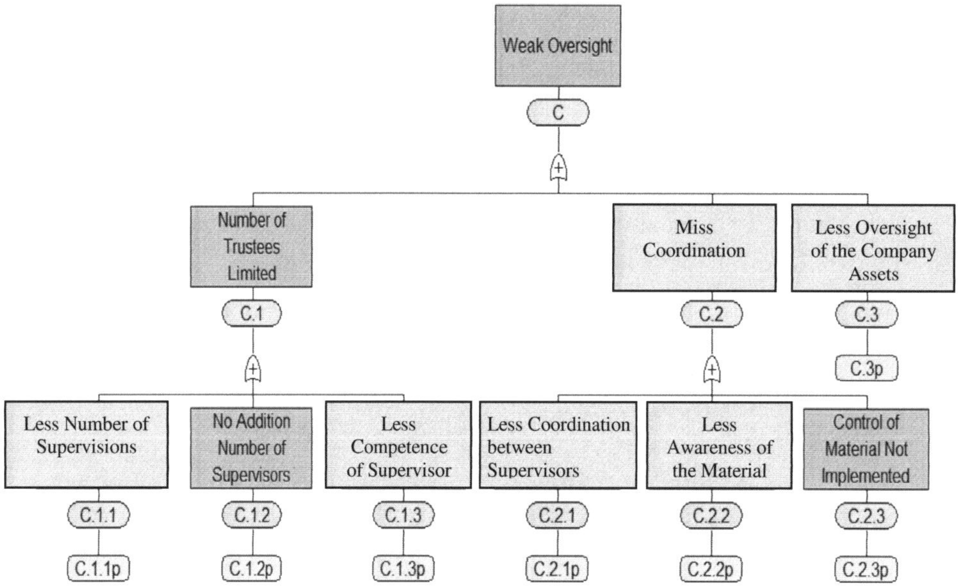

Figure 4. FT Model Weak Oversight.

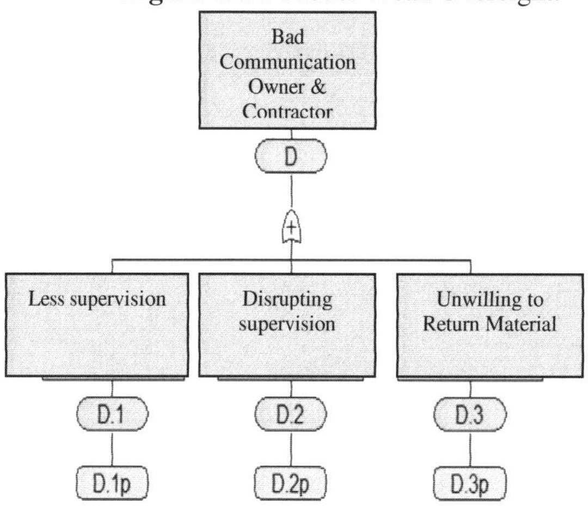

Figure 5. FT Model Bad Communication Owner & Contractor

Step 2: Determining Frequency of Occurrence

The frequency of occurrence in fault tree is obtained from the twenty experts based on questionnaire. The frequency index used in this paper is based on Table 1. The FTA result for support material is not returned is shown in Table 2. Total probability of support material is not returned is 0.7783. The frequency of occurrence derives from the result of FTA which is then used as the frequency of initiating event in ETA.

Table 1. Frequency Index

Rating	Qualitative	Quantitative
Frequent	Likely to occur in each production of platform	10^{-1}
Reasonably Probable	Likely to occur once in the range of 5 times production of platform	10^{-2}
Remote	Likely to occur once in the range of 25 times production of platform	10^{-3}
Extremely Remote	Likely to occur once in the range of 75 times production of platform	10^{-4}
Extremely Improbable	Likely to occur once in the range of 100 times production of platform	10^{-5}

Table 2. Top Event Probability

Rank	Intermediate Event	Probability
1	The contract agreement is not good	0,2317
2	Weak Oversight	0,2257
3	Communication between the contractor and subcontractor is not good	0,1679
4	Bad management system	0,1530
	Total	**0,7783**

Step 3: Building the event tree

The frequency of initiating events multiplied by the pivotal events will result the frequency of occurrence of the outcomes. Generally, there are several possible outcomes deriving from the initiating event followed by the pivotal events which include the success or failure possibility. Once the event tree diagram is constructed, the frequency of occurrence can be applied to the diagram for each path.

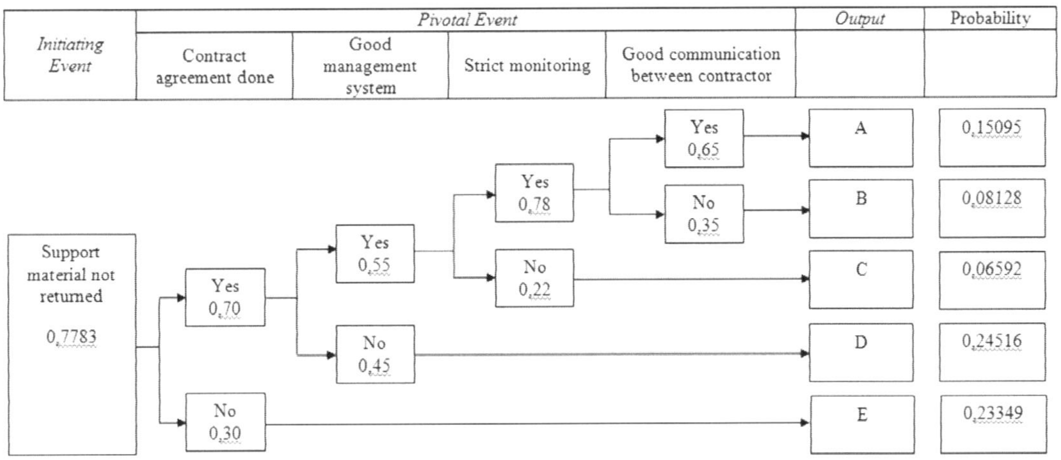

Figure 6. Event Tree Diagram

Event tree diagram for support material is not returned can be seen in Figure 6. Event tree is binary analysis with the initiating event support material not return. Initiating event consist of several pivotal events namely contract agreement done, good management system, strict monitoring and good communication between contractor. Each of the pivotal event have two path yes and no, yes path showing that the pivotal events is succeed and no path showing that the pivotal events is not succeed. There are five outputs based on the event tree, namely:

a. Output A is financial loss more than Rp. 125.000.000,- Less than Rp. 300.000.000,-
b. Output B is financial loss more than Rp. 350.000.000,- Less than Rp. 500.000.000,-
c. Output C is financial loss more than Rp. 600.000.000,- Less than Rp. 2.500.000.000,-
d. Output D is financial loss more than Rp. 2.500.000.000,- Less than Rp. 5.000.000.000,-
e. Output E is financial loss more than Rp. 10.000.000.000,-

Table 3 shows the frequency index, severity index and risk index for determining the risk level. The probability of each output is determined based on frequency index as shown in Table 1. Then the respondents give their judgment for the severity index. Once the frequency index and severity index have been determined, the next is to calculate the risk index.

Table 3. Frequency Index

No	Output	Frequency Index (FI)		Severity Index (SI)		Risk Index (RI)	
1.	Output A	5	Frequent	1	Minor	5	Moderate
2.	Output B	4	Reasonably Probable	1	Minor	4	Moderate
3.	Output C	4	Reasonably Probable	2	Moderate	8	Moderate
4.	Output D	5	Frequent	2	Moderate	10	Moderate
5.	Output E	5	Frequent	3	Serious	15	High

Table 4 describes the risk matrix of each output based on event tree diagram. Risk matrix consist of two axis. The x axis shows the severity index while the y axis shows the frequency index. From the risk index it can be seen that output A,B,C,D are in the Moderate risk level and output E is in the High risk level.

Table 4. Risk Matrix of Each Output

FI	Rating	Severity Index (SI)			
		1 Minor	2 Moderate	3 Serious	4 Catastrophic
5	Frequent	5	10	15	20
4	Reasonably Probable	4	8	12	16
3	Remote	3	6	9	12
2	Extremely Remote	2	4	6	8
1	Extremely Improbable	1	2	3	4

A B C D E

3. Conclusion

The main objectives of this paper have been accomplished and the conclusions were shown below:

1. The root causes of support material not returned is due to the contract agreement not good, bad management system, weak oversight and leak communication between contractor and owner.
2. The consequence is the contractor losses up to Rp.10.000.000.000,-. Beside financial losses, the other consequence is the company's reputation declined and bad publicity from the media
3. The output A,B,C,D are in the Moderate risk level and output E is in the High risk level.

4. References

[1] Barlian D A 2012 *Analisis Waktu Dan Pembiayaan Untuk Proses Loadout Jacket Structure Menggunakan Dolly Dan Skidway, (Surabaya: Tugas Akhir S1-Teknik Kelautan, Institut Teknologi Sepuluh Nopember)*

[2] A Suroso 2003 *Aspek Perancangan Platform Migas Laut Dalam (Oil Deep Sea Platform) Untuk Perairan Laut Indonesia* (Surabaya: Jurusan Teknik Kelautan ITS)

[3] F F Clifford, W L Erik 2007 *Manajemen Proyek: Proses Manajeria* (Yogyakarta: Andi Publisher)

[4] J Sitorus 2008 *Faktor-faktor Risiko Yang Berpengaruh Terhadap Kinerja Waktu Proyek EPC Gas di Indonesia* in *Teknik Sipil Universitas Indonesia, Jakarta.*

[5] Lestari A N 2012 *Studi Pembiayaan Pembangunan Jacket Structure Dengan Menggunakan Metode QFD (Quality Function Deployment)* TugasAkhir S1-Teknik Kelautan, Institut Teknologi Sepuluh Nopember, Surabaya.

[6] C S Howat 2008 *Introduction to Fault Tree Analysis in Risk Assessment.*

[7] PT PAL Indonesia 2014 *Jacket Fabrication and Erection Procedure* (Surabaya: Madura BD Field Development – EPCI Project Management Document)

[8] PT PAL Indonesia 2014 *Analisa Kajian Risiko Permintaan Pengembalian Material Dan Kepastian Pembayaran.* (Surabaya: Banuwati Project Management Document)

[9] Ericson A Clifton 2005 *Hazard Analysis Techniques for System Safety* (Hoboken, New Jersey: John Wiley & Sons, Inc.)

[10] Kurniawan Reza 2015 *Studi Keterlambatan Pembangunan Proyek Kapal Kargo Dengan Menggunakan Metode Bow Tie Analysis* (Surabaya: TugasAkhir S1-Teknik Kelautan, Institut Teknologi Sepuluh Nopember)

[11] Rosyid, D M 2007 *Pengantar Rekayasa Keandalan* (Surabaya: Airlangga University Press)

[12] Sukoroto, Suparno 2012. *Manajemen Risiko Usaha pada Tender Lokomotif di PT. INKA (Persero)* Prosiding Seminar Nasional Manajemen Teknologi XV (Surabaya: Program Studi Magister Manajemen – Institut Teknologi Sepuluh Nopember)

Carrying Capacity Model Applied to Coastal Ecotourism of Baluran National Park, Indonesia

H D Armono[1], D M Rosyid[1] and N I Nuzula[1]
Institut Teknologi Sepuluh Nopember, Surabaya 60111, Indonesia

Tel.: +62 31 5928105; fax: +62 31 5928105.
E-Mail : nike.prasade@gmail.com

Abstract. The resources of Baluran National Park have been used for marine and coastal ecotourism. The increasing number of visitors has led to the increasing of tourists and its related activities. This condition will cause the degradation of resources and the welfare of local communities. This research aims to determine the sustainability of coastal ecotourism management by calculating the effective number of tourists who can be accepted. The study uses the concept of tourism carrying capacity, consists the ecological environment, economic, social and physical carrying capacity. The results of the combined carrying capacity analysis in Baluran National Park ecotourism shows that the number of 3.288 people per day (151.248 tourists per year) is the maximum number of accepted tourists. The current number of tourist arrivals is only 241 people per day (87.990 tourists per year) which is far below the carrying capacity.

1. Introduction

One of the tourist destinations who has been and developed in Indonesia is a Baluran National Park that not only functions as a conservation for various types of plants and animals, but also have potential for ecotourism. This region has a physical landscape structure that can be used for tourism activities. Currently, many coastal tourism activities were already in progress, such as beach tourism, snorkeling, diving and mangrove tours. Various community groups such as family group, students, government and foreign tourists usually spend their long holiday to travel in Baluran National Park. In the recent two years, the number of tourist visiting to Baluran National Park has increased. In 2013, the number of tourists is 39,866 persons, and in 2014 reached to 58,169 persons. The increasing number of visitor absolutely has an impact, both positive and negative effect, especially for coastal natural resources condition in Baluran National Park. Coastal tourism activities in Baluran National Parks essentially contribute to economic growth and prosperity of local communities. On the other hand, these activities also had a negative impact on the preservation of coastal resources like coral reefs and mangroves. For this reason, restrictions on tourists' visit are necessary in considering natural resource support in Baluran National Park.

The concept of tourism carrying capacity arises from the perception that tourism cannot grow continuously in a particular region without causing irreversible damage to the local [1]. The concept of carrying capacity in tourism has its origins in the 1960's, when it was developed to place limits on the numbers of visitors that a tourist attraction (such as a natural reserve) or destination could cope with [1]. The concept of carrying capacity in tourism was very much in vogue in the 1970s and has encountered a fall in interest since then [2]. Many definitions of tourism carrying capacity have been proposed [3,4] but the best is the World Tourism Organization's definition; i.e: "the maximum number of people that may visit a tourist destination at the same time without causing destruction of the physical, economic or socio-cultural environment and an unacceptable decrease in the quality of the tourist satisfaction'' [3]. This definition implies various capacities: physical, economic, perceptual, social, ecological, and political [5]. This research is aimed to determine the sustainability of coastal ecotourism management based on the carrying capacity of the ecological, economic, social and physical environment.

2. Methods

This research was conducted in Baluran National Park which is administratively located in the District Banyuputih, Situbondo, East Java, Indonesia. This area is restrained by Madura Strait in the North, Bali Strait in the East, and the River Bajulmati, Wonorejo village in South, next to the West River Klokoran, Sumberanyar village. The Baluran National Park has an area of 25,000 ha with a coastline of 42 km [6]. This study is focused on coastal areas. Baluran National Park is located in the Latitude of 7°50'10.3" and Longitude 114°24'19.4". Figure 1 shows the map of research sites.

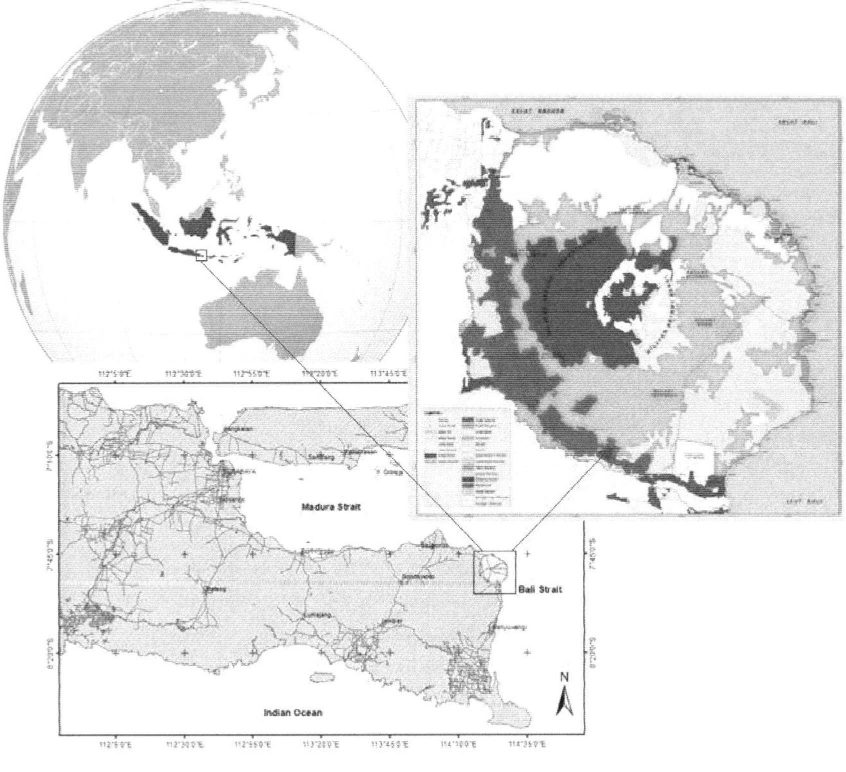

Figure 1. Study Area

The data collected in the study were classified into primary data and secondary data. The object of the research is natural resources associated with the coastal tourist activities include coral reefs, mangrove forests, beaches, water quality, rating, community, tourism entrepreneurs, supporting individual infrastructure and institutions related to the management of the park. Collected data were tabulated and grouped by interest analysis to address concerns and research purposes.

2.1 *Carrying capacity of ecotourism coastal ecology*

The ecological carrying capacity of ecotourism is analysed by the area attractions that are susceptible to direct damage and the maximum load of waste.

2.1.1. *Utilization of tourist areas.*

Estimation of the carrying capacity activities for utilization conservation area is following the government regulation (Government Regulation No. 18/1994) about the utilization of natural tourism attraction in the zone of national parks and Natural Park which is 10% of the utilization zone. Based on these considerations, A formulation to calculate the carrying capacity of the area for coastal tourism activities in conservation areas [7] is expressed as:

$$\text{Carrying capacity} = 0.1\left[K\frac{L_p W_t}{L_t W_p}\right] \tag{1}$$

K is a maximum number of tourists per unit area, Lp is the area or length of the area that can be used, Lt is a unit area for particular category, Wt is the time reserved area for tourism activities per day, Wp is the time spent by a visitor to any particular activities. Space requirements for coastal tourism refers to [8], [9], [10].

Table 1. The value of K, Lt, Wp and Wt for coastal tourism activities [5],[3],[6].

Number	Activity	K	Lt	Wp	Wt
1	Diving	2	1000 m^2	2	8
2	Snorkelling	1	300 m^2	3	6
3	Beach Tourism	1	20 m^2	3	6
4	Mangrove Excursion	1	100 m^2	2	8

2.1.2. Sea water pollution. The number of human population who move to coastal areas influence the maximum limit of allowed water usage to maintaining standar water quality for coastal tourism. Standards of coastal tourism water quality is the range of values that can sustain water quality for aquatic biota. It is assumed that tourists and residents who move in the area of the park produce waste that possibly pollute the waters.

2.2. Physical carrying capability

World tourism organization, in [11] provides a standard development of tourism facilities (resort and supporters) in coastal areas and small islands in order to limit the number of tourist visits. It is designed that natural resources in the region socially sustainable (not interfere with the local community). Physical carrying capacity indicates the amount of area that can be used for infrastructure / tourist facilities without disturbing the comfort of locals or other travelers. Standard space requirements for coastal tourist facilities can be seen in Table 2.

Table 2. Standard Space Requirement for Coastal Tourism Facilities.

Number	Commentary	Unit	Information
1	Coast capacity	m^2	Max.Quantity of person per 20-50 m^2 coast
	Low Class	10	2.0-5.0
	Middle Class	15	1.5-3.5
	High Class	20	1.0-3.0
	Luxurious Class	30	0.7-1.5
2	Coast Facilities	Hygienic Facilityies : 5 unit of toilets. 2 bathrooms and 4 showers for each 500 people	
3	Capacity of accommodation	60-100 bedroom/ha	

2.3. Social and economic carying capacity

The social carrying capacity express that the increment of time and the number of people, resulted in the increment of the need for human interaction and competition among human beings in the occupied area [12]. Consequently, discomfort (dissatisfaction) arise between one human being to another and disturbed (unsustainable). Some of the parameters needed to analyze the social carrying capacity are the public perception of tourism, feelings and reactions to the arrival of tourists, lifestyle changes related to tourism, and tourists as well as local communities' perceptions associated with comfort in interacting and conducting respectively. The method used to assess the carrying capacity is descriptive analysis and agreement [12].

Economic carrying capacity analysis in this study is approached by supply and demand, derived from the balance between supply and demand functions which generate coastal tourism product prices and the optimum number of tourists throughout the year. Based on these two values, the magnitude of the economic value of coastal ecotourism resource can be observed. Economic carrying capacity with the bidding approach is an approach used to analyze the economic potential of coastal resources developed as coastal ecotourism products. The calculation of the economic carrying capacity to micro bidding approach related to activities of travel services by companies that have consequences on production costs. The total cost (TC) issued by travel company is a function of supply whose value depends on the number of tourist arrivals (V) or mathematically written TC = f (V). Based on this, the economic capacity analysis is supported by the analysis of the costs arising marginal. Marginal cost (MC) is the ratio of change in the total production cost of coastal ecotourism to the changes in the number of tourists or can be written:

$$MC=\partial TC/\partial V \quad (2)$$

Travel demand approach is an approach used to analyze the magnitude of coastal tourism product demand by tourists is limited by the cost of travel, tourist income, changes in prices and other factors. Approach this request was analyzed by measuring the magnitude of the ability to pay (Willingness to Pay, WTP) by travelers in conducting coastal tourism. The method used to measure the WTP is the method of travel expenses (Travel Cost Method TCM) in order to obtain the value of consumer surplus. The maximum amount (the carrying capacity of the economy) tourists visiting the Baluran National Park and the maximum price that can be paid is obtained by balance between function related to coastal ecotourism product demand function (Supply = Demand).

2.4. *Composite carrying capability*

The combined carrying capacity analysis performed to obtain the value of the carrying capacity of a standard that will be presented as the basis for the management of coastal ecotourism in the National Park Baluran. By taking into account the four dimensions of ecological, social, economic and physical [3] in [8]. Operational approach in determining the carrying capacity that integrated from the fourth dimension is optimizing the values of the technical parameters and the carrying capacity that has produced by four dimensions. The fourth optimization carrying capacity by using linear goal programming, on the basis of the following formula [8]:

$$Z = \sum_{i=1}^{5} DU_i \text{ atau } DO_i \quad (3)$$

$$\sum_{i=1}^{5} a_i X + DU_i - DO_i = b_i \approx X, a_i, b_i \geq 0 \approx Du, DO = 0 \quad (4)$$

Where:

DU and DO = bearing capacity has not been reached and exceeded the carrying capacity of the target b_i
a_i = coefficient of each parameter function carrying capacity constraints (*i*)
X = optimal carrying capacity (combined)
b_i = target each parameter carrying capacity

3. Result and Discussions

Generally, environmental carrying capacity is a maximum capacity to support an organism to grow. The amounts of capacity in an area are influenced by the condition and characteristics of existing resources in an area. The capacity of the environment and resources in Baluran National Park will become a limiting factor in determining the appropriate utilization of space. A geobiophysics factor in coastal tourism affect the fragility of an ecosystem on the carrying capacity of coastal tourism [13]. Coastal ecotourism carrying capacity comprises of physical, ecological carrying capacity, social carrying capacity and economic carrying capacity.

3.1. Ecological carrying capacity

Ecological carrying capacity analysis is based on the maximum number of visitors who physically can fit in the area provided at a certain time without disturbing the natural and society. In this research, is based on the necessary space that can be accommodated in the region is very suitable for coastal tourism at a certain time. Due to the study is in the area of conservation (Baluran National Park), tourism activities are not mass tourism and visitor space is very limited, the determination of the carrying capacity of the region should consider aspects of environmental sustainability. So, the authorized area for tourist utilization is 10% of the total area. There are several values used in the calculation of the carrying capacity to the conditions and perceptions of tourism players in Baluran National Park, such as average time of diving, snorkeling, mangrove and beach tourism. The results of the analysis of ecological carrying capacity in Baluran National Park are presented in Table 3.

Table 3. Carrying Capacity Number Region Coastal Tourism in NTB.

No.	*Coastal tourism activities*	*DDK (people/day)*	*Condition Existing (people/day)*
1	*Mangrove Excursion*	*1813*	*50-70*
2	*Beach tourism*	*3680*	*50-100*
3	*Snorkeling*	*756*	*10-30*
4	*Diving*	*404*	*5-10*
Total		*6653*	*Maximum 210*

Table above shows that total carrying capacity for the coastal tourism activities is 6,653 visitors a day and 306,038 annually (assuming the effective day of activities 46 travel days per year). Snorkeling and diving capacity have a smaller amount than the other, because of the limitations of rental equipment and diving guide. Considering the current utilization and enthusiast, then the possibility of snorkeling will reach to maximum carrying capacity, especially in peak season. In the area of tourism, carrying capacity of mangrove which is usually used as a tourist attraction tracking only in Resort Bama, while other areas are still undeveloped. Attempts to increase the visitor carrying capacity should be supported by the aquatic condition in the Baluran National Park. The results of the aquatic analysis compared to the quality standard for marine tourism indicate below the standards (in accordance with the Environment Decree No. 51 2004). The results of the analysis show the maximum linear probity the number of tourists as many as 14 585 people per day.

3.2. Physical carrying capacity

Physical carrying capacity is the maximum utilization of a resource or ecosystem that can be absorbed by an area without causes a decrease in physical quality. Method to analyzing the physical carrying capacity was compared between standard requirements space [11] with the reality of the use of coastal areas to travel at this time. Some indicators used are; beach capacity, room capacity, density and the presence of boat rentals presented in Table 4.

Table 4. Space Needs Analysis for Coastal Tourism Facilities.

No.	Description	Reality in Baluran National Park	Standard Facility of coastal eco-tourism	informati on
1	Coast facility	Each room: 2 people, 1 toilet, 1 bathroom, 1 shower. There are 4 showers out room in homestay for 20 people.	Facility cleanliness: 5-unit toilet, 2 bathrooms, 4 showers for 500 people	suitable

2	Capacity of accommodation	80 bedrooms	60-100 bedroom	suitable
3	Facility of traditional fishing boat	3 boats, 1glass boat,75 traditional fishing boat	50-200 traditional fishing boat	suitable

Table 4. Shows that physically, coastal tourism facilities in each coastal tourist business location not exceed the carrying capacity (standard requirement amenities). There are some areas that still do not meet the infrastructure facilities and infrastructure. Thus, when the peak season there are several coastal tourist activities used with limited facilities. In peak season or certain activities, room needs sometimes exceed capacity (80 rooms), which reaches 160 people per day, so travelers forget reduced.

3.3. *Social carrying capacity*

Social carrying capacity assessment in this study is the level of public acceptance of local (host) with the arrival of the visitors (tourist) without disturbing the existing comfort. It is assumed that there is the maximum number of tourists visiting the Baluran National Park, so that local people do not feel disturbed by the arrival of tourists. Results showed 70% of local people give the perception that there is no change in the behavior of the local community since the tourists especially foreign tourists and 10% of people expressed no behavioral change in local communities. The behavioral changes regarding everything related to the activity which is always valued by money or there is a trend shift in the value of individuality (lack of help each other), and change the way they dress. In addition, the presence of tourists has not been a significant influence in the economy and changes in the quality of life of local communities so that presence to be addressed by ordinary travelers.

Associated with the convenience of local communities to the presence of tourists, the results showed that a variety of opinions and ratings of local people and tourists about the optimum ratio between tourists and local communities. Generally, local people claimed that in addition to the increase in the number of tourist visits, discomfort can be disturbed society mainly due to the way social interaction. But if people given the freedom to choose the ratio travelers with local people, then as many as 90% of respondents expressed a 1-30 rating compared to the local population (10% choose one versus 20). This shows that there is a possibility where a traveler can interfere with the 1 or the 30 local communities, depending on how the interaction and behavior among tourists with the local community. If we assume the total population Wonorejo village in 2014 was 6576 people, and the interaction between tourists and local communities, the maximum number of tourists visiting in Baluran National Park is 220 persons per day (this number is still smaller than the ecological carrying capacity 6653 people). As explained by [12], that inconvenience a person can choose to receive when someone else got to interact (Social Carrying Capacity), although ecologically (Biological Carrying Capacity) is still available space for people to interact [12].

3.4. *Economical carrying capacity*

Carrying capacity of the economy in this study using the approach of supply and demand, derived from the balance between supply and demand functions which generate coastal tourism product prices and the optimum number of tourists throughout the year. Based on these two can be seen also the magnitude of the economic resources of coastal ecotourism. The results of the economic analysis of the carrying capacity of tourist demand side will be obtained coastal ecotourism products consumer surplus per year amounting to Rp 189 328 thousand per traveler or USD 87.790 16 659 million per traveler. The average cost incurred by tourists today is Rp 300 thousand per day. Economic resources coastal tourism is still relatively small. This could be due to low accessibility to the coastal ecotourism locations in the park, and the lack of promotion by the central and local governments. Consumer surplus could increase if carried out improvements in support services and facilities [8].

Carrying capacity analysis in this study is a travel product on the balanced position between demand and supply of coastal ecotourism products. The results of analysis of the carrying capacity in the economy equilibrium position is 131 975 people per year with optimal price coastal ecotourism products that can be applied to any visitor to US $ 1,444.82 per visit. Maximum economy which can be obtained in the use of coastal tourism in Baluran National Park Rp. 34.33 billion per year. Based on the carrying capacity of this economy, the maximum number of tourists who can visit the tourist business is 2869 people per day. The carrying capacity is smaller than the ecological carrying capacity. The Economic capacity of a conservation area can be improved through effective and optimum [14] management (in terms of conservation management) and an increase in tourist knowledge. Improved knowledge of marine ecotourism activities especially snorkeling, diving tourism, distribution and rotation of each dive, setting space and time for snorkeling and underwater photographers is expected to provide a relatively small impact damage [8].

3.5. Combined analysis carrying capability

Carrying capacity can be measured in a level of tourist arrivals, which is associated with individual activities, amenities and special services [8]. Combined fourth-carrying capacity is intended to obtain an optimal value of the carrying capacity of coastal ecotourism that has considered the fourth aspect. Operationally means that the number of tourists coming into Baluran National Park by a combined carrying capacity, able to minimize the degradation of coastal resources, are economically profitable, in accordance with the capacity of lodging, tourists and people feel comfortable with the coastal ecotourism activities.

The considering the four aspects of carrying capacity, the maximum number of tourists who can go into the coastal ecotourism in Baluran National Park is 3288 people per day (151 248 tourists per year). The combined carrying capacity is an optimal value that accommodate the interests of ecological, social, economic and physical management of coastal ecotourism. Value effectiveness of coastal ecotourism management and carrying capacity generated is expected to be used as guidance in the management of coastal ecotourism including local understanding of ecotourism. The management of ecotourism always consider the ecological, social and economic aspects[15], therefore ecologically highly vulnerable marine tourism activities will damage marine resources and socially may bring inconvenience to the local community.

4. Conclusions

The results of the combined carrying capacity analysis in ecotourism Baluran National Park shows that the number of 3.288 people per day (151.248 tourists per year) is the maximum number of accepted tourists. The current number of tourist arrivals is only 241 people per day (87.990 tourists per year) which is far below the carrying capacity. The condition shows the high possibility to increase the number for tourists. Based on the carrying capacity per dimension per day, it shows that the ecological carrying capacity is 6.653 people per day, while the maximum number of tourist which is able to be accepted considering physical carrying capacity is 160 people per day. In terms of Social carrying capacity, Baluran National Park can accept 220 people per day, and in terms of Economical aspect, 2.869 people per day can be accommodated. Based on the analysis, the carrying capacity of the area can be improved if the magnitude of the extent of coral reefs and mangrove forests is well maintained and increased its presence.

References

[1]. Coccossis H and Mexa A 2004 *The challenge of tourism carrying capacity assessment : Theory and practice* (Aldershot: Ashgate)
[2]. Butler RW 2010 Carrying capacity In DG Pearce & RW Butler (Eds.) Tourism research A 0-20 vision (pp. 53–64) Oxford: Goodfellow
[3]. Coccossis H, Mexa A and Collovini A 2002 Defining, measuring and evaluating carrying capacity in european tourism destinations final report B4–3040/2000/294577/MAR/D2 Athens (Greece): European Union.
[4]. Saarinen J 2006 Traditions of sustainability in tourism studies *Annals of Tourism Research*

33(4) 1121–40
[5]. Getz D 1983 Capacity to absorb tourism. Concepts and implications for strategic planning *Annals of Tourism Research* 10, 239–63
[6]. Muryono, Mukhammad. 2011. Analisis tata ruang (zonasi) pengembangan ekowisata di kawasan taman nasional baluran jawa timur. Berk Penel Hayati Volume 17 halaman 115-117
[7]. Hutabarat A, Yulianda F, Fahrudin A, Harteti S and Kusharjani 2009 Pengelolaan pesisir dan laut secara terpadu. Bogor : Edisi I Pusdiklat Kehutanan, Deptan, SECEN-KOREA International Cooperation agency.
[8]. Laapo A 2010 *Optimasi Pengelolaan Ekowisata Pulau-Pulau Kecil (Kasus Gugus Pulau Togean Taman Nasional Kepulauan Togean)* (Disertasi. Bogor: Sekolah Pascasarjana. Institut Pertanian Bogor)
[9]. Dahuri R, Ginting SP, Rais J and Mj. Sitepu 1996 *Pengelolaan Sumberdaya Wilayah Pesisir dan Lautan Secara Terpadu.* (Jakarta: PT. Pradnya Paramita)
[10]. Davis D and Tisdell C 1995 Recreational scuba-diving and carrying capacity in marine protected areas. Ocean and coastal Management, 26 (1): 19-40. *Elsevier*. Northern Ireland.
[11]. Wong PP 1991 Coastal tourism in Southeast Asia *ICLARM Education Series* 13 40p Manila
[12]. Saveriades A 2000 Establishing the social tourism carrying capacity for the tourist resorts of the east coast of the Republic of Cyprus *Journal of Tourism Management* 21 147-56
[13]. Soemarwoto O 2004 *Ekologi, Lingkungan Hidup dan Pembangunan Edisi ke-10* (Jakarta : Penerbit Djambatan)
[14]. Davis D and Tisdell C 1996 Economic management of recreational scuba diving and the environment *Journal of Environmental Management* 48 229-248 in. Tisdell C *Tourism economics, the environment and development: analysis and policy* (Brisbane : Department of Economics University of Queensland)
[15]. Tisdell G 1998 Carrying capacity of coastal tourism sites : A mhetodological aapproach. Paper presented at the First Annual Sessions of the Faculty of Graduate Studies, University of Sri Jayewardenepura, 27th March, 1998.

IOP Conf. Series: Earth and Environmental Science **79** (2017) 012005 doi:10.1088/1755-1315/79/1/012005

Modelling of Phenol Contamination in Wonorejo Estuary, Indonesia

Suntoyo, and W Sakinah
Insitute of Technology Sepuluh Nopember

E-Mail: suntoyo@gmail.com

Abstract. Phenol as a part of marine pollution in Wonorejo estuary has been a concern today. Mass death of clams and decreasing growth of shrimps often occur in this area. The citizens made assumption that source of marine pollution from Sidoarjo mud disaster. This paper investigates the distribution of phenol from modelling with MIKE 21, Eco Lab module. Measurement were done in Wonorejo estuary to get a validation of modelling. Modelling result shows phenol concentration in northern Wonorejo estuary is higher than other areas. Hence, it was concluded that the phenol source is not from Sidoarjo mud disaster.

1. Introduction

Phenol is often sourced from chemicals that are generally derived from industrial waste and from different disinfectant in household cleaning products and products that are consumed like toothpaste and mouthwash. Phenol is also called carbolic acid which is an organic compound with alcohol structure C_6H_5OH. Phenol can be readily decomposed in the air, but the phenol contaminated water takes longer to be decomposed [1]. Phenolic compound is also a pollutant derived from oil spills or disposal of waste oil into the sea [2]. Phenol concentration that contaminated water became toxic to aquatic life and can cause a bad taste in the meat of fish [3].Other impacts are decreasing antibody and corrosive to the skin (irritation), and even can cause interference on human health [2]. Clams mass death in Wonorejo estuary made people to assume there are phenol contamination from Lapindo mud disaster waste (oil and gas well blow out) in Sidoarjo.

2. Methods

The research was performed at Wonorejo estuary, Surabaya, East Java province, Indonesia, see figure. 1. Coordinate range of Wonorejo estuary is 7°15'19,60" - 7 °17'13,25" S latitude and 112°48'35,69" - 112°48'40,72" E longitude.

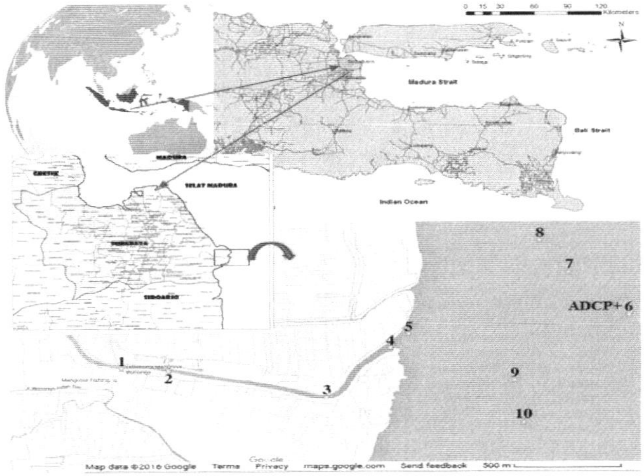

Figure. 1. Sampling location at Wonorejo estuary

Phenol measurement in ten sites as modelling validation was explained in Sakinah, *et al.* [4] (*waiting for publication*). Sampling was performed in dry season. A hydrodynamic and water quality model with DHI's Mike 21 Hydrodynamics and ECO Lab module are used to examine distribution of phenol in Wonorejo river estuary. Hydrodynamics module (HD) as basic computational component to simulate hydrodynamics and ECO Lab module simulates ecological parameter coupled in HD module.

2.1 Modelling methodology

From modelling reference with Mike 21 [5], the water level was prescribed along the open boundary for flow computation i.e. Northern, Southern, and Eastern boundary located in the Madura Strait. The western boundary of the model is the Wonorejo river with water discharge (observed data from Jasa Tirta Surabaya in dry season in the year of 2015 is 0.15 kg/m^3) as the boundary condition. Hydrodynamics and ECO Lab modelling was done with the input parameter of modelling consist of hydrodynamics parameters, such as, bathymetry, river flow discharge, tides, wind speed and wind direction, as well as the five boundary conditions, see figure 2. and data of phenol as water quality parameter. Model simulation was carried out for one month (720 hours) with the time step interval = 3600s, number of time steps = 720s, from 23rd August 2015 until 22nd September 2015.

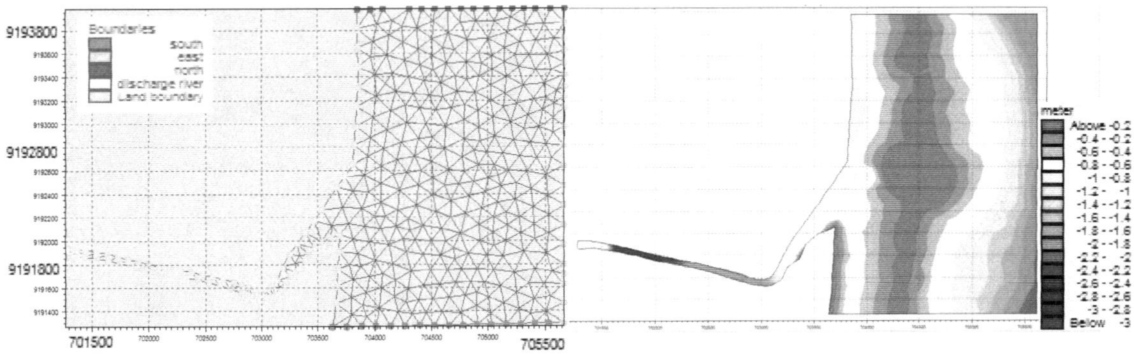

Figure 2. Boundary condition of modelling and bathymetrical data

2.2 Model validation

Phenol measurement at site 1, 6, 7, and 10 [4] are used to validate the modeling result. Site 1 represent a western side, site 6 for eastern side, site 7 for northern side, and site 10 for southern side. From the results of modeling and measurement are presented in table 1, which has a relatively small value difference. It could be concluded that the simulation result of phenol distribution in Wonorejo estuary was valid and can be used to predict the source of phenol contamination.

Location site	Measurement data (mg/L)	Modelling result (mg/L)	Error (%)
1	0.021	0.021	0.92
6	0.086	0.076	11.35
7	0.065	0.053	18.75
10	0.023	0.023	0.09

Table 1. Validation results of the phenol from measurement data

The calibration of hydrodynamics model was conducted with examining the water surface elevation from Dishidros-AL (2015) data and the simulation results at site 6, the root mean square error (*RMSE*) is 0.01 and absolute percentage error is 0.04%. Another calibration of hydrodynamics model was conducted the sea water current from measurement with acoustic Doppler current propeller (*ADCP*) and the simulation result at site 6, the root mean square error (*RMSE*) is 0.015 and absolute percentage error is 27.02%.

3. Result and Discussions

3.1 Hydrodynamics simulation results

Figure. 3 show the surface currents distribution during ebb and tide with condition as shown on point (a) and (b) in the upper part of figure 3. The distribution of high tide (figure. 3(a)), the current enter the estuarine mouth with the water level 2.53 m and maximum speed of 0.003 m/s occur in the river body and 0.017 in the river mouth for the simulation step of 203. During the ebb current (figure. 3 (b)) is with the water level of 0.32 m and maximum speed of 0.017 m/s occur in the river body and 0.02 in the river mouth for the simulation step of 160, the current moving to the downstream direction or toward the river mouth of Wonorejo estuary. At the highest tides, currents in the Wonorejo estuary move toward upstream and at the lowest tide currents moving in the opposite direction.

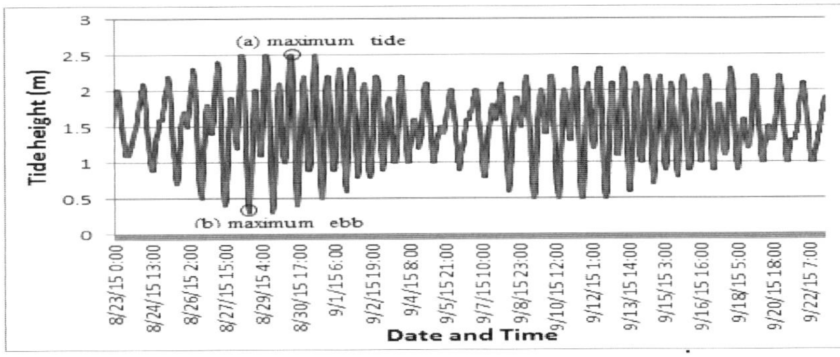

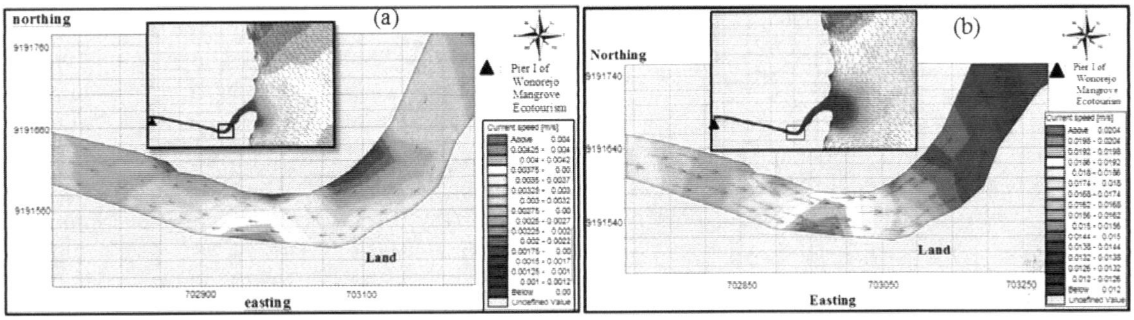

Figure 3. (a) Tide current pattern; (b) Ebb current pattern in the river body of Wonorejo estuary

3.2 The distribution pattern of phenol contamination in Wonorejo estuary based on measurement data and numerical modeling result

The distribution of phenol at the tide condition as shown in figure. 4(a), is large enough in the sea and smaller along the river. The minimum value of phenol is -7x10^{-6} mg/L occurs in river body and the maximum value is 0.89 mg/L occur in the northern sea. Fig. 4(b) shows the distribution of phenol at ebb condition with the minimum value is 1x10-5 mg/L occurs in river body, was marked with the purple colour at the modelling result and the maximum value is 0.88 mg/L occur in the eastern sea. Phenol concentration in the tide condition was higher than ebb condition.

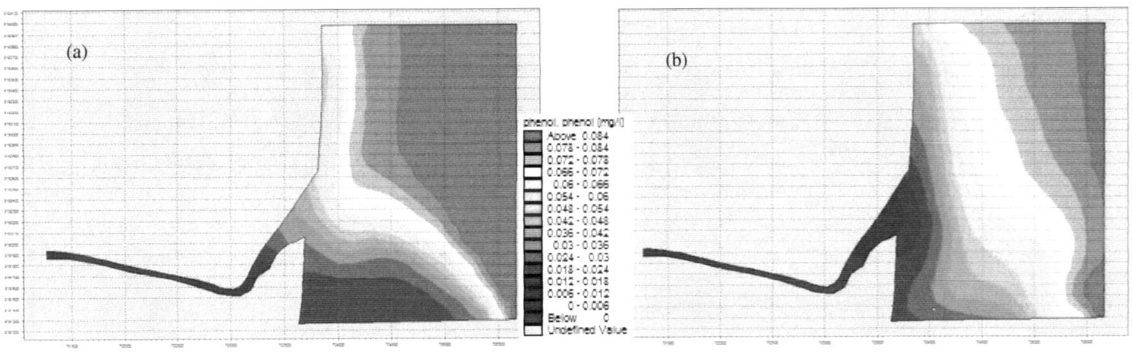

Figure. 4. Phenol distribution pattern (a) during tide condition; (b) during ebb condition

Figure 5 gave the information obtained from the numerical modelling results for the distribution of phenol after one month. The minimum value of phenol is -1.2x10^{-5} mg/L occurs in river body and the maximum value is 0.9 mg/L occur in the northern sea.

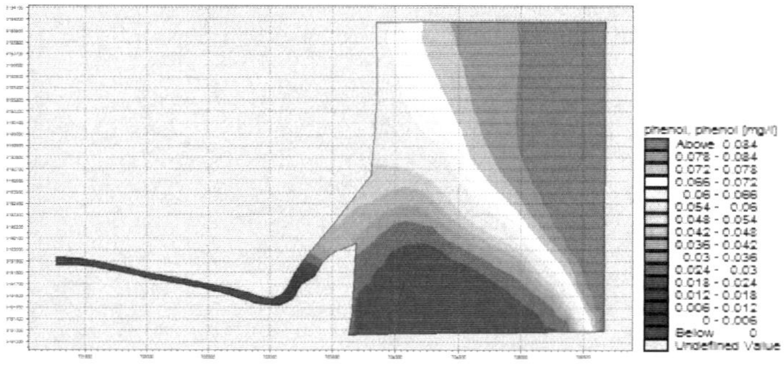

Figure. 5. Phenol distribution pattern after one month based on numerical modelling result

The phenol concentration become higher from time to time and the phenol concentration pattern almost same as surface elevation pattern, see figure. 6. The highest concentrations of phenols were at sea while the lowest concentrations are in the area of the river. This is because the sources of pollution are not on the river, but in the environment or from the sea. This phenomenon is reinforced by the concentration on modelling with a negative value. Output appears on modelling using Mike 21 with a negative value means the sources of pollution comes from the environment, negative values also indicate the direction which means the source of the waste water is not towards the environment, but rather away from the environment [6].

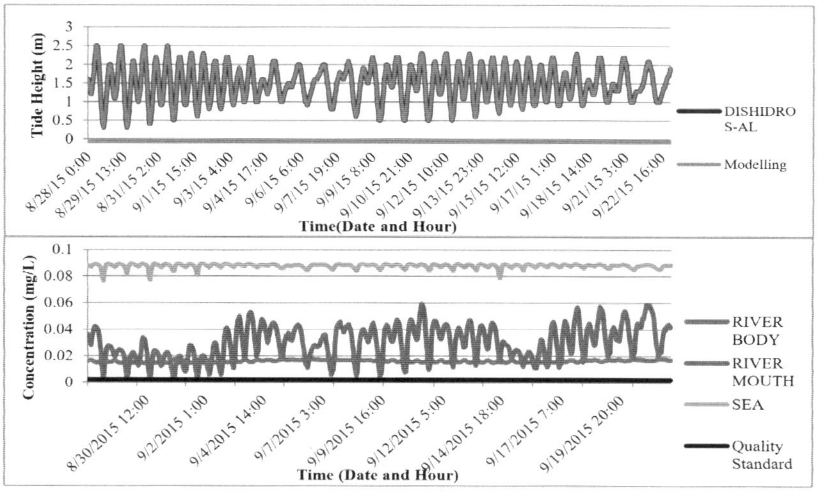

Figure. 6. Comparison of phenol distribution between river body, river mouth, and sea in Wonorejo estuary

The last simulation results in figure. 5 showed that phenol contamination in the estuary area becomes higher, it can be seen from the color in modelling result becomes more green, which means the concentration was intense. But for the southern of the sea area, the blue color becomes wider than at the time of high tide and low tide, see figure. 4. This proves that the assumption of the public about the contamination of Lapindo mud waste, Sidoarjo cannot be substantiated. Laboratory test results from Airlangga University also showed the phenol concentration of Lapindo mud waste is still below the quality standard. This is reinforced by another research [5], which shows the distribution and concentration parameters of water quality in estuaries Porong, Sidoarjo as a result of the Lapindo mud waste. The study showed in Porong estuary area relatively close to the source of the sludge are still in the safe criteria and does not exceed from the quality standard of the Government Regulation Number 82/2001 on the Management of Water Quality and Water Pollution Control. The simulation result shows the phenol contamination actually comes from the northern sea, the direction of the island of Madura. This explains the source of phenols derived from Madura strait because of the many ports in the strait so that there is a possibility of oil spills on ships sailing in the straits of Madura. East Surabaya seas, including the modelling area are shipping lines that are highly vulnerable by oil pollution from waste or disposal of the ships in the harbor and across the waters, and mining activities [7].

4. Conclusions

Numerical modelling with Mike 21 results phenol distribution is higher from time to time and already exceed from quality standard. The maximum concentration of phenol is 0.89 mg/L occur in northern sea and the minimum concentration is 7×10^{-6} occur in river body. Phenol concentration during tide condition is higher than ebb condition. All simulation show the highest concentration of phenol occurs in northern sea. But, Lapindo mud disaster occurs in southern sea. This proves that the assumption of the public about the contamination of Lapindo mud waste, Sidoarjo cannot be substantiated.

Acknowledgements

Authors thank to the Centre of Research and Development of Marine and Coastal Resources, the Ministry of Marine Affairs and Fisheries Republic of Indonesia for providing the facilities of DHI's Mike 21/3 model.

References

[1] U.S. Department of Health and Human Services. (2008). *Toxicological Profile for Phenol.* Public Health Service, Georgia.

[2] Dewilda, Y., Afrianita, R., Iman, F.F. (2012). Degradasi Senyawa Fenol oleh Mikroorganisme Laut. *Jurnal Teknik Lingkunngan UNAND*, **9**, No. 1, 59-73.

[3] Water, Air and Climate Change Branch. (2002). Ambient Interim Water Quality, *Guidelines for Phenols-Summary Report. Ministry of Water, Land and Air Protection.* Canada.

[4] Sakinah, W., Suntoyo, Mukhtasor. (2016). *Impact Identification of Estuarine Water Quality to Marine Biota: A Case Study in Wonorejo Estuary*, Indonesia, AMM Journal.

[5] Suntoyo, Ikhwani, H., Zikra, M., Sukmasari, N.A., Angraeni, G., Tanaka, H., Umeda, M., Kure, S. (2015). Modelling of the COD, TSS, Phospate, and Nitrate Distribution due to the Sidoardjo Mud Flow into Parang River Estuary, *Procedia Earth and Planetary Science,* Vol. -,146-153.

[6] DHI. (2012b). *Mike 21 Flow Model-ECO Lab Module-User Guide.* DHI. Denmark.

[7] Wardhani, M.K., Sulistiono, Siregar, V.P. (2011). Tingkat Kerentanan Lingkungan Pesisir Selatan Kabupaten Bangkalan terhadap Potensi Tumpahan Minyak (Oil Spill), *Jurnal Ilmiah Perikanan dan Kelautan*, **Vol. 3**, No.1, 13-19.

IOP Conf. Series: Earth and Environmental Science **79** (2017) 012006 doi:10.1088/1755-1315/79/1/012006

East Java Maritime Connectivity and Its Regional Development Support

H Purboyo and M Z Ibad
Department of Regional and City Planning
Institute of Technology Bandung
Indonesia

Abstract. The study presents an evolution of maritime connectivity index of East Java which is associated with accessibility and mobility index of regions in East Java. The findings show that East Java increased connectivity more than three times from 1996 to 2011. Initially, the East Java is importer but then become exporter to national territory. For accessibility, the inland regions of East Java in general is higher than the coastal areas. And for mobility, inland regions initially have a small index, but in subsequent years its index is greater than the coastal areas.

1. Introduction

A region in need of other regions to develop [1]. Interaction between the regions will result in specialization, economic growth, efficiency, and lead to the development of the region. Christaller stated in [2] that there are three elements in the development of the region, namely the central nodal, sphere of influence, and transportation networks. The transportation network serves to connect or meet dependency between the central and the area surrounding them. This interaction may be the movement of people, goods and factors of production one region to another. Road infrastructure is important in regional development because it is considered to be a key prerequisite of social and economic development [3].

Seaport is one link of the transport process from the point of origin to point of destination for movement of goods, people, or other production factors [4]. Economic theory considers the port as an important factor in economic development. As evidence, many large cities in the world which is a port city [5]. Marine transportation led to the expansion of the market involving various sectors of the economy, including the extractive sector of natural materials (primary), manufacturing (secondary), and trade (tertiary). Seaports can be categorized as a mouthpiece of economic development due to its nature as a catalyst and driving the development of other economic sectors. The ports support economic development by enhancing competition through market expansion [6], so as to stabilize prices for consumers. More broadly, the port is the point of the central node relationships on a support area (hinterland) and liaison with the remaining areas. By doing so, it can be called a maritime connectivity (connectivity territory by sea) is one indicator of economic growth in the region.

East Java is the province with the highest number of main port (*pelabuhan utama*) in Indonesia, and its port production is the third in Indonesia ports production. East Java's role in the national maritime connectivity is quite important. To maintain or even to increase its role, good regional development in East Java province is needed. For this research, regional development will be associated with the road infrastructure. So, this paper will reveal how the accessibility and mobility index both in the coastal and inland areas are related with the maritime connectivity index. The districts/city in East Java province are:

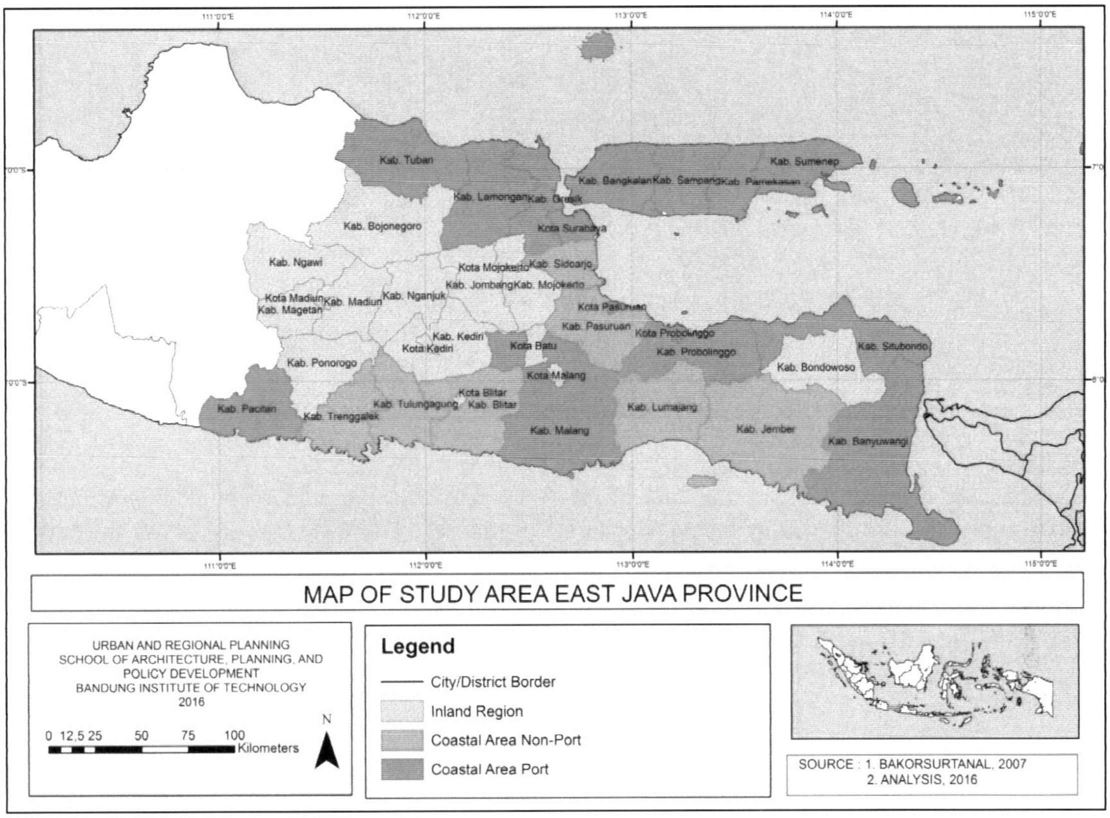

Figure 1. Map of Study Area East Java Province.

The paper will answer the questions below:

1. How much change the maritime connectivity index of East Java Province?
2. How is the growth of accessibility and mobility index in the coastal areas without ports, the coastal areas with port, and in the inland areas in East Java province?

2. Methods and Data

Analysis is divided into two scopes as summarized in the table below

Table 1. Methods and Data

No	Question on	Method	Data
1	Maritime connectivity of East Java province	Analysis of maritime connectivity index	- Origin-Destination of goods between provinces, 1996 (Ministry of Transportation, 1998) - Origin-Destination of good between seaports 2011 (Ministry of Transportation, 2013) - Distance between provinces.
2	Regional Accessibility and mobility in East Java province	-Region of East Java province is divided into 3 areas/	-Length of roads for each class category of *kabupaten* (regency)/ city, 2010 and 2014 (East Java in Number ((*Jawa Timur dalam Angka*), 2016)

	regions, namely: inland region, coastal areas non-port, and coastal areas with port. -Analysis of accessibility and mobility index for each region.	-Number of population of *kabupaten* (regency)/ city, 2010 and 2014 (East Java in Number ((Jawa Timur dalam Angka), 2016) -Area of kabupaten (regency)/ city, (East Java Bureau Central Statistic, 2016)

- Analysis-1

Analysis of Connectivity Index is an analysis that calculates the number of links between the various points (nodes) divided by the number of nodes that can reflect the connectivity of the region. This analysis is useful to know how good the transportation network connecting points of origin and destination [7]. Link is a relationship between 2 nodes, while the node is a point that can be either terminal, ports, airports, stations, cities, and so on. Here is the formula connectivity index [8]:

$$CI = \sum_{j}^{n} c_{ij} \quad (1)$$

Note: CI = Connectivity Index; cij = Presence link = 0 or 1; n = number of nodes

In this study, link is coupled the number of movement of goods between the two nodes, and the distance between nodes is given by a bird's eye distance, so that the formula in maritime connectivity index is as follows:

$$CI_k = \frac{\sum_{i=0}^{n}(a \times q_{kn})}{(\sum_{i=0}^{n}(a \times d_{kn}))n} \quad (2)$$

Modified from Oxera, 2015

Note:
CI_k = Connectivity Index node-k
a = 0 or 1 (depends on the existence of link/);
q_{kn} = The movement of goods from node k to node n (Ton);
d_{kn} = distance from node k to node n (Km);
n = number of nodes

Additional information: Analysis of connectivity index requires data connectivity matrix (link) between nodes; matrix of goods movements between nodes; matrix of distance between nodes; and the number of nodes in the region. Calculations using Microsoft Excel. The index does not have a notation.

- Analysis -2

Accessibility Index Analysis is an analysis by calculating the length of a divided by road-square kilometre area. Analysis of accessibility index indicates the size of the ease of an area to be accessed [9]. This index represents the number of available road network system in the region. In addition, there are other variables to determine the level of accessibility of a region that is the number of means of transport, road capacity, and quality of roads [10].

In this study, the accessibility index is approached by weighting for each class of road. For the National Road (11 m) = 1.1; Provincial Roads (9.5 m) = 0.95; Distric/ City Roads District / City (7.5 m) = 0.75; and Rural Roads (*Jalan Desa*) / Local (6.5 m) = 0.65. Weighting is based on a minimum

IOP Conf. Series: Earth and Environmental Science **79** (2017) 012006 doi:10.1088/1755-1315/79/1/012006

width of the road on the technical requirements of the Government Regulation No. 36 of 2006 on the road. Calculations using Microsoft Excel. The index does not have a notation. Here is the formula to calculate the accessibility index districts / cities in East Java province:

$$Accesibility_{it} = \frac{(1{,}1)\ L_{it}\ Nas+(0{,}95)L_{\ it}\ Prov+(0{,}75)L_{\ it}Kab\ Kota+(0{,}65)\ L_{it}\ Desa/Lokal}{A_{it}} \quad (3)$$

Modified from Miro, 2010

Note:
$Accesibility_{it}$ = Accessibility Index on District/City in year t
$L1_{it}$ = Length of National Class Road on District/City i in year t (Km)
$L2_{it}$ = Length of Province Class Road on District/City i in year t (Km)
$L3_{it}$ = Length of District/City Class Road on District/City i in year t (Km)
$L4_{it}$ = Length of Local/Rural Class Road on District/City i in year t (Km)
A_{it} = Area on District/City i in year t (Km^2)

Analysis Mobility Index is an analysis to assess a person's ability to move in the region [11]. Mobility can be determined by dividing the length of the road with a total population of the region. The index does not have a notation. Formula to determine the mobility index is as follows:

$$Mobility_{it} = \frac{(1{,}1)L_{it}Nas+(0{,}95)L_{it}Prov+(0{,}75)L3_{it}Kab\ Kota\ +(0{,}65)\ L_{it}\ Desa/Lokal}{N_{it}} \quad (4)$$

Modified from Syahban, 2005

Note:
$Mobility_{it}$ = Mobility Index on District/City in year t
$L1_{it}$ = Length of National Class Road on District/City i in year t (Km)
$L2_{it}$ = Length of Province Class Road on District/City i in year t (Km)
$L3_{it}$ = Length of District/City Class Road on District/City i in year t (Km)
$L4_{it}$ = Length of Local/Rural Class Road on District/City i in year t (Km)
N_{it} = Population on District/City i in year t

3. Result and Discussion

3.1. Maritime Connectivity of East Java Province

Maritime connectivity of East Java province is represented by maritime freight movement and connectedness of the link between the East Java province with other provinces. Here is the data maritime movement of goods of East Java province in 1996 and 2011:

Table 2. Maritime movement of goods of East Java Province with other provinces, 1996 and 2011

1996				2011				96 - 2011
PROVINCE	Import	Export	Total	PROVINCE	Import	Export	Total	
ACEH	0	0	0	ACEH	26378	48920	75298	+
NORTH SUMATERA	0	0	0	NORTH SUMATERA	363764	910400	1274164	+
WEST SUMATERA	403995		403995	WEST SUMATERA	12570	148666	161236	-
				RIAU	178799	166850	345649	+
RIAU	0	0	0	RIAU ISLANDS	86894	7118	94012	+
				BANGKA BELITUNG	12860	20005	32865	+
JAMBI	9716	23	9739	JAMBI	11952141	16267	11968408	+

1996				2011				96 - 20 11
PROVINCE	Import	Export	Total	PROVINCE	Import	Export	Total	
SOUTH SUMATERA	0	0	0	SOUTH SUMATERA	30568	103551	134119	+
BENGKULU	0	0	0	BENGKULU	5353	26774	32127	+
LAMPUNG	0	0	0	LAMPUNG	105254	66004	171258	+
JAKARTA	279875	0	279875	DKI JAKARTA	25822589	20751544	46574133	+
WEST JAVA	0	0	0	BANTEN	552711	7346526	7899237	+
				WEST JAVA	2298912	150534	2449446	+
CENTRAL JAVA	4660	0	4660	CENTRAL JAVA	65747	86572	152319	+
EAST JAVA	0	0	0	EAST JAVA	341118,5	342218	683336	+
BALI	0	3089	3089	BALI	85380	13608	98988	+
WEST NUSATENGGARA	3890	455746	459636	WEST NUSATENGGARA	46515	11649	58164	-
EAST NUSATENGGARA	113817	166959	280776	EAST NUSATENGGARA	85213	130846	216059	-
WEST KALIMANTAN	14473	0	14473	WEST KALIMANTAN	1685220	41052	1726272	+
CENTRAL KALIMANTAN	0	196471	196471	CENTRAL KALIMANTAN	4095204	364078	4459282	+
SOUTH KALIMANTAN	1165622	385558	1551180	SOUTH KALIMANTAN	739252	451648	1190900	-
EAST KALIMANTAN	1640367	0	1640367	EAST KALIMANTAN	599492	15147957	15747449	+
NORTH SULAWESI	1254773	0	1254773	NORTH SULAWESI	363764	372447	736211	-
				GORONTALO	131838	132557	264395	+
CENTRAL SULAWESI	3064	0	3064	CENTRAL SULAWESI	282931	76468	359399	+
				WEST SULAWESI	424236	7660	431896	+
SOUTH SULAWESI	165174	5338	170512	SOUTH SULAWESI	130345	4428133	4558479	+
SOUTH EAST SULAWESI	1317		1317	SOUTH EAST SULAWESI	30568	4296299	4326867	+
MALUKU	0	0	0	MALUKU	163437	238237	401674	+
				NORTH MALUKU	122082	7390572	7512654	+
PAPUA	740774	0	740774	PAPUA	125670	98861	224532	-
				WEST PAPUA	497851	314042	811893	+
TOTAL	5801517	1213184	7014701	TOTAL	51464661	63708066	115172727	+
MARITIME CONNECTIVITY INDEX		17,95		MARITIME CONNECTIVITY INDEX		65,34		

3.2. Accessibility and Mobility of District/City in East Java Province

Table 3. Accessibility and Mobility of District/ City in East Java Province

City/District	Accessibility Index			Mobility Index		
	2010	2014	2010-2014	2010	2014	2010-2014
Ngawi	0,345	0,345	+	0,000587	0,000580	-
Bojonegoro	0,613	0,678	+	0,001174	0,001275	+
Mojokerto	0,621	0,757	+	0,000590	0,000689	+
Jombang	0,724	0,870	+	0,000671	0,000785	+
Nganjuk	1,095	1,094	=	0,001382	0,001354	-

IOP Conf. Series: Earth and Environmental Science **79** (2017) 012006 doi:10.1088/1755-1315/79/1/012006

City/District	Accessibility Index			Mobility Index		
	2010	2014	2010-2014	2010	2014	2010-2014
Ponorogo	0,880	1,115	+	0,001455	0,001821	+
Kediri	1,082	1,196	+	0,001098	0,001183	+
Batu City	1,595	1,5947	=	0,001694	0,001622	-
Magetan	1,747	1,7467	=	0,001988	0,001968	-
Madiun	1,862	1,8617	=	0,003146	0,003091	-
Bondowoso	0,801	2,808	+	0,001708	0,005828	+
Kediri City	3,503	3,530	+	0,000900	0,000876	-
Mojokerto City	4,588	4,9558	+	0,000763	0,000795	+
Malang City	7,341	7,480	+	0,000985	0,000973	-
Madiun City	7,395	8,9257	+	0,001471	0,001740	+
Blitar City	20,893	20,8932	=	0,005225	0,004988	-
1. Inland Region (Total)	**3,443**	**3,741**	**+**	**0,001552**	**0,001848**	**+**
Lumajang	0,490	0,490	=	0,000879	0,000862	-
Jember	0,095	0,565	+	0,000136	0,000783	+
Blitar	0,212	0,560	+	0,000334	0,000916	+
Trenggalek	0,617	0,617	=	0,001139	0,001119	-
Pacitan	0,808	0,808	=	0,002121	0,002088	-
Situbondo	0,891	0,894	+	0,002272	0,002216	-
Sidoarjo	1,063	1,109	+	0,000394	0,000383	-
Tulungagung	1,274	1,274	=	0,001481	0,001443	-
Pasuruan	1,192	1,633	+	0,001171	0,001547	+
2-1 Coastal Area Non-Port	**0,738**	**0,887**	**+**	**0,001103**	**0,001262**	**+**
Banyuwangi	0,102	0,102	=	0,000236	0,000232	-
Lamongan	0,211	0,274	+	0,000315	0,000406	+
Tuban	0,489	0,489	=	0,000865	0,000844	-
Bangkalan	0,474	0,519	+	0,000681	0,000716	+
Gresik	0,159	0,533	+	0,000167	0,000531	+
Sumenep	0,600	0,604	+	0,001200	0,00118	-
Sampang	0,630	0,633	+	0,000886	0,000844	-
Probolinggo	0,309	0,777	+	0,000482	0,001171	+
Pamekasan	0,569	0,818	+	0,000570	0,000779	+
Malang	1,703	1,703	+	0,002406	0,002329	-
Pasuruan City	2,100	2,100	=	0,000428	0,000413	-
Probolinggo City	3,060	3,084	+	0,000887	0,000734	-
Surabaya City	3,097	6,759	+	0,005391	0,00079	-
2-2 Coastal Area with Port	**1,039**	**1,415**	**+**	**0,001117**	**0,000844**	**-**
2. Coastal Area Total	**0,916**	**1,199**	**+**	**0,001111**	**0,001015**	**-**

The growth of maritime conectivity index reflects the increase of seaport performance to distribute the freight from internal area to the external one and vice-versa. Ideally the increase of external connectivity should be supported by internal province accesibility and mobility to balance the freight movement. Therefore, it is important to measure the road availability using index of accesibility and index of mobility in internal province area. To know better internal condition of East Java province in term of those two indexes, the province region divided into coastal area and inland area. The coastal area provide the location of seaport. Accordingly, to support export and import activities through seaport, this area should have good accessibility and mobility indexes.

In 1996, the index of maritime connectivity of East Java province to the other provinces in Indonesia is 17,95, while in 2011 significantly increase to 65,34. It means, there are 364 percent growth of sea transport between East Java to other provinces. In line with the fact that the East Java province during those periods was getting connected with other provinces in Indonesia, the total number in tonnage of export-import activities via its seaports increases from 7.014.701 to 115.172.727 or more than 15 times. So, the interdependency of East Java province in maritime connectivity was increase.

From the internal province view from 2010 to 2014, accessibility index is increased in all areas. Accessibility index in inland areas increases (from 3,443 to 3,741) more than in coastal area non-port (from 0,738 to 0,887) and also in coastal area with port (1,039 to 1,415). It indicates that during 2010-2014 there were the construction of new roads in all areas with different total length of the road per area. In term of mobility index, it grows differently. Mobility index in inland area increases from 0,001552 to 0.001848, in coastal area non-port increases from 0,001103 to 0,001262, and in coastal area with port decreases from 0,0011117 to 0,000844. It indicates that the ratio between population growth and new road constructed in coastal area with port is higher comparing with other areas. So, it can be said that the movement of traffic in coastal area with port experienced more disruption caused by the movement of its population comparing to other areas. Therefore, movement to the port to be less supported in the coastal area with port and also in the coastal area in general. In other word, the increase of maritime connectivity relates to the increase of accessibility, but is less supported by the increase in the number of population that are not balanced with the addition of new road.

4. Conclusions

This study shows that from the external conditions, by using maritime connectivity index, the development of East Java province increases during the period of 1996-2011. The growth of maritime connectivity ideally should be supported by regional infrastructure to balance the freight movement. The coastal area serves the freight transport from all East Java area including its inland area to the external area via seaport. But from the internal conditions, based on accessibility and mobility index, the coastal areas where the seaports are located, in general, have road infrastructure supply lower than inland areas. Whereas, the road availability in coastal area has vital support to a whole freight transport and economic activity. Broadly speaking, it can be said that East Java province has poor support for regional development. Thus, a recommendation that can be issued is increasing number of road infrastructure for both the length and the capacity of the road, especially in the coastal area of East Java province.

References

[1] Sullivan A 2009 *Urban Economics* (New York: McGraw-Hill/Irwin)
[2] Adisasmita S A 2011 *Transportasi dan Pengembangan Wilayah (*Yogyakarta: Graha Ilmu)
[3] Ivanova, Eva and Masarova, Jana. *Importance Of Road Infrastructure In The Economic Development And Competitivenes.* Economics and Management: 2013. 18 (2)
[4] Maritime World. *Fungsi Umum Pelabuhan.* http://birulautku.blogspot.com/2008/12/sekilas-tentang-kepelabuhanan.html (accesssed May 25, 2016)
[5] Rodrigue, Jean-Paul, dan Joseph Schulman (2008). *The Economic Impacts of Port Investments.2013.* http://people.hofstra.edu/geotrans/eng/ch7en/appl7en/ch7a5en.html (accessed May 25, 2016).

[6] Ferrari C 2011 Ports and Regional Economic Development. *Global Ports and Urban Development: Challenges and Opportunities (*Paris: OECD)
[7] Litman T 2015 *Evaluating Accessibility for Transportation Planning* (Victoria: Victoria Transport Policy Institute)
[8] Oxera 2010 *Understanding The Theory of International Connectivity (*United Kingdom: Oxford).
[9] Black J A 1981 Urban Transport Planning: Theory and Practice (London: Cromm Helm)
[10] Miro 2010 Aksesibilitas Lokasi (Universitas Sumatera Utara)
[11] Syahban A 2012 *Aksesibilitas dan Mobilitas Transportasi* http://azissyahban2005.blogspot.co.id/ 2012/12/ aksesibilitas-dan-mobilitas-transportasi.html (accessed June 20, 2016)
[12] Rahardjo A 2014 *Pertumbuhan Wilayah & Wilayah Pertumbuhan. (*Yogyakarta: Graha Ilmu)
[13] Agustina C, Roberto P, John M, Elizabeth J, and Aurora R-R Understanding connectivity to international markets: a systematic review *Journal: Transport Reviews* pp 1-24
[14] Earl B 2007 *The Practice of Social Research: Eleventh Edition (*Belmont: Thomson Wadsworth)
[15] Bhattacharyay B 2012 Seamless sustainable transport connectivity in Asia and the Pacific: Prospects and challenges *International Economics and Economic Policy* 9 pp 147–189.
[16] Burghouwt G and Redondi R 2013 Connectivity in air transport networks: Models, measures and applications *Journal of Transport Economics and Policy* 47 pp 35–53
[17] Deng P, Lu S, and Xiao H 2013 Evaluation of the relevance measure between ports and regional economy *Transport Policy 27* pp 123-133
[18] Kachigan S K 1986 *Statistical Analysis (*New York: Radius Press)
[19] Logistic Shipping. *Indeks Konektivitas Maritim Antar Negara dan Poros Maritim Dunia.* Oktober 2015. http://jurnalmaritim.com/2015/10/poros-maritim-dunia-indeks-konektivitas-maritim-bilateral-indonesia/ (accessed March 7, 2016)
[20] Jung, B 2011 Economic Contribution of Ports to the Local Economies in Korea *The Asian Journal of Shipping and Logistics 27*(1) pp 1-30
[21] Latin-American research maritime networks, services structure and maritime trade *Networks and Spatial Economics* 11 pp 555–576
[22] Lin J and Ban Y 2013 Complex network topology of transportation systems *Transport Reviews* 33 pp 658–685
[23] Li-zhuo L. 2012 Analysis of the Relationship Between QinHuangDao Port Logistics and Economic Growth *Advances in information Sciences and Service Sciences AISS 4*(4) pp 105-114
[24] Moreno R and Lopez E 2007 Returns to local and transport infrastructure under regional spillovers *International Regional Science Review* 30 pp 47–71
[25] Perroux F 1964 Economic Space: Theory and Applications Friedmann J and Alonso W (ed). *Regional Development and Planning.: A Reader (*Cambridge, Massachusetts, The M.I.T. Press)
[26] Jean-Paul R et. al 2006 *The Geography of Transport System* (Tylor & Fracis e-Library)
[27] Pritta A W 2015 *Pengaruh Pelabuhan Pengumpan Regional Terhadap Perekonomian Wilayah Kabupaten/Kota.(* Bandung: Institut Teknologi Bandung)
[28] Pius Honggo W 2012 *Terminal Penumpang Kapal Laut Pelabuhan Harbor Bay* Tesis (Yogyakarta: Universitas Atma Jaya Yogyakarta)
[29] Wilmsmeier G and Hoffmann, J 2008 Liner shipping connectivity and port infrastructure as determinants of freight rates in the Caribbean *Maritime Economics & Logistics* 10 pp 130–151
[30] Eveline W 2015 The Relalion between Port Performance and Port Development. *Bachelor Thesis. Erasmus School of Economics* (Rotterdam: Erasmus University

IOP Conf. Series: Earth and Environmental Science **79** (2017) 012007 doi:10.1088/1755-1315/79/1/012007

Towards Sustainable Ambon Bay: Evaluation of Solid Waste Management in Ambon City

S Maryati, M Miharja, A F Iscahyono, S Arsallia, AN S Humaira
Research Cluster for Regional and City Infrastructure System, School of Architecture, Planning, and Policy Development, Institute of Technology Bandung
Jalan Ganesha 10 Bandung 40132, Indonesia

E-mail: smaryati@pl.itb.ac.id

Abstract. Ambon Bay is a strategic area in the context of regional economic development, however it also faced environmental problems due to economic development and the growth of population. One of the environmental problems in the Ambon Bay is the growing solid waste which in turn lowers the quality of the water. The purpose of this study is to evaluate solid waste management in the Ambon City and propose recommendation in order to reduce solid waste in the Ambon Bay. The analytical method used is descriptive analysis by comparing a number of criteria based on the concept of solid waste management in coastal region with the current conditions of solid waste management in Ambon City. Criteria for waste management are divided into generation, storage, collection, transport, transfer and disposal. From the results of analysis, it can be concluded that the components of solid waste management at transport, transfer, and disposal level are generally still adequate, but solid waste management at source, storage and collection level have to be improved.

1. Introduction

Ambon Bay is a strategic area in the context of regional economic development. Ambon Bay administratively consists of Ambon City and Central Maluku Region. Based on Spatial Plan of Maluku Province 2007-2027 Ambon Bay is Strategic Area of Maluku Province due to its environmental carrying capacity. In addition, based on the Spatial Plan of Ambon City 2011 - 2031, Ambon Bay is stated as strategic area regarding its environmental condition. Ambon Bay is a coral reef ecosystem, conservation area and water infiltration area. Ambon Bay is also a riparian area that serves to resist erosion and stability of the riverbank. Beside its environmental function, Ambon bay also a strategic area related to its economic function, with the main function in fisheries and marine tourism.

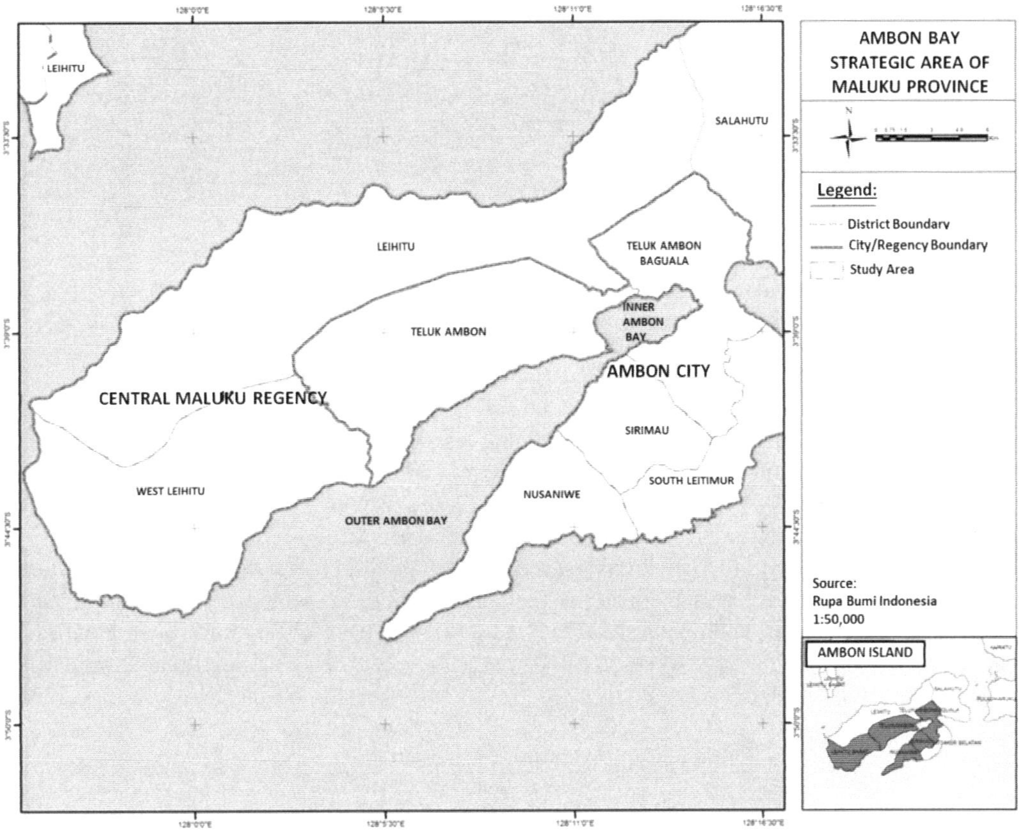

Figure 1. Study Area

As mentioned above, Ambon Bay is strategic area in term of environmental aspect. Ambon bay has the conserved natural resources, such as coral reefs. Nevertheless the existence of the natural resources is threatened because of high level of pollution. One of the reasons for high level of pollution is solid waste. Solid waste is often found in the Ambon Bay. It is estimated that solid waste comes from households or facilities which is not handled properly. The waste enters the streams, which ultimately ends up in the Ambon Bay. The purpose of this study is to evaluate solid waste management in the Ambon City and propose recommendation in order to reduce solid waste in the Ambon Bay. Increasing number of waste generation will give negative impact to environment, especially pollution in air, water, and land, due to the improper waste maangement [1].

The study will be focused in Ambon City, which is one of the administrative region in Ambon Bay. The selection of this city is based on the condition that the number of population in the city is much higher compared to another administrative region, that is Central Maluku Region. Ambon City has developed intensively compared to Central Maluku Region and give more impact to the Ambon Bay compared to Central Maluku Region. The higher the number of population, without good solid waste management, the higher the level of the problem.

The city of Ambon consist of 5 Districts namely Nusaniwe, Sirimau, Teluk Ambon, Teluk Ambon Baguala, and Leitimur Selatan. Ambon City is a coastal region. There are five districts in Ambon City and all of the districts have connection to the bay and sea. Total length of coastline is 102.7 km. There are four districts with direct connection to the Ambon Bay. In this study, the analysis was done in districts which have direct connection to the Ambon Bay. The districts are Nusaniwe, Sirimau, Teluk Ambon, Teluk Ambon Baguala. Nusaniwe is the most densely populated area in Ambon City. The number of population, area, and density of each district in study area are presented in Table 1.

Table 1. The number of population, area, and density of each district in study area

District	Area (km^2)	Population	Density (person/km^2)
Nusaniwe	88.35	111,671	1,264
Sirimau	86.81	174,045	2,005
Teluk Ambon	93.68	47,777	510
Teluk Ambon Baguala	40.11	66,442	1,656

2. Methods

The analytical method used in this study is descriptive and comparative analysis approach. Descriptive analysis is used to describe normative criteria and existing condition in Ambon Bay, whereas comparative analysis is used to compare the normative criteria or standard with existing condition. The study begins by identifying the normative criteria in the provision of waste infrastructure, based on the concepts and rules that exist in Indonesia. The criteria are shown in Table 2. Furthermore, these criteria compared with existing conditions in the city of Ambon. Descriptive analysis is used to identify the phases of waste management in the city of Ambon, which are generation, storage, collection, transport, transfer and disposal. After the problem of each phases have been identified, the type, quantity, and quality of infrastructure within each phase are being explored and compared with the planning standard in Indonesia.

Table 2. Criteria of solid waste management

Criteria	Indicator
Waste Generation	Availability of solid waste bin
	Availability of reduce, reuse, recycle in source level
Waste Collection	Availability of waste collection system and equipment (1m3 of volume for 640 people or 128 households)
Waste Transfer and Transport	Availability of temporary waste disposal station with service coverage 500 m)
	Availability of waste transport mechanism done by government or private sector
	Availability of waste transport equipment (eg. dump truck, arm roll truck, compactor truck, trailer truck)
Waste Disposal	Availability of sanitary landfill or at least controlled landfill

3. Result and Discussion

3.1. Waste Generation

In line with the increasing number of urban population, amount of generated waste is also increasing. The increasing number of population causes an increase in the activity of the population, which directly affects the increase in the amount of waste. Presented below is the amount of the generated waste in the city of Ambon from year 2010-2014.

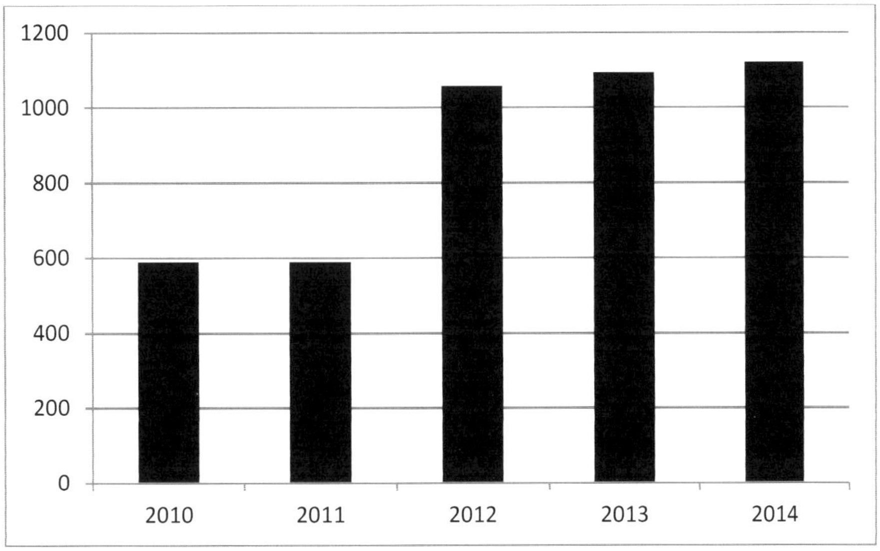

Figure 2. Waste generated per day (m3)
Source: Central Bureau of Statistics, Ambon City 2015

Based on the chart above, it can be seen that the production of waste per day increases from 2010 to 2014. In addition, an average waste of 2.8% from the amount of total waste produced, are not collected at the transfer station. Meanwhile, from all the waste collected at transfer station, an average of 19.7% are not transported to the landfill.

Besides the government, the society also participates in waste management. Based on existing condition, society has participated in the provision of waste management facilities, such as providing trash bin at the source of waste and providing means of collection from the source of waste to the transfer station. In addition, there are groups of people who have provided a means of waste composting and recycling at home. Several schools have done composting and recycling of waste in the school environment. In addition, society also has a role in waste management, starting from collecting waste from the source to the depot transfer, and managing waste at the depot transfer.

Waste generation is not only produced from household activities, but also generated from the tourism sector. Therefore, it can be projected that waste generation will be increased if there is no action to reduce waste. Based on Government Regulation No. 81/2012 about the Management of Household Waste, waste reduction activities consist of restrictions of waste generation, waste recycling, and waste reuse. The three activities are the principles of environmentally sustainable waste management called 3R (reduce, reuse, recycle). In addition, for waste management on the waste production sources such as homes, restaurants, shops, schools, offices and others, waste containers needs to be provided, at least two pieces for each location, one for organic and one for inorganic waste.

If we compare to the criteria that have already defined above, in waste generation aspect in some place there have been reduce, reuse, and recycle system. Government and community also have participated in provision of waste generation facilities. The problems are the facilities and also reduce, reuse, and recycle system have not been equally distributed.

3.2. Waste Collection

Waste collection is the activity of transporting the waste from waste sources to the transfer station or 3R transfer station. Waste collection is defined as waste management from its origin location to the transfer station before heading to the next stage. Based on SNI 3242-2008, waste collection can be classified as individual direct, individual indirect, communal direct, and indirect communal. The terms direct and indirect refer to the process whether the waste is transported directly to the final disposal or collected in a transfer station before the waste is transported to final disposal. The collection process can be done by the use of garbage motorcycles, garbage carts, garbage bicycle, and others.

Based on an interview with the District Head of Teluk Ambon, Teluk Ambon Baguala, and Nusaniwe, garbage in the residential area is transported by the people to the transfer station. In Teluk Ambon and Teluk Ambon Baguala, each housing area has its own initiative in the collection of waste to the transfer station. People in the same neighborhood collectively use garbage carts, and then these carts haul the waste from each house to the transfer station. However, only in the sub-district of Nusaniwe, Benteng, Wainitu, Mangga Dua, Udameti, and Nusaniwe, where the waste collection is served by waste carts. Meanwhile, for the people who are living at the sub-districts where are not served with the garbage carts, they must bring their own waste to the depot transfer. The size of the depot transfer generally is 2 x 2 x 2 meters, with a capacity of 8 m^3. In District of Nusaniwe, people are only allowed to collect their garbage in transfer station between 10.00 pm to 05.00 am.

If we compared with the indicator have been specified above, the waste collection system and equipment in Ambon City is already available. The same as the problem in waste generation, the distribution of the system and equipment is still in question.

3.3. Waste Transfer and Transport

The operational of waste transfer and transport in the Ambon City done by The *Dinas Kebersihan Kota Ambon* (Ambon City Sanitation Department). Waste transportation facilities that are currently available are shown in the table below. Waste that is transported into the landfill is the waste that is produced at household area, offices, schools, and public facilities (shops, markets, and terminals). In addition, medical waste, which is produced from health facilities, is usually handled by the medical personnel.

All the waste vehicles operate daily, from 05.00 am until afternoon when most of the waste has been transported. Wastes, which are transported by waste motorcycles, are the waste that are collected at the garbage bins along the roads and waste from the roads itself. Then, the wastes that are transported by motorcycle are collected in the trash containers that are located in the city of Ambon.
Waste in the City of Ambon, are transported from the transfer station to landfill by the Sanitation Department's garbage truck. There is no transfer station that is located at Laha Sub-District in Teluk Ambon District. While in the District of Nusaniwe and Teluk Ambon Baguala, all sub-districts are served by the transfer station. However, in the District of Teluk Ambon Baguala, people requested to add more numbers of transfer station to increase the service coverage. The sub-districts that ask for additional transfer station are Negeri Lama Sub-district and Halong Sub-district. Moreover, several districts request to provide carts in order to haul garbage from residential area to the transfer station, such as Passo Sub-district.

Table 3. Number of transportation waste facility in Ambon City 2016

No	Type of Facility	Total	Capacity (m3)	Note
1	Dump Truck	17	10	3 broken
2	Arm Roll Car	8	5	2 broken
3	Container	10	8	-
4	Pick Up Car	8	5	3 broken
5	Speed Boat	3	-	-
6	Garbage Carts	40	2	-
7	Tossa Motorcycle	9	2	-

Table 3 shows that capacity of transportation facilities in Ambon City (Dump Truck, Arm Roll Car, Pick Up Car, Garbage Car, and Tossa Motorcycle) is 293 m3, whereas the number of solid waste generation by using 2014 data is 1120.2 m3. It means that in average each transportation waste facility have to work 4 times a day for transporting waste.

Other than land waste, marine debris is also generated by the people's activities at the Ambon Bay. Therefore, Sanitation Department has a regular schedule to collect marine debris that is done by speedboat. In addition to marine waste transportation schedule, the department also has regular schedule to pick up waste, which are located at the transfer station, apparent transfer station, road waste, and *Kliko* garbage bins.

Currently in Ambon City there are 146 transfer stations, spreading at the four districts, which the 59 units are located in Sirimau District, 37 units in the Nusaniwe Distric, 24 units in the Teluk Ambon Baguala District, 26 units in the Teluk Ambon District and none in Leitimur Selatan District. In addition, to accommodate the increasing volume of waste, there are 18 units of garbage container at the 18 points of garbage disposal. In addition, there are 76 apparent transfer stations, which is a point of storage or accumulation of garbage that is not formed as a transfer station. This apparent transfer stations generally found in densely populated urban areas in the central city of Ambon, such as in the Sub-district of Ahusen, Honipopu, Uritetu, Rijali, Wainitu, Waihaong, Kuda Mati, and Batu Merah. Moreover, there are also garbage bins "Kliko", which separates among organic, non-organic, and dangerous waste.

When compared to SNI 3242-2008, waste transport by carts or motorcycles with a pickup or sectional pickups, needed to be done at least two days and transport to the transfer station or integrated transfer station. Additionally, streets and parks litter sweeping in the neighborhoods area, needed to be done by the environmental waste management according to the schedule that has been set. Transportation of waste from transfer station or integrated transfer station to the landfill, needed to be done when the container has been filled and in accordance with the schedule that has been confirmed with the transport of municipal waste management.

3.4. Waste Disposal

Final waste processing is done at the landfill. In the City of Ambon, there is a waste processing field (namely *Tempat Pengolahan Akhir Sampah* in Indonesian term), which is located in the hamlet of Toisapu Villate, Hutumuri Sub-district, Leitimur Selatan District. The figure below shows the landfill site in the city of Ambon.

The waste processing system at Toisapu Landfill is controlled landfill. Currently the area covers with about 9 hectares, and the 5 hectares are full-filled with waste. With this type of processing system, the waste compacting is done once a week. The landfill area comprises of 3 landfill blocks, the garbage block 1 and 2 are currently full, and the garbage block 3 is now reaching a height 2 meters, with a maximum capacity height of 5 to 7 meters. The landfill has an estimated maximum of the existing capacity of 47.450 tons. In addition to the garbage blocks, the landfill area also has a leachate pool for water treatment that is generated from the waste. In addition, the landfill has a development plan to add an area of 2 hectares to as the garbage block 4. Moreover, there is also an Integrated Waste Treatment Installation (namely *Instalasi Pengeolahan Sampah Terpadu* in Indonesian term) located at the landfill area. Thus, some types of waste are able to be recycled. The landfill serves City of Ambon, but District of Leitimur Selatan is not served due to the mountainous area that is difficult to reach.

Figure 5 explained the flow of waste management in the City of Ambon. Starting from household waste that is transported individually or collectively to the depot transfer near to their household area. Furthermore, wastes that are collected at the transfer depot are then transported to the landfill area by the garbage truck. In the landfill, the wastes are sorted into two, which are organic and non-organic waste, by the scavengers. Organic waste is further processed in the waste treatment installation to become compost with an average production of 20 m^3 per month. Meanwhile, non-organic waste that is unable to be processed will be store at the landfill, and non-organic waste that is able to process, is processed in the waste treatment installation. Currently the non-organic waste that can be processed is processed by plastic chopping system with production as much as 3 tons per day. Then they are sent to Surabaya for further processing. In addition, there is also non-organic waste that is used to make plastic craft, which then sold at the Waste Bank. Organic and non-organic wastes that can be processed then contribute to the income of the city of Ambon.

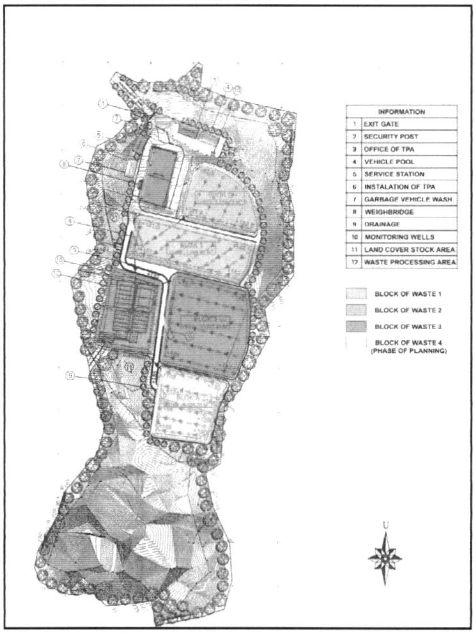

Figure 3. Plan of the landfill in the City of Ambon
Source: IPST Kota Ambon, 2016

When compared to SNI 3242-2008, waste management in the integrated depot transfer, should be done by sorting the organic and inorganic waste. Additionally, organic waste needed to be composted of organic waste with environmental scale. Inorganic wastes also needed to be sorted according to the type of waste such as Inorganic waste that can be recycled, waste from stalls that can be sold, dangerous household waste, and residual waste. After that, selling the waste that has economic value to the city, needed to be managed in accordance with applicable regulations, and collect residual waste into the container to be transported to the landfill.

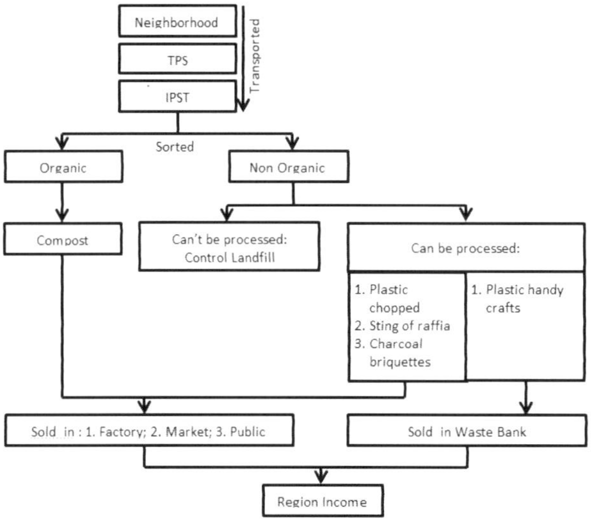

Figure 4. Waste processing flow in City of Ambon
Source: Result of Observation and Interview with Head of IPST City of Ambon, 2016

4. Conclusions

Based on the analysis, the completeness of waste management in the city of Ambon can be considered as a good waste management system. It is because of the stages of waste management already run most likely according to the applicable standards and regulations. However, several points remain to be improved, such as the quantity and quality of personnel and infrastructure which are still limited, the waste transport operation which can be more optimal, the public which can be more supportive to the waste management and the lack of action against violations of regulations on waste disposal. In addition, expansion of the landfill area and improvement of the waste processing tools are needed, so that all kinds of process-able inorganic waste can be processed in Ambon City.

Acknowledgement

The research, on which this article is based, is funded by Institut Teknologi Bandung under the scheme of 'Riset Inovasi Kelompok Keahlian 2016'. The title of the research is 'Model Konseptual Penyediaan Infrastruktur berdasarkan Persepsi dan Preferensi Stakeholder'. The authors would like to thank all the stakeholders involved in this research. However, the authors alone are responsible for any mistakes and shortcomings

References

[1] Uz Zaman A and Mohammad S 2016 Performance evaluation and benchmarking of global waste management systems *Resources, Conservation and Recycling* 114 pp 32-41
[2] Government of Ambon City 2016 *Installation of Integrated Waste Management (IPST) in Ambon City*
[3] Government of Ambon City 2011 *Spatial Plan of Ambon City 2011-2031*
[4] Government of Maluku Province. 2007 *Spatial Plan of Maluku Province 2007-2027.*
[5] Central Bureau of Statistics Ambon City 2016 *Ambon Municipality in Figures 2016.*
[6] Central Bureau of Statistics Ambon City 2015 *Ambon Municipality in Figures 2015.*
[7] Government Regulation of Indonesia Number 81/2012 regarding Domestic Waste Management.
[8] National Standardization Agency of Indonesia (BSN). Indonesian National Standard (SNI) Number 3242:2008 regarding Waste Management in the Settlement Area.

Integrated Coastal Zone Planning Based on Environment Carrying Capacity Analysis

M Miharja and S Arsallia
Regional and City Infrastructure Systems Research Group
School of Architecture, Planning, and Policy Development
Bandung Institute of Technology
Jalan Ganesha no. 10, Bandung 40116, Indonesia

E-mail: mimingm@pusat.itb.ac.id

Abstract. Coastal zone is a crucial area in terms of planning development. It holds high economic value, which affect to increasing number of inhabitants living in the area. As a result, this condition influences environmental degradation. Thus, in every attempt towards coastal zone development, it is crucial to always refer to environment carrying capacity. Carrying capacity is the limit of a certain coastal zone capability to support all human created activities, in which all ecological performances are maintained at sustainable level. The failure to establish strong and clear method and regulation on carrying capacity analysis will lead to a very risky coastal zone development, which in turn would threat the area's sustainability. This paper discusses method for analysing carrying capacity of coastal zone as important input for the area development plan. Two key parameters, i.e. land and clean water carrying capacities are discussed to form carrying capacity analytical method. Furthermore, an empirical data of Ambon Bay, Moluccas Province, is used to illustrate the operationalization of the method.

1. Introduction

As one of the largest archipelagic state, Indonesia owns 17.504 islands [1] and coastal line of 99.093 km^2 [2]. Territorially, Indonesia area includes around 65% of waters area and the rest of land areas (berjubel.net, 2015). Moreover, Indonesia is a habitat of a highly diverse marine life. As one of the proof, Indonesia has been included to the Coral Triangle Initiative (CTI), which is a multilateral partnership of six countries to address the urgent threats facing the coastal and marine resources of one of the most biologically diverse and ecologically rich regions on earth (CTI, 2007). With high potential on marine life it is no longer a question that Indonesia owns high economy value on its coastline. Therefore, around 60% of Indonesian lives along the coastline [3].

There are variety of definitions have been used to describe the area of coastal zone. Coastal zone is the interface where the land meets the ocean, encompassing shoreline environments as well as adjacent coastal waters [4]. Its components can include river deltas, coastal plains, wetlands, beaches and dunes, reefs, mangrove forests, lagoons, other coastal features. Moreover, another source defines coastal area as a junction between terrestrial ecosystems, marine ecosystems and ecosystems meet each other in a balance that is susceptible [5]. The transition between land and sea in the coastal region has formed a diverse ecosystem and is very productive; therefore, provide tremendous economic value to humans. In line with population growth and increased socio-economic development activities, the "value" of coastal areas continues to grow. With the unique potential and economic value was then faced with the highly threatened of coastal areas, it should be handled specially the coastal areas so that these areas can be managed in a sustainable manner.

The notion of carrying capacity or sustainability yield has become a basic criterion of sustainability. The concept of carrying capacity is rooted in a notion of "limits to growth". Ecosystems and populations have a limited capacity to cope with environmental stress; above a certain amount of stress there may be detrimental effects for the ecosystems [3]. Carrying capacity is defined as "the growth limits an area can accommodate without violating environmental capacity goals" [6]. In recognition of the diverse nature of carrying capacity as a concept, a variety of types of carrying capacity have been identified. Most of these fall into the following categories: physical, ecological, social, and economic. Among others, this paper focuses on physical carrying capacity. This is a measure of the spatial limitations of an area and is often expressed as the number of units that an area can physically accommodate, for example, the number of berths in a marina [7].

Ambon Bay, which is located in the eastern part of Indonesia, has a fairly rapid economic development. The increasing number of people who settled on the Ambon Bay as well as the number of tourists, both domestic and foreign, derive the increasing number activities that have an impact on the increasing need for land and natural resources, especially water resources to support the activities. In addition, land development also causes an increase in the pollution of Ambon Bay. This raise concerns on environmental degradation, particularly on Ambon Bay in which owns a variety of coral reefs, sea grass and mangrove areas, where these ecosystems need to be protected. Therefore, carrying capacity of the environment must be analyzed to prevent or decrease the environmental degradation, which caused by Ambon Bay development.

2. Methods

In order to do their activities human needs several aspects in their lives, including land, water, and supporting infrastructures, such as transportation infrastructure, waste management infrastructure, electrical infrastructures, and others. Human's needs may increase or decrease time to time, depends on the number of population. In the case of developing country, normally the cities experience increasing number of population. Thus, the needs of resources also increase in line with the increasing number of population. Several aspects of the human's needs may adjust according to the increasing number of demand. For example, due to the increasing number of population, number of waste generation is increasing. Thus, the waste management infrastructure needs to cope with this problem. It is possible that this infrastructure to be adjusted to meet people needs, the question may be only the technology improvement. This aspect of needs is called derived needs.

On the other hand, there are also several aspects of needs that are given. This means that, the supply of this type of human needs is nonadjustable to the increasing number of demand. Land and water are the example of the given needs. Even though up-scale development could be possible, in the end this type of development will find its maximum capacity. When there is no more land left to do any development, this means that land has meet its maximum carrying capacity. The same case also happens for clean water. Clean water is not adjustable to the people's increasing demand. When there is no more clean water left from the sources, this means that clean water has meet its maximum carrying capacity.

Based on the discussion above, it can be concluded that the supply amount of land and clean water are classified as given, nonadjustable, or not able or hardly to be improved. While, land and clean water is the main needs of human to live and to do their activities. Thus, these two criteria are chosen to be the threshold of environment to cope with the increasing number of population or increasing development. The time when clean water or land is not able to deal with the number of population or activities, it is a sign that the environment carrying capacity has been reached. Thus, this criterion is crucial to be applied, as the basis for the prevention of the other ecological risks to rise.

Additionally, the analysis method also refers to the Regulation of Environment Ministry Number 17 Year 2009 on the guidelines on determining the environment carrying capacity in spatial planning. Knowing the capacity of the natural environment and resources to support human activities does the identification of environment carrying capacity. Environment carrying capacity is divided into two components, which are supportive capacity and assimilative capacity. In this guideline, the environmental carrying capacity assessment is limited to the supplying capacity of the natural resources, particularly on the capability analysis of land and clean water. The capability analysis looks at how the supply meets the demand. As the natural resources capacity depends on the land and water capability, therefore the determination of environmental carrying capacity in this guideline is based on four approaches, namely:

- The capability of land to allocate people's activities.
- The capability of the potential clean water supply to fulfill people activities.
- Comparison between the supply and demand of land.
- Comparison between the supply and demand of clean water.

In order to know the land availability, analysing the land capability is the first step to it. By overlaying the slope map, soil map, erosion potential area map, and other disaster potential area map, the land capability map can be produced. In this kind of map, the land characteristic is identified and then divided into several different classes. Thus, this land classes have their own role to address the growing development. In this research, the land needs that are calculated are the land for settlements and its supporting facilities (e.g. housing area, health facilities, education facilities, office and trade facilities, tourism facilities).

To calculate land area needs for housing area, a standard from SNI 03-1733-2004 on Procedures for Urban Housing Environment Planning is used. The assumption is that one family consist of five members (i.e. father, mother, and three children. Following calculation is using the previously mentioned standard.

The main floor area	: (2 x 9,6) + (3 x 4,8) m^2 = 33,6 m^2
Floor area of service	: 50% x 33,6 m^2 = 16,8 m^2
Total Floor Area	: 50,4 m^2

Assuming the Building Coverage Ratio (BCR) is 50%, then the minimum parcel of housing area for families with five members is 100,8 m^2. Moreover, other supporting facilities have different area standards.

The calculation of water supply is based on the current water source discharge located in Ambon Bay area. While as the water requirement is calculated based on the number of population and the standard of clean water needs per person. This formula is used in the calculation of water needs:

$$D_A = N \times KHL_A$$

D_A	: Total clean water needs (litre per day)
N	: Number of population (people)
KHL_A	: Standard of clean water needs (litre per people per day)

In the Moluccas Province Spatial Plan 2013 – 2033, Ambon Bay is appointed as a Provincial Strategic Area from the aspect of environment support. It has a strong impact in the provincial scope in terms of its environment quality. Thus, Ambon Bay's development should be prioritized, and requires an immediate support of spatial planning and management. Ambon Bay has natural resources that need to be protected, such as coral reefs, sea grass beds and mangrove forests. However, the pollution level in this area is at a point that is very alarming. Several issues cause the pollution, such as land use shift from green areas into settlement areas that may lead to a huge level sedimentation and hardly-decomposed waste that is disposed on the water surfaces (e.g. rivers, drainage channels), and then carried into the Ambon Bay which makes marine ecosystems are threatened.

In addition, based on Ambon City Master Plan Year 2011 – 2031, Ambon Bay is appointed to be a conservation area of coral reef ecosystems. In addition to its role as a provincial strategic area from the aspect of environment support, Ambon Bay also has a role as a provincial strategic area from the aspect of economic support. Thus, it is also planned as a regional flagship fishery product and marine tourism. Those both roles might be contradictory, as economic activities usually lead to more development, while environment needs to be protected. Due to this condition, the coastal zone planning must facilitate the growing economic activities without harming the natural environment condition.

The area of coastal zone planning consists of 46 villages in five districts on the island of Ambon, which equals to 374,86 km^2 with a population of 52.354 people. The area consists of chosen sub-district that are adjacent to Ambon Bay or sub-district that are adjacent to a river that empties at the Ambon Bay.

3. Result and Discussion

3.1. The capability of land to allocate people's activity

Comparison between the land needs and land availability is done to see the ability of land to accommodate the settlement and its activities in the future. Before comparing those two, firstly land capability must be analyzed. The land capability analysis is conducted to see the capacity of the land based on existing criteria such as slope, soil type, and other factors. Factors that are used to analyze land capability in Ambon Bay area is the data of slope and soil type. In addition, based on the Ambon City Master Plan and the Moluccas Province Master Plan, there are several areas, which are classified as protected zone. Thus, this zone is not available to be developed. These factors are then processed spatially in ArcGIS.

3.2. Land carrying capacity

Land needs is divided into two scales, which are land needs for sub-district (namely kelurahan in Indonesian term) and district (namely kecamatan in Indonesian term). The calculation of sub-district scale land adequacy is conducted to see the comparison between the land availability and the land needs in sub-district unit. The land needs include land area for settlements and the supporting facilities to support activities on the sub-district unit. The calculation of this land needs refers to SNI 03-1733-2004 on procedures for housing planning in urban area. The standard used in order to calculate the housing area requirement is that, each household, which consists of 5 people, requires 100 m^2 as land to be designated as their settlement. Moreover, supporting facilities which supposed to serve in a sub-district unit is included to the calculation, such as educational facilities, medical facilities, religious facilities, cultural and recreational facilities, open spaces, parks and athletic fields. As for the calculation of land needs for district unit is done to get the number of the land area, which are used to serve the district scale facilities and tourism facilities. The needs for district scale facilities refers to SNI 03-1733-2004 on procedures for housing planning in urban area, while for the needs for tourism facilities refers to DG of Tourism, Post and Telecommunications No. 22/U/VI/ 1978 dated June 12, 1978.

Table 1. Projection of Population Number and Population Density in 2027 and 2037

No.	Sub-district	Area (km^2)	Population (people)		Population Density (people/km^2)	
			2027	2037	2027	2037
1	Teluk Ambon	93,68	81.188	125.916	867	1.344
2	T.A. Baguala	40,11	118.405	191.288	2.952	4.769
3	Sirimau	86,81	297.715	464.031	3.429	5.345
4	Nusaniwe	88,35	191.568	299.106	2.168	3.385
5	Leihitu Barat	65,91	13.387	14.027	203	213
	Total	374,86	702.263	1.094.369	9.620	15.057

Firstly, before we are able to calculate the land needs, calculating population projection is needed. Population projections up to 2037 carried out by the geometric projection model. The geometric projection model is chosen other than arithmetic and exponential projection model because it generates the smallest standard deviation value. Geometric projection model is performed by the following formula.

$$P_n = (1 + r)^n$$

Pn : number of tourists in year n
Po : the number of tourists in the base year
r : growth rate of tourists
n : number of intervals

Figure 1. Land Availability Calculation

Land area that can be developed minus the land area needed equals to land area availability. The land availability is divided into two categories, which are surplus and deficit. Surplus of land availability indicates that the sub-district still having an area to be further developed, while the deficit shows shortage of land area and it is impossible to carry out further development. Thus, from this analysis, it can be indicated that necessary anticipation in the increasing number of population in a sub-district with a future land deficit is needed. Moreover, future development can be done in the sub-district that is still having surplus of land in the future.

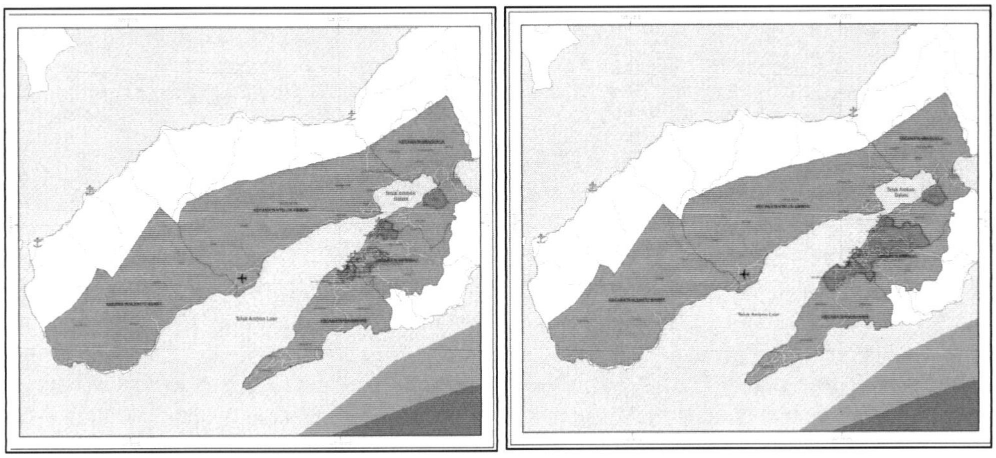

Figure 2. (a) Comparison Between Land Supply and Demand in 2027 (b) Comparison Between Land Supply and Demand in 2037

On the figure, it is shown that the unit of analysis is sub-district area. The figure above explains that the land availability is divided into two classifications, which are surplus (shown by the green color on the map) and deficit (shown by the red color on the map). If the land availability is surplus, it means that the available land is able to serve the land needs on the specific year. However, deficit of land availability means that the available land is no longer able to serve the land needs on the specific year. From the figure, it is shown that in 2027, there are number of sub-district which are having deficit land availability. Moreover, on the year 2037, more sub-districts are having the land availability status of deficit. Thus, it is concluded that the land carrying capacity in Ambon Bay, is decreasing time to time, which is shown by the increasing land deficit in the sub-district units.

3.3. Clean water carrying capacity

This analysis part shows the calculation of the water carrying capacity in the area, taking into account the supply and demand of water resources for the people who live in the area. Thus, it will be known whether the water resources in the area in a state of surplus or deficit.

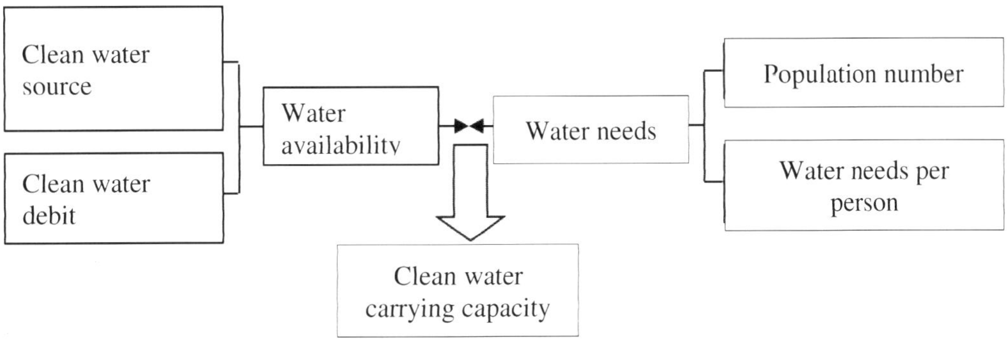

Figure 3. Comparison Between Supply and Demand of Clean Water

Water supply is calculated based on the water discharge produced from the water sources. There are several water sources, such as:

- Surface water, is water on the surface of the earth, such as river, lake, and others.
- Groundwater is water present beneath the earth's surface.
- Spring water, is water from a spring, which is a natural situation where water flows from aquifer to the earth's surface.

Water demand is calculated based on number of population and the standard water requirement per person. Based on the water supply standards from Directorate General of Human Settlements on Year 2000, there are two types of water demand, which are:

- Domestic water, is household water demand
- Non-domestic water is non-household water demand, such as commercial and industrial area.

The domestic and non-domestic water demand differs for every type of cities, ranging from metropolitan, big scale city, medium scale city, small city, and village.

Other than domestic and non-domestic water needs, to include water needs of every activity is necessary. In Ambon Bay, water needs of tourist and livestock activity are included to the water demand calculation. For tourist, it is assumed that their water needs is used for bathing, washing, and toilet activities, which is *mandi, cuci, kakus (MCK)* in Indonesian term. The standard requirement refers to the SNI 03-2399-2002 on the supply specification for MCK. Standards clean water demand for MCK is 45 liter/person/day. Moreover, an addition of 2 liter/person/day is required for tourist for drinking water purpose. Thus, the total need for clean water for tourist is 47 litre/person/day. In addition, the calculation of livestock water needs refers to SNI 19-6728.1-2002 on balancing water resources. Different types of livestock have different water needs. As a case study, data on Ambon Bay clean water carrying capacity analysis is presented.

In the Ambon Island, PDAM (Regional Clean Water Company) and another private company distribute the clean water. Clean water sources used by both of them come from spring water and groundwater. In total, the current spring water discharge reaches up to 20.380.800 liter/day and the groundwater discharge is 9.288.000 liter/day (Ambon City Clean Water Master Plan, 2012). After the clean water demand and the clean water supply data are obtained, the difference between them can be counted. If the difference shows a negative value, then it means that Ambon Bay is on a water deficit condition. Vice versa, if the value is positive, then it means that Ambon Bay is having a water surplus. In Table 2, it is shown that Ambon Bay is having water deficit starting 2017. Thus, it is necessary for Ambon Bay to search for more water resources. Otherwise, the number population and activities must be restricted to a number which water resource could handle. From this finding, it is concluded the water carrying capacity is not sufficient to serve current Ambon Bay population and activities. In the end this condition may lead to environment degradation.

Table 2. Comparison Between Supply and Demand of Clean Water in Ambon Bay

Year	Demand (liter/day)	Supply (liter/day)	Difference Between Supply and Demand (liter/day)
2017	116.853.543,64	29.668.800,00	-87.184.743,64
2022	144.603.896,85	29.668.800,00	-114.935.096,85
2027	179.164.581,53	29.668.800,00	-149.495.781,53
2032	222.268.647,82	29.668.800,00	-192.599.847,82
2037	276.094.020,51	29.668.800,00	-246.425.220,51

4. Conclusion

Environmental carrying capacity is a method of analysis, which is necessary to be done in order to see how the environment is able to cope with the area development. Since coastal zone owns high economical value, then it is vulnerable to uncontrolled development. Clean water and land area are the two main environment factors that are also the main needs for living. Thus, without the adequate number of clean water and land area, it is impossible for people to live and to do their activities. Moreover, if development is still going with adequate number of clean water and land, environmental degradation will be the case. This study analyses the land and clean water carrying capacity in Ambon Bay. From the land carrying capacity, it is shown that number of areas which are unable to meet the needs increase time to time. As for the clean water carrying capacity, it is shown that since the year of 2017 Ambon Bay is having a water deficit problem. From these two reasons, it can be said that the development in Ambon Bay must be adjusted in order to meet the balance between the environment carrying capacity and the development itself, particularly to anticipate possible problems due to the great scale of marine tourism development in this area.

From the case of Ambon Bay, it can be seen that the land utilization has not been spread out. Only several sub-districts that are experiencing land deficit up until 2037. As a solution, the land in other sub-district, which is still having a surplus quantity of land, could be more utilized. This can be done by equally distributing the number of population. As for clean water, the solution is to look for other clean water potential sources. Ambon Bay is a home of several wide rivers. It can be the case that not all rivers have been utilized. Thus, searching for more clean water potential might be the answer. However, in the end, after every population or development adjustment are done, but the environment carrying capacity has not been met, that is the time when number of population or development needs to be restricted. Otherwise, environment degradation will be the case. The goal of ICZM is to find the resolution of land use conflicts for the development and conservation of coastal and marine environment. Thus, by being aware of the environment carrying capacity and also to include the solutions into the ICZM, will contribute into the conservation of coastal and marine environment.

References

[1] National Security Council (DKN) 2016 Number of Islands in Indonesia

[2] Geospatial Information Agency (BIG) 2015 Importance of Geospatial Information to Organize Indonesian Sea

[3] Mayoka F 2009 Coastal Population *Carrying Capacity Analysis* (Alexandra: Mexa)

[4] United Nations Environment Programme (UNEP) 2010 Coastal Zone Definition and Geographic Coverage of the ICZM Protocol.

[5] Ortolano L 1984 Environmental Planning and Decision Making (New York: John Wiley and Sons)

[6] Beatley et al 2002 An Introduction to Coastal Zone Management *Coral Triangle Initiative on Coral Reefs, Fisheries, and Food Security* (Washington: Island Press Coral Triangle Initiative)

[7] MacLeod M and Cooper J A G 2005 Carrying Capacity in Coastal Areas *The Encyclopedia of Coastal Science* M.L. Schwartz, Springer (Dordrecht, The Netherlands) pp 226-227.

[8] National Standardization Agency (BSN) 2002. SNI 03-2399-2002 on The Supply Specification for bathing, washing, and toilet activities (MCK)

[9] National Standardization Agency (BSN) 2002 SNI 19-6728.1-2002 on Balancing Water Sources

Ecosystem Approach Studying Development of North Coast Jakarta

P Rahardjo
Department of Urban and Regional Planning
Faculty of Engineering
University of Tarumanegara, Indonesia

E-mail: parinor19@gmail.com

Abstract. The Inhabitant population of the city of Jakarta continues to increase, the extent of land is limited, and the increasing demand for housing and support facilities, to encourage the conversion of land in Jakarta, for example, the conversion of mangrove forests and reclamation on the north coast of Jakarta in a bid to increase the area of land for the construction of housing and business and commercial centre. Changes in land cover caused by changes in the ecosystem affect the North Beach neighbourhood of Jakarta. The purpose of this paper examines the impact of changes in land cover and environmental development of coastal areas of Jakarta. The qualitative research approach used. Analysis of changes in land uses the space, temporal analysis instruments, to study the change of land use in 1995, 2000, 2005, 2010, 2015. The scope of the study is limited to environmental issues, which include the natural and built environment of the Environment. The study results get, the development of the north coast city of Jakarta ignores ecosystems, such as coastal ecosystems, open green spaces, river ecosystems, resulting in a puddle of water at high tide.

1. Introduction

Jakarta has a diverse role as the capital of the State, business and commerce, and municipal services, promote economic growth, increased more than in the other cities in Indonesia and drivers of migration, resulting in increased population. According to National Statistic Bureau In 1980 the population of the town amounted to 6.503 million inhabitants and a population census in 2010 increased to 9.607 million. The Jakarta Capital City has a land area of 661.52 km2 and sea area of 6977.5 km2 and recorded ± 110 islands scattered in the Thousand Islands, with the growing population, and breadth of the land is limited encourage the conversion of mangroves and swamps (wetlands) for settlement and region business and commerce, as well as the reclamation along the coast to add to the breadth of the land. Waters of Jakarta Bay with a long coastline reaches 50.5 km and reached 39,238.5 ha of water area[1]. The addition of the land area of Jakarta town reclamation area includes the addition of breadth Taman Impian Jaya Ancol and Pantai Mutiara carried out by private companies, while the conversion of mangrove forests and wetlands (swamps) occurs in Kapuk area become residential areas, business, and recreation. The addition of the land area of the town of

Jakarta is reclaimed to create artificial islands. With the pattern follows the river which empties into the Bay of Jakarta. With the 13 estuaries, then at least will form 17 islands reclaimed.

According to Law No. 1 of 2014 on the amendment of Law No. 27, article 1, paragraph (2) of Coastal is the transition between terrestrial and marine ecosystems are affected by changes in land and sea, whereas in paragraph (21) controls the coastal border, mainland along the banks of a width proportional to the shape and physical condition of the beach, a minimum of one hundred (100) meters from the highest tide landward.

Climate change can be caused by processes internally or eexternal strength, or the presence of anthropogenic changes that are constantly in the composition of the atmosphere tau in land use, [2]. The definition of climate change at the top shows how the climate system change, adaptation is the adjustment effort lifestyle and ingredients to climate change, while mitigation is basically mitigation efforts to prevent climate change is getting worse. Mitigation can also be interpreted as a variety of active measures to prevent / slow down climate change.

According to Macintosh and Ashton (2004), "Mangrove greenbelts can provide significant coastal protection from erosion and should be established along erosion-prone coastlines and riverbanks and in areas which experience significant damage from typhoons, tidal surges, cyclones, and geomorphic erosion (Macintosh and Ashton 2004). Greenbelts should be a minimum of 100 m, but preferably up to 500 m or 1 km (advocated in Mekong Delta, which is subject to typhoons) at the open coast and 30-50 m along riverbanks and lagoons, and >10 m on islands, creeks, and channels (Macintosh and Ashton 2002; Macintosh and Ashton 2004). In the Red River Delta in Vietnam, engineers estimate that an earthen sea dyke with rock facing will last about five years before it requires repair from wave damage, whereas the same sea dyke with a 100 m wide protective mangrove belt will last up to 50 years! [3]

"Strategies for adaptation to sea level rise, there are three main categories: retreat, accommodate and protect. In the Pacific islands, this could mean leaving low-area islands or, if available, on the same island, leaving the area of low and move to higher ground. Accommodation suggests changes in land use when the water level rises, such as improving the building or changing to crops that are more tolerant to salt. According to Nicholls and Leatherman (1996) often use a protection embankment built to keep the sea from the shoreline could use hard structures such as seawalls and breakwaters, while Watson et al. (1996) chose the use of vegetation to stabilize the beach. Adding sand and rocks to the beach there or raise the height of the coastal village location may be useful in some places (Nicholls and Leatherman, 1996), according to Watson et al (1996) the precautionary approach, including enforcement of land use regulations and building regulations, increases the range of buildings on the beach [4]

According to the researchers the United States, "the best thing to protect the property from sea level rise is a sand dune, or swamp mangroves, or coral reefs, or seagrass [5] and Bruttomesso (2001) "Identifies three types of activity to waterfronts, normally require: 1)'recomposition': giving a common unitary sense to the different parts, both physical and functional, of the waterfront; 2) 'regeneration': revitalizing urban areas which can be of considerable size and often centrally located; and 3) 'recovery': the restructuring and restoration of existing buildings and structures. Typically, these are linked to initiatives aiming to 're-join' the city and the waterfront physically and functionally [6].

Procurement of materials to improve the land by creating artificial islands and reclaimed northern coast of Jakarta City posed a problem, especially when the need to dredge the seabed. This will lead to the occurrence of landslides in the area of the seabed around the location of the backfill and threatened sinking small islands around the dredging. Reclamation can have adverse effects on the marine environment, especially coastal ecosystems, such as coral reefs, mangroves, and biota including humans. The purpose of this study examines the development of the northern coast of Jakarta City with the ecosystem approach. The scope of this study is limited to environmental issues, which include the natural environment and the built environment

IOP Conf. Series: Earth and Environmental Science **79** (2017) 012009 doi:10.1088/1755-1315/79/1/012009

2. Methods

The research method uses a qualitative approach, analytical instruments using descriptive methods, to assess the changes in land use analysis of space, temporal, using satellite imagery data from 1995, 2000, 2005, 2010, 2015. Data was collected using a purposive sampling method to interesting phenomena and influence the research by conducting field observations. This research took place at 6^0 07 'South latitude until $6^0$12' and $106^0$72 'East longitude to $106^0$82'. An administrative research site located in the Penjaringan District and Pademangan District, the area explained in the picture.

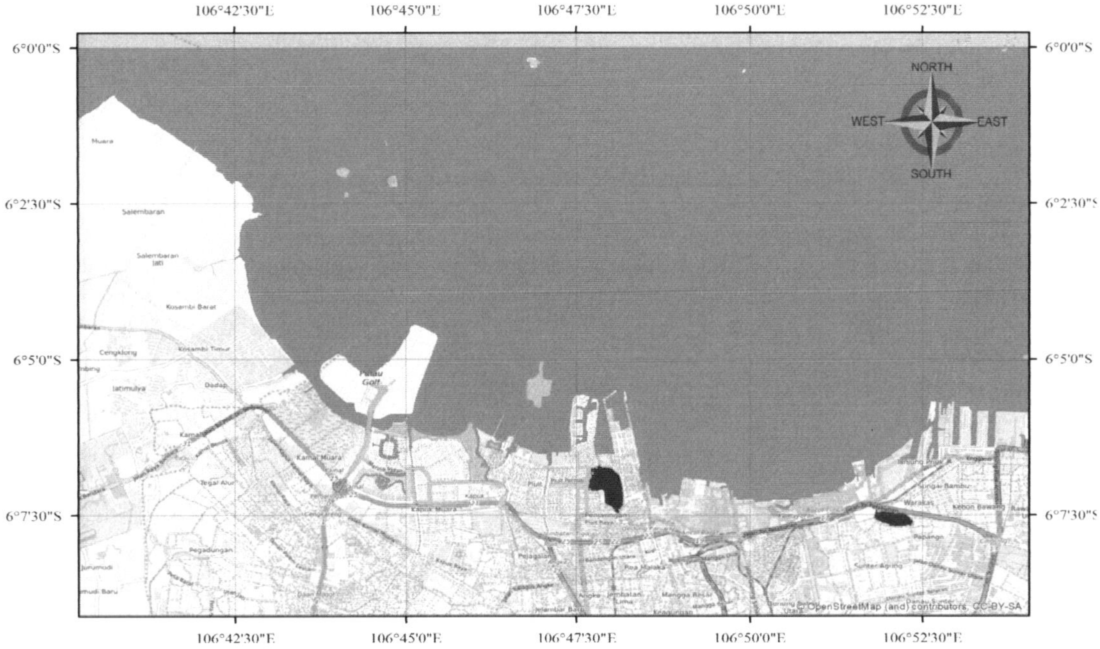

Figures 1. Research Location
Source: ArcGIS 10.2

3. Result

3.1 Natural Environment

Northern coastal ecosystem of Jakarta, among others: 13 rivers flowing through Jakarta is Kali Mookevart (on the side of Jalan Daan, mostly across the West of Jakarta), Kali Angke (across the South And West of Jakarta), Pesangrahan (across the south and west), and Grogol (South And West), Krukut (South, Central), Kali Baru Barat (Jakarta of Center), Ciliwung (South, East, And Central), Kali Baru Timur (Central, South and East), Cipinang (East and Middle), Sunter (Middle East), Buaran (East), Jati Kramat (East), and Kali Cakung (East). Kali or the rivers that eventually empties into the Bay of Jakarta. There is a direct way to the sea, there is also a need to connect with the West Flood Canal, the East Flood Canal, Cengkareng Drain, and Cakung Drain.

Angke Kapuk Forest Preserve is a formal conservation area owned by Jakarta in the mainland, is Penjaringan subdistrict, North Jakarta, extending from Angke river estuary in the east to the border of Jakarta to Banten in the west. Based on the Decree of the Directorate General of Forest Inventory and Land Use. No.08 / KPPS / VII 4/94 protected forest has an area of 44.76 ha and designated as a conservation area, with Mangrove trees as dominant vegetation.

The existence of the types of wildlife is intimately associated with the type of vegetation in the region. An area of the beach with mangrove vegetation types and special habitat types, so that wildlife that is found in protected forests are species of mammals and aves. In addition to the type of long-tailed macaque (Macaca fascicularis) discovered the protected species legislation or migratory, among others: kingfisher blue (Alcedo coerulescens), Pecuk snake asia (Anhinga melanogaster), herons small (Egretta garzetta) ibis rokoroko (Plegadis falcinellus) and turns mustache (Chilidonias hybridus). Angke Kapuk protected forest as well as a conservation area is one potential object for the purpose of travel / ecotourism. In order to improve the development of tourism sector in Indonesia is also supporting education and research system.

Muara Angke Wildlife Reserve set in 1998 by the Government through the Minister of Forestry and Plantations No. 097 / kept-II / 1998 as a Muara Angke Wildlife Reserve with a total area of 25.02 ha., This area includes the Village of Kapuk Muara, Penjaringan, North Jakarta. The area is adjacent to the Housing Muara Karang and fishing settlement Muara Angke Angke restricted. On the north side, the Angke-Kapuk protected the forest that is within the authority of the Department of Forestry in Jakarta. Vegetation in Muara Angke Wildlife Reserve is a mangrove forest, with a fairly high species diversity. Found the swamp overgrown with grass, reed (Saccharum spontaneum) and water hyacinth (Eichchornia crassipes), while the fauna encountered in the form of lizards, monkeys, and various species of birds. Overall the mangrove forests around Pantai Indah Kapuk, among others: the protected forest (mangrove) 49.25 ha, Muara Angke Nature Reserve (Mangrove) 21,45 ha, 91.37 ha Forest Tourism, Forestry nursery 10.47 ha, 29.05 ha Cengkareng Drain, Transmission Line PLN 29.99 ha. Sedyatmo toll road and green track 91,37ha (Hadi Alokodra, 1996).

Location PIK originally inhabited mangrove forest fauna, among other things: monkeys, snakes, crocodiles. Coastal marsh as a puddle, converted as residential and business district (an area of 831 ha), is estimated as the cause of the flood that caused the motorway Sedyatmo as well as the surrounding residential areas submerged when it rains (http://www.menlh.go.id).

Taman Impian Jaya Ancol and inaugurated on June 25, 1967, was opened to the public. Widely used initially only 137 ha land area of 552 ha available. Taman Impian Jaya Ancol is situated between the international port of Tanjung Priok and between domestic ports, Sunda Kelapa. Built of swamp and scrub, marshy ground in 1960 is still full of bushes. Through PP 5111960, Ancol belongs to the central government and will be used as an industrial area (Ancol Barat), housing (Ancol West and East), and the designation as the East Ancol recreational area (http://www.jakarta.go.id/encyclopedia)

3.2 Built Environment

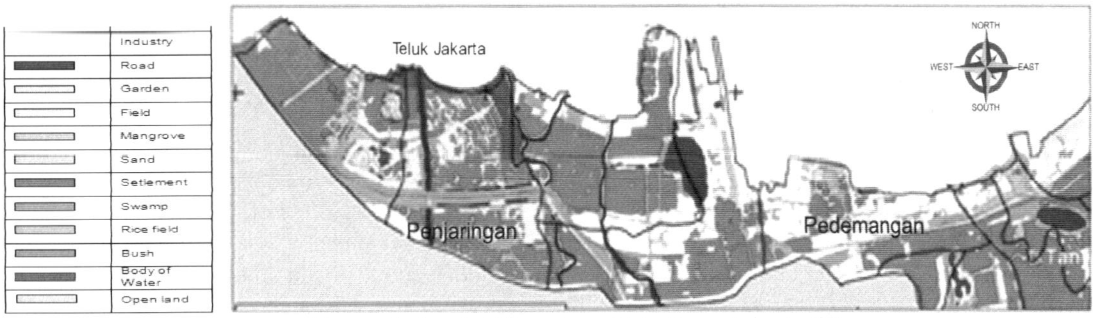

Figure 2. Spatial Planning of Northern Jakarta City
Source: Angraeni, 2012

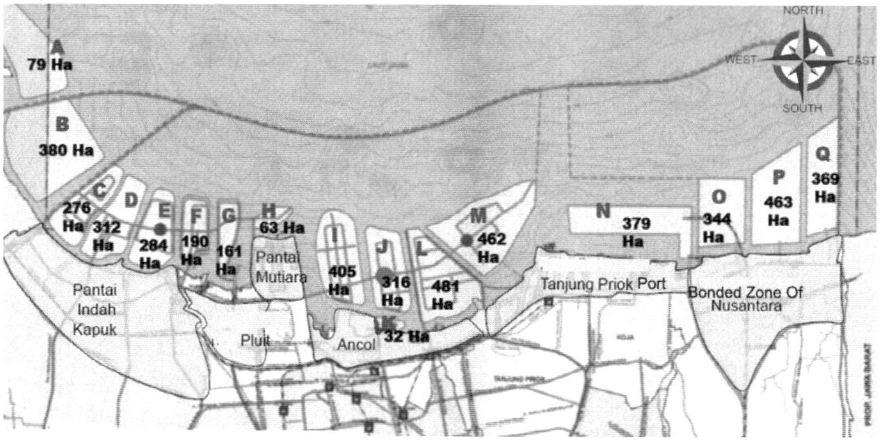

Figure 3. Reclamation Plan Region Pantai Indah Kapuk
Source: Pemprov DKI, 2014

Reclamation construction of North Coast Jakarta, Pantai Mutiara conducted since 1985, this development until 1999. First the reclamation in Jakarta with a total area of 110 ha. PT Pembangunan Jaya Ancol reclamation of the north side of the industrial area and recreation around the year 1981. Ten years later, turn Pantai Indah Kapuk conversion of mangrove forests, swamps, ponds for the luxury residential area which is now known as Pantai Indah Kapuk. In 1995, after the reclamation is used for industry, namely Bonded Marunda http://nationalgeographic.co.id/ 15:09, 12 September 2016).

After converting the coastal region Acol, extensive additions Beach Ancol Dreamland carried back by reclaiming the beach. As the expansion of recreational areas as well as for industrial and apartments, this is also done at the Mutiara Beach neighborhood and Pluit.

Ocean Ecopark is located in Taman Impian Jaya Ancol has a land area of nearly 34 hectares of the area of the resulting from the conversion Golf Course Ancol Become recreational facilities with the values of education (edutainment) and (adventure) approaches adventure for green lifestyle, into an open space with a composition of open space green and blue. Blue open space is a channel with a linear curve shape (organic) with a length of 400 meters with +/- 6 meters wide and the middle forming ring-shaped island, the which is used as a place of educational tours http://swa.co.id/swa/trends/

Toll Road Airport (Prof Sedyatmo) on adding lanes to the elevation of the road surface is higher in comparison with the toll lanes other existing, this is done to facilitate the flow of traffic, because the toll lanes have never experienced submerged due to heavy rain, elevation of the face is done also on local roads area of Pantai Indah Kapuk and Pluit. "The volume of rainwater in a measuring Salama 24 hours (starting clock 07OO GMT until at 0600 GMT, in the Territory of Jakarta, North Jakarta area, is the location with the largest volume of water that occurs on the 9-10 February 2015, with a total volume of water in North Jakarta on February 10, 2015 reached 36,670,726 m3, with a range of 100-400 mm rainfall / day (Kadarsah, et al). "Penjaringan and Pademangan sub-district is an area that is often experienced as a result of the floods and waterlogging caused by rain runoff. Cause ponding of water is partly due to the extent of land-proofed compared with green open spaces. This is due to the widespread development of residential and business district and trade. From 2010 Until 2030 the vast puddles on Penjaringan and district Pademangan will reach 1045 ha (settlements) and 463 ha (Industry), this condition occurs when not in use change in both these districts [7].

Currently in North Jakarta area, there are 26 locations prone to tidal flooding, among others: Penjaringan, Pluit, Kamal estuary, Kapuk estuary, Tanjung Priok, Kalibaru, Ancol, Pademangan, Marunda, Koja, Lagoa, Sunter Karya South, Papanggo, Sunter Agung, Warakas, Kebon Bawang,

Sungai Bambu, Jampea, Kramat Jaya, Kelapa Gading, KBN Cakung, Sunter Jaya, and Yos Sudarso (http://bnpb.go.id/berita).

4. Discussion

In the reclamation plan and manufacture, the artificial island would require the formation material that could reach millions m3. Large volumes needed considering the plan of the artificial island will reach an area of 2700 ha. With this comprehensive revitalization of artificial islands and the northern coastal city of Jakarta will form a new town requires infrastructure support Supplies electrical energy, water, waste management, transport, road network, green open spaces and public facilities, and social.

To expand the land in the northern coastal city of Jakarta to reclamation by using materials obtained from the surrounding waters of Jakarta Bay by way of dredging the seabed, seabed mining for the purpose of hoarding with a large area can damage coral reefs in the Gulf of Jakarta and surrounding areas, and can submerge small islands at the mine site as a result of the loss of coral reefs and landslides on the seabed in the dredger, because "the coral reefs in addition to beauty is also a convenient place for marine life and have the physical capability reduces the pressure waves of the sea [5].

Artificial islands reclaimed should be able to produce an ecologically built environment, the development of new islands as a city should be in accordance with its carrying capacity. "The carrying capacity is defined as the maximum continuous load that can be imposed by humans on the environment. (The Independent Commission on Population and Quality of Life, 2000)."

Of the seven urban ecosystems on an artificial island, just agricultural land cannot be implemented, while six other ecosystems can be applied.

Urban ecosystems consist of several subsystems interrelated, each representing a complex system and interplay with each other at the level of structure and function differently. This the ecosystem can be understood and studied in various sizes. Is it a pond, river, lake, or a piece of the pond, forest, or landscape. Even the laboratory is a unit of the ecosystem can be observed. As long as the essential components and interact to form a partnership to achieve a functional stability, even if only a short time, the force can be considered as an ecosystem. Urban ecosystems are clearly different from natural ecosystems, in particular on the features that contain a variety of artificial facilities.

Urban Ecosystem consists of seven elements identified as 1) street trees; 2) lawn & garden 3) the urban forest 4) agricultural land 5) wetlands 6) lake / sea, and 7) rivers (Bolund and Sven 1999). Of the seven urban ecosystems on an artificial island, just agricultural land cannot be implemented, while six other ecosystems can be applied. At 17 artificial islands being developed will generate 13 channels which is a continuation of the 13 rivers in Jakarta and upstream Jakarta Bay, otherwise it will form a channel with a channel width of 200 meters that separates the coastal plain north of the city of Jakarta with 17 artificial islands. Around the edge of the canal, the author of the category as a wetland of arable Mangrove. Mangrove planters can also be done on an artificial island beach overlooking the open sea. Mangrove plants around the artificial island will form a green belt that has a function as a barrier waves and attempt to restore the coastal ecosystem strengthen the structure of the artificial islands and barrier of wave, in addition to Mangrove also must cultivate coral reefs, the breeding of coral reefs, in addition, to withstand waves can also serve as a residence of marine life, area of land which formed naturally, slowly but surely made island will have the same ecosystem with the urban ecosystem

Ideally, the whole artificial islands are equipped with urban forests, to control the micro law that lowers the temperature of the air on artificial islands. The Urban forest is a collection of trees with different species of trees with dense plantings, around the entire city, with a continuous pattern and form connecting all the ecosystems on the artificial island. Besides reducing micro air temperature, the plant can reduce the kinetic energy of rainwater use, until the pressure of rainwater on the surface to be reduced, because the rainwater pressure eased after touching the trees and other vegetation. The rainwater that hits the ground is channelled into the ground through the trunk and the rest flowed on the surface soil. Besides works to increase the absorption of rainwater, and reduce the pressure on the surface of the soil, vegetation has other functions, among others, can reduce air temperature due to the radiant heat of the sun, absorbed by trees and vegetation, trees can also reduce wind pressure, thus reducing the air temperature around the plant grows. The existence and preservation of vegetation in a place, then the quality of life in the neighbourhood of the vegetation is to be good. Vegetation has a role in the ecosystem, among others: As a patron, binding energy for the entire ecosystem, and as a source of mineral nutrients.

Vegetation actually the most crucial species in the ecosystem because it has a role as a producer in the food chain system. Breadth of green open space, vegetation density impacts the ability of the soil to absorb rainwater (permeability), thus increasing infiltration capacity, so as to control the flow of surface water (runoff), so that the process of the hydrological cycle can occur that result in groundwater levels swell, which in turn intrusion sea water is inhibited.

The layout of open green space in this artificial island could be centralized or spread across multiple locations. "Green open space is an elongated area / line and / or in groups as a place to grow crops, whether naturally grown or planted (Act No. 26 of 2007, Article 1)." Then in chapter 29, the green open spaces can be public open spaces and private spaces, (verse 1). Spacious green open space in the urban area of at least 30 (thirty) percent of the area of the town, (paragraph 2). Spacious green open spaces in urban public least 20 (twenty) percent of the total area of the town. The condition of at least 30% area RTH expected of artificial islands. Ecosystem-made island in the form of an urban forest or a body of water, such as lakes or ponds may be stabilizing, with the function of maintaining the balance of hydrology and the availability of clean air, lowering micro air temperature, and as a place of recreation that is accessible to the entire population of the town and surrounding areas, resulting in interaction between people who are on an artificial island and locals outside of the island. The function of the ecosystem is the earth's ability to sustain life for a long period of time [8], other than that of ecosystems, according to Sarukhan (2005), provide Four services provided by ecosystems (table 1)

Table 1. Ecosystem Services

Services	The benefits
Provisioning	Food ingredients Clean water supply
Culture	Recreation Education And Knowledge Spiritual
Regulation	Resisting Abrasion Withstand Wave Lowering The Temperature Of Micro Air
Supporting	Biodiversity Nutrient Cycling Strengthening The Structure Of The Soil

On the development of the town in this artificial island should not be any exclusive island with certain class citizens. As in the natural surroundings of biodiversity will be better in comparison with monocultures that are sensitive to external threats the species concerned. In social life, diversity in terms of ethnicity (ethnic), culture, and social class will provide beauty and harmony when it is going well. In environmental science, there are five interrelated elements, which is building a life and can be used on the social system, the five elements, diversity, harmony, the interaction, interdependence, and sustainability. By the very nature of life in the world of interdependence is something absolute, human beings need each other while running their lives.

Reclamation by creating an artificial island will result in settlement of fishermen in the estuary angle fish market and harbour of fishing vessels in the back, it should be planned well how out access fishing boats, and how to access the resident fishing settlement to a new island. This needs the attention of the local government should not be a gap in treatment between resident fishing settlement with town residents on an artificial island later.

Sea-level rise due to climate change in urban development in artificial islands must be taken into consideration in spatial planning. There are four roles of spatial planning in disaster risk reduction [9] as follows:

1. Limiting the development of disaster-prone areas are dangerous.
2. Allocate land use different settings for disaster-prone areas.
3. The land-use plan or zoning has legally binding.
4. Modify the intensity and frequency of hazards.

Based on the four roles of spatial land-use planning should allow the restriction of the use of space, especially in areas that are sensitive to the threat of sea level rise, especially on the beach, using the equivalent line the river, as a green corridor to minimize losses due to the brunt of the waves.

The development of artificial islands, with the level of 2700 hectares and the revitalization of the old beach area of 2500 hectares, will be a new town development held private parties, namely developers and involve more than one developer. Master Plan prepared by the developer must constitute a single entity, which is the result of a compromise between the developer and the government. The Master Plan is the result of a compromise is used as a reference in the development of the island, so there is no overlap, and as an instrument of coordination of actions in the field. "Urban areas that have major non-agricultural activities with the regulatory function of the area as a place of urban settlements, concentration and distribution of government services, social services, and economic activities (Law No. 26 of 2007, chapter 1, verse 25)." Based on the formulation of a new town as a whole should have a structure that can support the activities of its citizens as well as support all existing activities. "The structure is the arrangement of residential space and network infrastructure, and tools that support the social economy in a society that a hierarchical structure. (Law No. 26 of 2007, article 1, paragraph 3)." Thus, the development of new towns includes several aspects, among others: the number of houses being built, infrastructure that includes public facilities, social facilities, and access to green open space (RTH), transportation networks, water treatment plants and solid waste.

The development of new towns must maintain a harmonious environment, integrated planning and can protect the region has unique characteristics, and the culture that needs to be protected planning so that the spatial development of new cities should consider the uniqueness and potential of natural landscape or areas that have unique value and culture, it must be supported by the spatial patterns to safeguard and maintain the uniqueness in this region where the new town was developed, as mandated by Law No. 26 of 2007. "Spatial distribution pattern allocation is in an area that includes spatial allocation and protection function space allocation for the function of culture (Article 1, paragraph 4) ", whereas, in article 1, paragraph 20, the region is the region whose main function is protection or cultivation. In the reclamation area in Penjaringan there Mangrove Forest Reserve area, and Wildlife must be protected.

Construction of a new town on an artificial island ideally has a large infiltration capacity, this condition can be achieved by building a low ratio ranged from 30% -40% so vast green open space larger than built-up areas, development of artificial islands covering an area of 2,700 ha, plus the revitalization of the old coast around the artificial islands makes the construction of an artificial island as a new town. With the ecosystem approach is expected to materialize an ecological town, a town that can repair and restore coastal ecosystems better, and the town that is friendly to citizens of the town.

5. Conclusion

Beaches in Jakarta Bay has a coastal ecosystem rich in marine life is diverse, reclamation to form artificial islands by bringing sand dredging the seabed may damage coastal ecosystems, among others: the damage and loss of marine organisms, such as coral reefs and cause abrasion finally submerge small islands on the location of the sand, other than that of mangrove forest ecosystems and wildlife in Muara Angke danger of being corrupted and can change the biota habitat is in it remembering the position of mangrove forests, and wildlife habitat change which was originally facing the sea into the back of the artificial islands. The ecosystem approach to the development of the city on an artificial island and the revitalization of the northern coast of the city can be as an alternative to improve the ecosystem of the north coast City of Jakarta, with shape of planting Mangrove as a green belt on the island Artificial, the urban forest and body of water, such as lakes or ponds on the artificial island.

The new town on artificial islands and the revitalization of the North Jakarta coastal cities will be occupied by residents of all ethnic groups (tribes), and social class, so the development of the city in this artificial island to be able to open up access to all social classes, and do not build up area exclusively for groups and particular classes, thus generating the harmonization of social systems in the new town of artificial islands and in the coastal plain old.

Acknowledgements
The chairman and students of the Department of Urban and Regional Planning, Faculty of Engineering of the University, which supports research. Thanks also to the informant who helped during the interview to get the data.

References

[1] Rositasari 2011 Karaktreistik Komunitas Foraminifera Di Perairan Teluk Jakarta *Jurnal Ilmu dan Teknologi Kelautan Tropis* **3** p 100-111
[2] Sutamihardja, Mari E M 2011 *Climate Change* (Bogor: Yayasan Pasir Luhur Bogor)
[3] Elizabeth M, Rodney V S 2006 Managing Mangroves for Resilience to Climate Change (Switzerland: IUCN)
[4] Kristie L E, Nancy D L, Carlos C 2006 Climate variability and change and their health effects in small island states : information for adaptation planning in the health sector *Environ Jealth Perspect* **114**
[5] Tim R 2013 Natural Barriers Protect Our Coasts Best, Say Researcher *Climate central*
[6] Smith H, Ferrari M S G 2012 *Waterfront Regeneration Experiences in City-building* (New York: Routledge)
[7] Anggraini N, Trisakti, Soesilo 2012 Pemanfaatan Data Satellite Untuk Analisis Potensi Genangan dan Dampak Kerusakan Akibat Kenaikan Muka Air Laut (Application of Sattelite Data to Analisys Inundation Potential and The Impact of Sea Level Rise) *Jurnal Penginderaan Jauh* **9** 149
[8] Alberti M 2005 *The effect of Urban Patterns on Ecosystem Function* International Regional Science Review **28** 2: 168–192
[9] Heri S, Abbas R, Ian B 2010 Studying Spatial Plan in Coastal Urban Cities – facing global threat adapting to local condition *FIG Congress 2010 Facing the Challenges Building the Capacity*
[10] Arsyad S 2010 *Konservasi Tanah dan Air* (Bogor : IPB Press)
[11] Badan Nasional Penanggulan Bencana (BNPB). (2016). Berita. http://bnpb.go.id/berita). 18 September, 6:25 PM.
[12] Girardet H, Mendonqa M 2009 *A Renewable World* (London: Green book Ltd.)
[13] Haughton G 1997 Developing Sustainable Urban development models *Cities* **14**
[14] Kadarsah et al. 2015 Kajian Curah Hujan Tinggi 9-10 February 2015 di DKI Jakarta (Jakarta: BMKG)
[15] Newman P, Jenning I 2008 *Cities as Sustainable Ecosystem* (Washington.D.C: Island Press)
[16] Soeryani M, Rofiq A, Rozy M 1987 *Lingkungan: Sumber Daya Alam dan Kependudukan dalam Pembangunan* (Jakarta: Penerbit Universitas Indonesia)
[17] Undang-Undang Republik Indonesia Nomor 1 Tahun 2014
[18] Tentang Perubahan Atas Undang-Undang Nomor 27 Tahun 2007 Tentang Pengelolaan Wilayah Pesisir Dan Pulau-Pulau Kecil. http://bk.menlh.go.id
[19] Undang-Undang Republik Indonesia Nomor 27 TAHUN 2007, Tentang Pengelolaan Wilayah

Pesisir Dan Pulau-Pulau Kecil. http://bk.menlh.go.id
[20] Geografis Jakarta. http://www.jakarta.go.id/v2/news/category/geografis-jakarta. 12 Sept 2-16, 23:34 PM.
[21] Sarukhan J, Whyte A 2005 *Ecosystems And Human Well-Being* (Washington DC: World Resources Institute)

IOP Conf. Series: Earth and Environmental Science **79** (2017) 012010 doi:10.1088/1755-1315/79/1/012010

Cellular Automata Modelling in Predicting the Development of Settlement Areas, A Case Study in The Eastern District of Pontianak Waterfront City

E Nurhidayati[1], I Buchori[2], Mussadun[3], T R Fariz[4]
[1,2,3]Diponegoro University, Jl. Prof. H. Soedarto Tembalang, Semarang 50275
[4]Semarang State University, Sekaran Gunung Pati Semarang 50229
Indonesia

E-mail: elmartptk@gmail.com

Abstract. Pontianak waterfront city as water-based urban has the potential of water resources, socio-economic, cultural, tourism and riverine settlements. Settlements areas in the eastern district of Pontianak waterfront city is located in the triangle of Kapuas river and Landak river. This study uses quantitative-GIS methods that integrates binary logistic regression and Cellular Automata-Markov models. The data used in this study such as satellite imagery Quickbird 2003, Ikonos 2008 and elevation contour interval 1 meter. This study aims to discover the settlement land use changes in 2003-2014 and to predict the settlements areas in 2020. This study results the accuracy in predicting of changes in settlements areas shows overall accuracy (79.74%) and the highest kappa index (0.55). The prediction results show that settlement areas (481.98 Ha) in 2020 and the increasingly of settlement areas (6.80 Ha/year) in 2014-2020. The development of settlement areas in 2020 shows the highest land expansion in Parit Mayor Village. The results of regression coefficient value (0) of flooding variable, so flooding did not influence to the development of settlement areas in the eastern district of Pontianak because the building's adaptation of rumah panggung's settlements was very good which have adjusted to the height of tidal flood.

1. Introduction

Cellular automata model is generally used to predict land development [1,2] which is a historical process depending on where the development was done in the past that might influence the future through local interactions among land parcels as studied by Wu and Webster 1998 [3]. explains that some unexpected features can possibly emerge during the simulation by properly defining transition rules. Within the GIS concept framework, spatial analysis focused more on investigating patterns, attributes, and regional overview using modeling to improve understanding and prediction [4]. In order to discover the relevant variables used in this research, we can refer to the spatial and ecological approach [5] or what is called as a complex regional approach combined with the cellular automata model.

In determining the transition rules as well as the determining of pull and push factors of the settlements development which can be determined by the causes of an area can develop. According to Hoyt in Syahar about the sectorization tendency theory [6], that the areas that have the potential to grow rapidly (High Quality Areas) there are 10 locations. But that is associated with this research theme was the settlements tend to thrive in the transportation / towards the centre of the trade, settlements tend to thrive on the existing transport lines and settlements tend to develop in the same direction for long periods.

In connection with the exposure to the above, this study use several variables as push factors of the settlements development such as; (1) Distance to the main road; (2) Distance to the non-main road (collector road); (3) Areas of potential flooding; (4) Distance to the Keraton Kadariyah, tourism and government centre; (5) Distance to the central business district and services; (6) The distance to facilities such as hospitals and universities; (7) The distance to utilities such as piers and bus terminals; and (8) Distance to the existing settlements.

2. Methods

This research is located in the eastern district of Pontianak, Pontianak City, West Kalimantan Province. This research object is a settlement which located in the waterfront areas of the Pontianak, cultural heritage areas (Masjid Jami' Sultan Syarif Abdurrahman Alkadrie and Keraton Kadariyah Pontianak), and the history of first settlement areas in Pontianak namely Kampung Beting which is the forerunner to the birth of Pontianak [7].

Based on Pontianak Government regulation policies in Spatial Planning which is oriented Pontianak as Waterfront City (A Critical Assessment Act No. 24/1992), describes Pontianak is one of region in province which is an autonomous region that has a typical water town. Settlement areas in this study refers of Law No. 4/1992 on the housing and settlement, settlements are part of the environment outside the protected areas, either in the form of urban and rural areas that serves as a living environment or settlement environment and the activities which support life and livelihood.

The data used in this study such as:

- Satellite imagery Quickbird 2003 from Google Earth
- Satellite imagery IKONOS 2008 from BAPPEDA Pontianak
- Elevation contour interval 1 meter from BAPPEDA Pontianak

This study using satellite imagery from Google Earth downloads, given the difficulty of high-resolution imagery for the study area. In the study Yuanita et al (2013) showed Quickbird image obtained from Google Earth has a low RMSe values in the relatively flat compared to areas that are relatively high, while the study area in this study is the relatively flat region as well.

Satellite imagery is then interpreted in order to obtain landuse cover information. Classification used the classification of Danoedoro (2012) & USGS (2015) which have been modified. Basic modification of the classification is to look at the goals and characteristics of the study area, the modification is done by dividing the cover into three classes, namely settlement, non-settlement can be converted, and non-settlement cannot be converted. Non-settlements can be converted could include landuse cover is vacant land and vegetation areas, while the non-settlement cannot be converted such as land cover in the form of a body of water plus the portion of land cover awakened namely industry and building.

Distance-related variables in this study using analysis Euclidean Distance, while variable potential flood areas constructed of DEM (Digital Elevation Model) created from the data contour interval of 1 meter. According to data from tidal Kapuas River of Meteorology and Climatology of Pontianak (2013) highest tide is 1.9 meters. Therefore, it can be assumed that the terrain with a height of 1.9 meters below is the potential flooding areas.

Moreover, the research also adds the variable density of the building. The building in question such as settlements, buildings and industry. This variable represents one of the factors inhibiting the development of settlement areas in the Eastern district of Pontianak namely fire disaster. Areas with dense buildings have a high vulnerability to fire risk disaster, in addition to density variable of buildings (Hoyt in Syahar, 2012) define settlement areas tends to develop in open area to further development of open country and there is no physical barrier means. Density variable of buildings constructed using Kernell Density analysis.

Geomorphological diversity and intensity of the settlements is the basis in determining changes in the settlement areas, so that each region has waterfront and coastal city characteristics are different, depends on the existing geological processes. A critical factor in the settlement areas is a distance from the water bodies. In modelling settlement patterns is needed to assess the effects of flooding in settlements on the riverbanks, it is due to human behaviour in managing water resources because they have the management capacity to water bodies [7, 8, 9, 10]

In making the prediction model settlement development, this study uses binary logistic regression integration with Cellular Automata-Markov models. This model is used by Wijaya and Susilo (2013) in predicting the development of non-built-up areas in Salatiga within overall accuracy (78.20%) and kappa index (0.48). This research differs from [11], this study did not consider the number of cells to be predictable change.

Cellular Automata-Markov model uses raster based data, therefore all the data used in this study is converted in raster form with a size of 10 meters x 10 meters. Data in 2003 and 2008 will be used as the basis of predicted development of settlement areas in 2014, the result of the prediction will be tested prediction precision, then it can be carried back in 2020 (see figure 1).

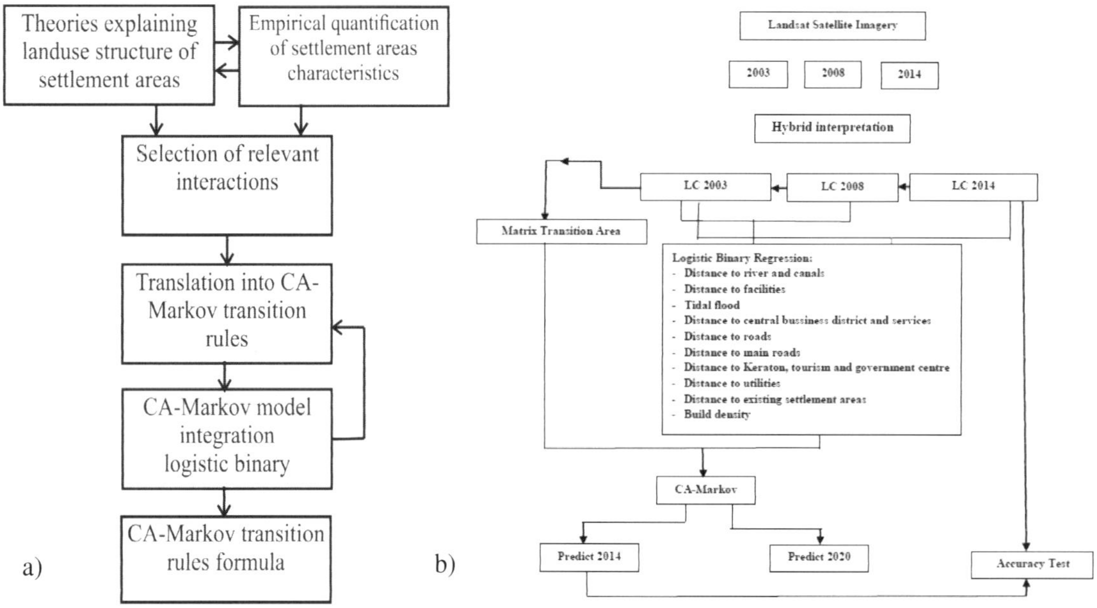

Figure 1. (a) The Procedure Followed for Defining Transition Rules (b)The Framework of Cellular Automata Modeling

3. Result and Discussion

3.1 Settlement Land Use Changes in the Eastern District of Pontianak

Based on the results of the analysis of changes in land use cover changes from 2003-2008, it is concluded that in this period the increasingly of sub-settlement areas (38.49 hectares) in The Eastern district of Pontianak. The settlements areas in the Eastern district of Pontianak (397.35 hectares) in 2008 while previously (358.86 hectares) in 2003.

The results of the analysis using interpolation spatial polynomial order of 3 to the location of land expansion of settlements in 2003 to 2008 in the District of Pontianak East, showing the centre of the development of non-built-up areas located in the central and the eastern part, more precisely in the the eastern part of Kampung Dalam Bugis Village, the southern part of Tanjung Hulu Village and the northern part of Saigon Village (see figure 4).

Analysis of push factors of landuse changes in binary logistic regression between land use changes for non-settlement into settlement areas in 2003-2008 resulted in the regression equation as follows:

IOP Conf. Series: Earth and Environmental Science **79** (2017) 012010 doi:10.1088/1755-1315/79/1/012010

$$Y = -3.6640 + 0.000877*X1 - 0.000000*X2 + 0.000358*X3 + 0.000843*X4 - 0.001819*X5 + 0.000382*X6 - 0.000505*X7 + 0.000684*X8 + 0.004674*X9 - 0.264610*X10 \quad (1)$$

Where: Y : Logit of land use change from non-settlement to settlement areas
X1 : Distance to rivers and canals
X2 : Distance to facilities
X3 : Tidal flood
X4 : Distance to central business district and services
X5 : Distance to collector roads
X6 : Distance to main roads
X7 : Distance to Keraton Kadariyah, tourism and government centre
X8 : Distance to utilities
X9 : Distance to existing settlement areas
X10: Buildings density

The equation (1) shows the highest regression coefficients are at distance variable to existing settlements (0.26). A positive coefficient indicates that the greater the coefficient of the independent variable, the greater possibility of land use changes from non-settlement into settlement areas, while a negative coefficient is the opposite which shows that the smaller the coefficient of the independent variable, the greater possibility of land use changes from non-settlements into settlement areas.

The existing settlements variable have the strongly influenced of the land use change in non-settlement into settlements areas which have a negative coefficient, so it can be concluded that the closer the existing settlements, the greater the likelihood that land use changes from non-settlement into settlements.

The smallest regression coefficient is variable of flooding (0). Based on this analysis, flooding variable did not influence at all to land use changes from non-settlement into settlements areas. Flooding that occurred in The Eastern district of Pontianak did not influenced the development of settlement areas, because the level of community's adaptation to the building structure on rumah panggung's settlements was very good, through rumah panggung's foundation have adjusted to the height of tidal flood on the Kapuas river (see figure 2 & 3).

a. b

Figure 2. (a) Tidal Flood of Settlement Area; (b) Perspective of The Settlement Area

a b

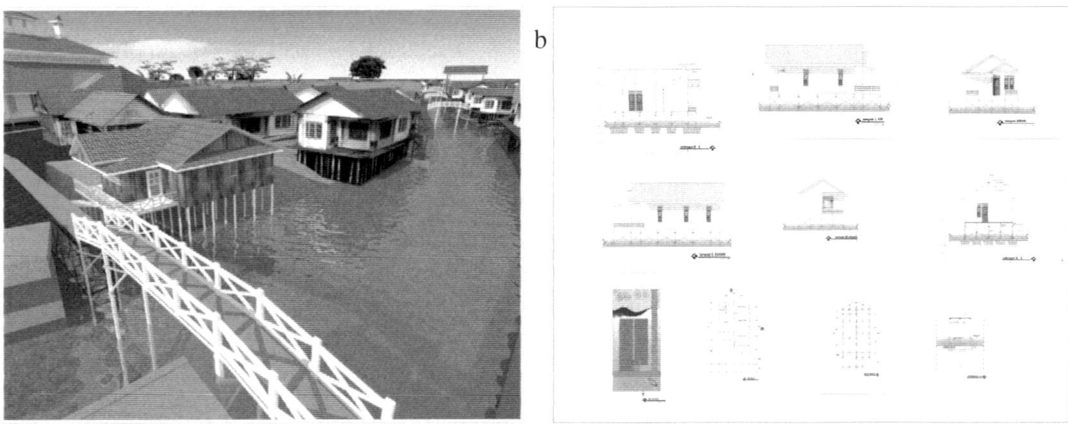

Figure 3. (a) Settlement Area of Rumah Panggung; (b) Sketch of Rumah Panggung

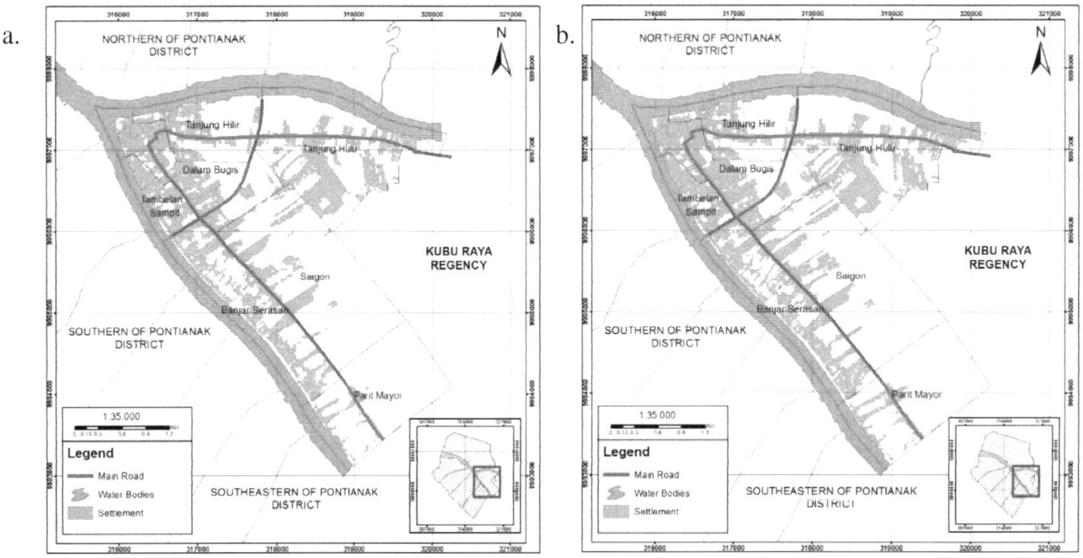

Figure 4. (a) Map of Existing Settlement Areas in 2003; (b) Map of Existing Settlement Areas in 2008

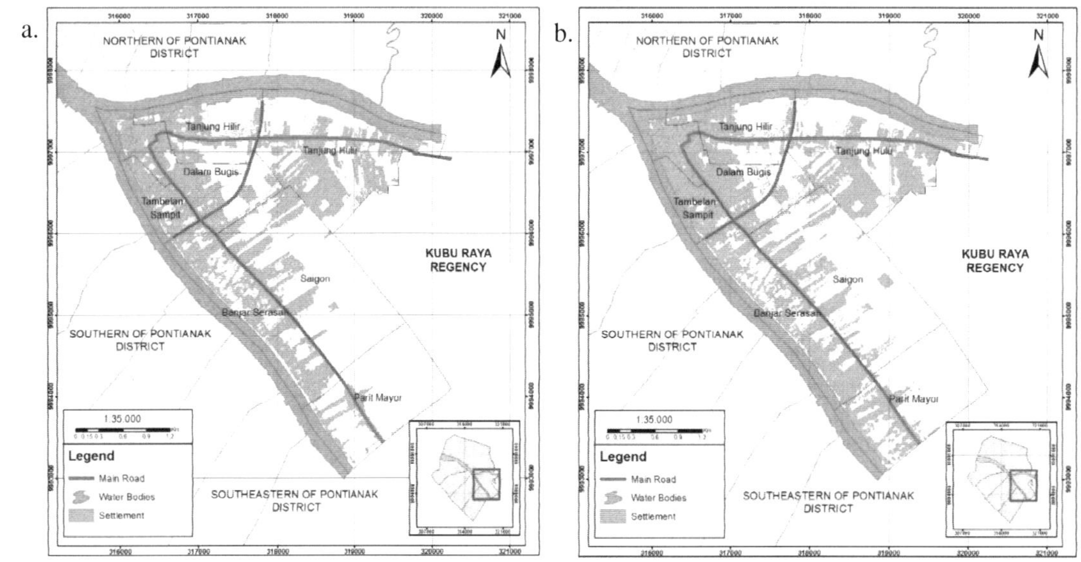

Figure 5. (a) Map of Existing Settlement Areas in 2014; (b) Map of Existing Settlement Areas in 2020

3.2 Prediction of Settlement Areas in the Eastern District of Pontianak

The settlement areas prediction results in 2014 (440.82 hectares), the predicted outcome is smaller (0.35 hectares) than the area of Pontianak urban settlements (441.17 hectares). The results show the accuracy in predicting settlements land use change in The Eastern district of Pontianak have been able to get the overall accuracy (79.74%) and the highest kappa index (0.55). Based on this analysis, the prediction of settlement areas using Cellular Automata-Markov modeling has been accurate and will be used to predict the development of settlements areas in 2020.

The prediction results show that in 2020 (481.98 hectares) of settlement areas. During a period of 6 years from 2014 to 2020 there was an increase of settlement areas (40.81 hectares) or 6.80 hectares/year. Based on the analysis using spatial interpolation polynomial order of 3 to the location of land expansion of settlements in 2020 in the eastern district of Pontianak, showing settlement areas are developing fairly evenly in Saigon Village, Tanjung Hulu Village, Banjar Serasan Village and the highest develops is in Parit Mayor Village (see figure 5).

Analysis of push factors of land use changes in binary logistic regression between land use changes for non-settlement into settlement areas in 2014 to 2020 resulted in the regression equation as follows:

$$Y = 4.4328 - 0.001154*X1 + 0.002352*X2 + 0.000000*X3 - 0.001400*X4 + 0.005696*X5 - 0.001250*X6 + 0.002758*X7 - 0.002166*X8 - 0.754588*X9 + 0.003258*X10 \qquad (2)$$

Where: Y : Logit of land use change from non-settlement to settlement areas
X1 : Distance to rivers and canals
X2 : Distance to facilities
X3 : Tidal flood
X4 : Distance to central business district and services
X5 : Distance to collector roads
X6 : Distance to main roads
X7 : Distance to Keraton Kadariyah, tourism and government centre
X8 : Distance to utilities
X9 : Distance to existing settlement areas
X10: Buildings density

The equation (2) shows the most significant regression coefficient is the distances to existing settlements variable (0.75). The distance to existing settlements have the most significant effect of changing in non-settlement into settlement areas which have a negative coefficient, so it can be concluded that the closer the existing settlements the greater the likelihood that land use change from non-settlement into settlement areas. Besides that, the distance to the main road and the density of the buildings are also highly enough to influence the land use changes from non-settlement to settlement areas. Just as logit changes in settlement areas in 2003-2008, the smallest regression coefficient is flooding variable (0). Based on this analysis, flooding variable did not influence at all to land use changes from non-settlement into settlement areas, it is similar to the percentage of flood exposure in a very small number (see figure 6). Based on the prediction results, Parit Mayor Village will be the centre of the highest development of settlements areas in the eastern district of Pontianak in 2020, because the Parit Mayor Village still has space of non-built-up areas, so that it can be developed into settlement areas (see figure 7). Moreover, in terms of accessibility factors in Parit Mayor Village has by passed the main roads to get crossing the Trans Kalimantan bridge.

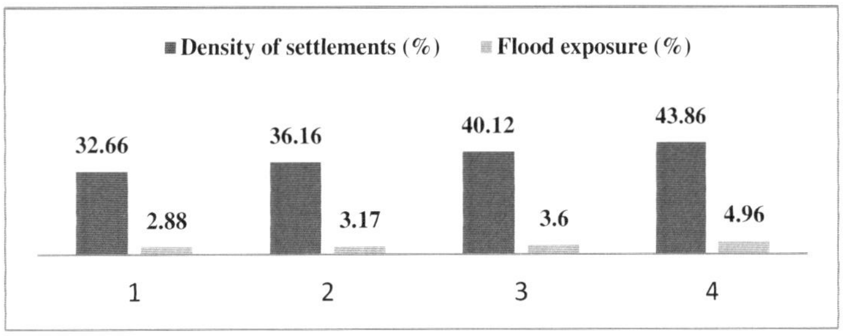

Figure 6. Chart of Density and Flood Exposures in Settlement Areas.

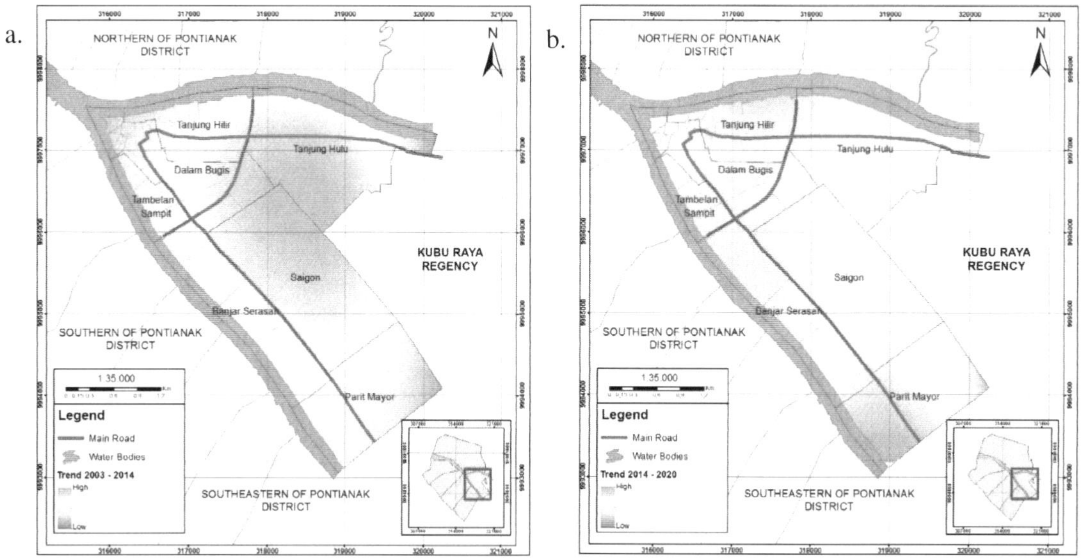

Figure 7. (a) Map of Trend Settlement Areas in 2003-2014; (b) Map of Trend Settlement Areas in 2014-2020

4. Conclusion

The settlement land use changes in the eastern district of Pontianak in 2003-2008 increased (38.49 Ha). The results of land expansion of settlements show the centre of the development of non-built-up areas located in the central and the eastern part, more precisely in the eastern part of Kampung Dalam Bugis Village, the southern part of Tanjung Hulu Village and the northern part of Saigon Village. Analysis of binary logistic regression resulted the strongly influenced is distance variable to existing settlements (0.26). Flooding variable did not influence to the development of settlement areas, because the community's adaptation to the building structure on rumah panggung's settlements was very good, through rumah panggung's foundation have adjusted to the height of tidal flood on the Kapuas river.

The prediction results of settlement areas in the eastern district of Pontianak in 2014-2020 increased (40.81 Ha). The results show the settlements land use change in the eastern district of Pontianak is overall accuracy (79.74%) and the highest kappa index (0.55). The prediction results show of settlement areas in 2020 (481.98 Ha) and 2014 (440.82 Ha). The increasingly of settlement areas (40.81 Ha). The results of land expansion of settlements show the settlement areas are developing fairly in Saigon Village, Tanjung Hulu Village, Banjar Serasan Village and the highest develops in Parit Mayor Village. Analysis of binary logistic regression resulted the most influential is distances to existing settlements variable (0.75). Flooding variable did not influence to land use changes from non-settlement into settlement areas, it is similar to the very small number of flood exposure percentage.

Acknowledgements

We would like to express our gratitude towards Prof. Dr. rer. nat. Imam Buchori, ST and Dr. Mussadun, ST, M.Si as a mentor in conducting this study and for giving us the opportunity to write this paper to support a research in Architecture and Urbanism Department of Diponegoro University.

References

[1] Hedge N P 2008 Settlement Growth Prediction Using Neural. Journal of theoritical and applied information technology *Journal of Theoretical and Applied Information Technology*

[2] Maria de Almeida C, Marinaldo G J 2005 Cellular automata and neural networks as a modelling framework for the simulation of urban land use change c *Anais XII Simposio Brasileiro de Sensoriamento Remoto, Goiania, Brasil* pp. 3697–3705

[3] F Wu 1998 An experiment on the generic polycentricity of urban growth in a cellular automatic city *Environment and Planning B-Planning & Design* **25(5)** pp. 731-752

[4] Rustiadi, Ernan et al. 2009 Perencanaan dan Pengembangan Wilayah (Jakarta: Yayasan Obor Indonesia)

[5] USGS 2015 Landsat 8 (L8) Data Users Handbook.

[6] Yunus, Hadi S 2010 Metodologi Penelitian Wilayah Kontemporer (Yogyakarta: Pustaka Pelajar)

[7] Syahar, Fitriana 2012 Pengaruh Faktor Artifisial terhadap Perkembangan Kota *Jurnal Skala* **2(4)** pp. 49-63

[8] Hasanuddin 2015 Pontianak Masa Kolonial (Yogyakarta: Ombak)

[9] Anunobi A I 2014 Informal Riverine Settlements and Flood Risk Management : A Study of Lokoja, Nigeria *Journal of Environmental and Earth Science* **4(12)** pp. 35-43

[10] Jaroslaw J, Hildebrandt-Radke I 2009 Using multivariate statistics and fuzzy logic system to analyse settlement preferences in lowland areas of the temperate zone: an example from the Polish Lowlands *Journal of Archaeological Science* **36(10)** pp. 2096–2107

[11] Nwilo P C, Olayinka D N, Adzandeh A E 2012 Flood Modelling and Vulnerability Assessment of Settlements in the Adamawa State Floodplain Using GIS and Cellular Framework Approach *Global Journal of Human Social Science* **12(3)** pp. 11–20.

[12] Watson D, Adams M 2011 Design For Flodding, Architecture, Landscape, and Urban Design for Resilience to Flooding and Climate Change (Canada: John Wiley & Sons Inc.)

[13] Wijaya, Muhammad S, Bowo S 2013 Integrasi Model Spasial Cellular Automata Dan Regresi Logistik Biner Untuk Pemodelan Dinamika Perkembangan Lahan Terbangun (Studi Kasus Kota Salatiga) *Jurnal Bumi Indonesia* **2(1)**

[14] Chander, Gyanesh, Markham, Brian 2003 Revised Landsat-5 TM Radiometric Calibration Procedures and Postcalibration Dynamic Ranges *IEEE Transaction on Geoscience and Remote Sensing* **41** pp. 2674-2677

[15] Projo D 2012 Pengantar Penginderaan Jauh Digital (Yogyakarta: Penerbit Andi)

Urban Agglomeration and Extension in Northern Coast of West Java: A Transformation into Mega Region

Y Octifanny[1], **D Hudalah**[1]
[1]Regional and Rural Planning Research Group
Institut Teknologi Bandung
Jl Ganeca No 10 40135
Bandung, Indonesia

E-mail: yoctifanny@gmail.com

Abstract. In the past decade, at least three neighbouring metropolitan areas in the north-western part of the island have been merging with each other: Jakarta Metropolitan Area (Jabodetabek), Bandung Metropolitan Area (Bandung Raya), and Cirebon Metropolitan Area (Ciayumajakuning). It is expected to be the first island-based mega-conurbation. This paper explores the potential emergence of mega region as a ground study, where mega region can be used for economic, logistic, transportation development. Authors use scoring analysis from economic and demographic indicators. The outcomes found a new and larger formation of city-region in the northern coast road networks (Pantura) – specifically western part of Java Island.

1. Introduction

The paper explores the latest urban dynamics across Java, the most developed and most densely populated main-island in Indonesia. It is where Indonesia's largest metropolitan areas located. According to McGee, extended metropolitan region like Jakarta-Bandung Corridor is dominating the urban system of Indonesia [1]. In Indonesia there is an Extended Mega Urban Region that combines two metropolitan region of Jakarta Mega Urban Region (JMUR) and Bandung Metropolitan Region that has been developed by Firman [2] and Firman [3] according to McGee in 2012 "In 2010 this Extended Metropolitan Region of JMUR and BMA had a population of 35 million which made it the second largest EMR's in the world second only to the Tokyo Metropolitan Area." [1]. However, in the past decade, there is a conjecture that at least three neighbouring metropolitan areas in the north-western part of the island have been merging with each other: Jakarta Metropolitan Area (Jabodetabek), Bandung Metropolitan Area (BMA) / Bandung Raya, and Cirebon Metropolitan Area (Ciayumajakuning). These three metropolitans are integrated based on Local Government Regulation Number 12 / 2014 about Managing Metropolitan Construction, Development, and Growth Centre in West Java. This policy is possible trigger for the extension of the JMUR and BMA in Indonesia to reach Cirebon Metropolitan Region with the population of more than 30 million and increasing, it is expected to be the first island-based mega-conurbation in Southeast Asia. This paper updates the previous research about mega region in Java.

This article is presented in four parts, following this introduction, the concept of mega region, discussion, and summary. The study discussed one major: delineating mega region. In delineating mega region, the authors use demography, economy, infrastructure, and socio politic analysis. This analysis is focus on deciphering spatial form of mega region in Indonesia that estimated in the area of northern coast of west java as the most populated, capital intensive, high performance economic concentration, socio-political importance, global role importance, and also most infrastructure importance settlement area in Indonesia

2. Theoretical Transition to Asian Background

2.1 Theoretical Evolution

The evolutions of mega urbanization agglomeration theory start from conurbation to mega region. Conurbation theory coined by Geddes (1915), it is a region that comprising a number of cities, large towns, and other urban areas that merge to form one continuous urban and industrially developed area as a result of population growth and physical expansion [4].

In the following years the theory evolve to a bigger more detailed explanation of settlement type, named Megalopolis coined by Gottmann (1957), defined as a region with national and international national and international significance because of the size and weight of megalopolis in economic and social matter [5]. In the following years, (Gottmann, 1961) elaborate the theory Megalopolis as a bigger region than metropolis – a vast area that encompasses many great cities, the most active crossroads on earth with the population close to 38 million [6].

In the early period of millennium, the theory evolves into new concept named Mega region. Prior from many forms of urbanization: city, metropolitan, megacities, and now mega region. Mega region also corresponds with several terminologies such as megalopolis, Global city Region, Mega-city region, Metro (politan) regions, and Mega Urban Region. Mega region is defined as in several perspectives. *First*, network of metropolitan centres and their surrounding areas that spatially and functionally linked through environmental, economic, and infrastructure interactions' (Ross et al. 2009). *Second*, this definition then adapted by America 2050 that defined this region as a region with interlocking economic systems; shared natural resources and ecosystem; and population centres that linked by transportation [7]. Different from previous perspective, Florida, Gulden, and Mellander (2008) define mega region with different perspective, differentiate urban mega region unlike urban in the past, that the city system has become global – compete in global terrain in the way that metropolitan regions relate to national economies. Furthermore, Mega region is defined not only as an area with a large population, but also possess large market, significant economic capacity, substantial innovative activity, and highly skilled talent [8]. Thus, there are many large megacities that do not meet the criteria for being a mega region.

2.2 Bridging to Asia

Urbanization in Asia and in Europe or America is inherently different. Asia urbanization has a lot of megacities; cities as the economic engine; the cities are built upon internal migration; high disparity; and non-environmentally friendly city growth are the urbanization trend in Asia based on study conducted by ESCAP [9]. Meanwhile, most of the theory that mentioned before was emergent from Euro-American cities which has different urbanization character. There has to be a recalibration and bridging of theory to urbanization in Asia cities with the theory rooted from Asia.

The similar theory emergent from Asia is Mega Urban Region (MUR), interaction of metropolitan centres to extend beyond administrative and traditional core-suburb agglomeration into rural areas along inter metropolitan corridors extending upwards of 100 kilometres from metropolitan centres because of the growth of vehicular transport and trunk highway system (McGee & Robinson, 1995). In 1995, there are studies which study the MUR in Indonesia. The merging of two metropolitan regions, Jakarta and Bandung, this Indonesian context of this concept has been developed by (Firman & Dharmapatni, 1995) that these two metropolitan and the corridor between them have become integrated into a fully functional MUR. However, as the time passes by the development of this megaregion after 20 years, after major changes like megaproject, foreign investment, decentralization, new regulations, and more. A new update of prior research need to be done.

Table 1 Theoretical Evolution Frame

Concept	Definition	Foundation Literature
Conurbation	A region that comprising a number of cities, large towns, and other urban areas that merge to form one continuous urban and industrially developed area as a result of population growth and physical expansion.	Geddes (1915)
Megalopolis	1. Megalopolis defined as a region with national and international national and international significance because of the size and weight of megalopolis in economic and social matter. 2. Megalopolis as a bigger region than metropolis – a vast area that encompasses many great cities, the most active crossroads on earth with the population close to 38 million.	1. Gottmann (1957) 2. Gottman (1961)
Mega Urban Region	interaction of metropolitan centres to extend beyond administrative and traditional core-suburb agglomeration into rural areas along inter metropolitan corridors extending upwards of 100 kilometres from metropolitan centres because of the growth of vehicular transport and trunk highway system	T. G. McGee and Robinson (1995)
Megaregion	Network of metropolitan centres and their surrounding areas that spatially and functionally linked through environmental, economic, and infrastructure interactions	Ross et al. (2009)
	A region with interlocking economic systems; shared natural resources and ecosystem; and population centres that linked by transportation	(Hagler, 2009)
	1. City system has become global – compete in global terrain in the way that metropolitan regions relate to national economies. 2. An area with a large population, but also possess large market, significant economic capacity, substantial innovative activity, and highly skilled talent	Florida, Gulden, and Mellander (2008)

Source: Authors Collection

3. Approach and Methods

The approach in this research is a deductive research, the research formulates the mega region indicators and see if the realization of the indicators. Here's the brief explanation:

Table 2 Summary of Method

Objectives	Analysis Method
Delineating the Distribution of Megaregion in Java Island	Demography: o growth rate o Population density
	Local economy: o Economic growth o GDP per capita
	Demography and local economic indicator score

The analyses in this article are literature review, statistic, and spatial analysis to show the development of mega region in Indonesia. The statistical data then developed into spatial map to illustrating distribution of demographic and economic activities in mega region.

4. Result and Discussion

4.1 Defining Structure and Pattern of Mega Region

In deciphering structure of mega region; demography, economy, infrastructure, and socio-politic aspect is used.

4.1.1 Demography and Urban Area

The pattern of population distribution can also explain the pattern of settlement that indicates the distribution of people in this area.

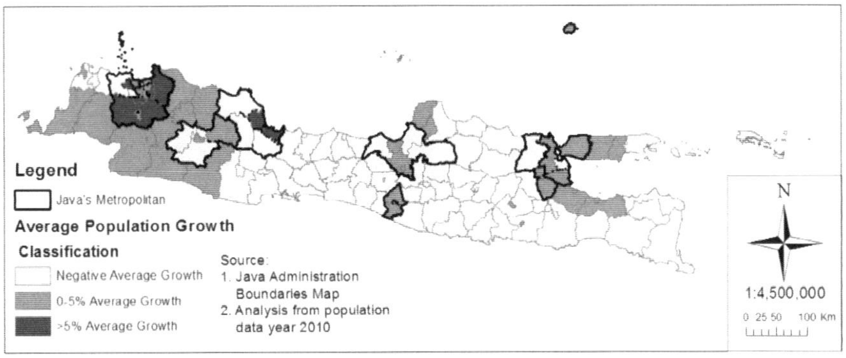

Figure 1. Average of 10 Years Population Growth in City and Regency of Java Island (Year 2000 – 2010)

It can be implied from the map above that most of Java Island experience medium range of population growth all across java. However, there are some parts that are having the most growth in Cirebon Metropolitan and also in Jakarta Metropolitan.

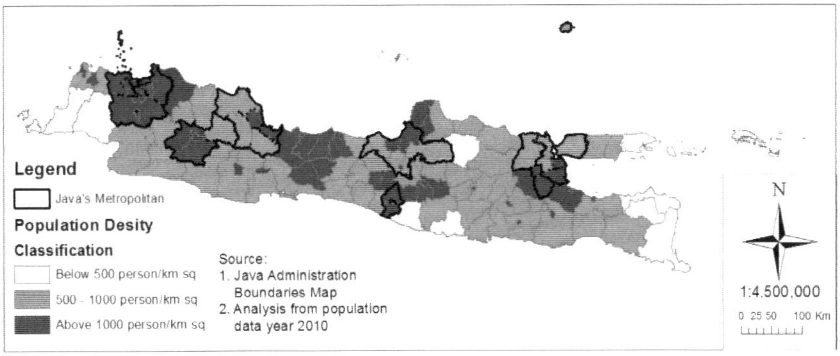

Figure 2 Population's Density in City and Regency of Java Island (Year 2010)

As for population density, it's clear that the high-density population is located around in metropolitans and their surrounding city or regency. This condition is happening in all metropolitan in Java Island.

4.1.2 Local Economy

New Formation of Economy is emergent in Java Island. Most of the city and regency in Java Island experience high GDP growth. However, when we take a closer look to GDP per capita, there is unequal GDP per capita. Inequality differs greatly between highest and lowest GDP per capita. The capital accumulation mainly located in major cities suitable to new formation of urban.

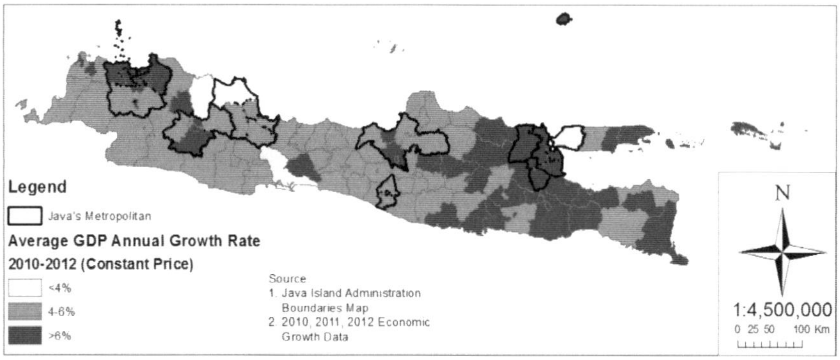

Figure 3. Average GDP Annual Growth Rate Year 2010 – 2012, Constant Price 2010 (Billion Rupiah)

The map above shows the area with GDP growth rate from 2010-2012 based on constant price 2010. We could see the growth rate is high in area inside metropolitan. The surprising condition is the most economic growth is happening inside and in the hinterland of Surabaya Metropolitan.

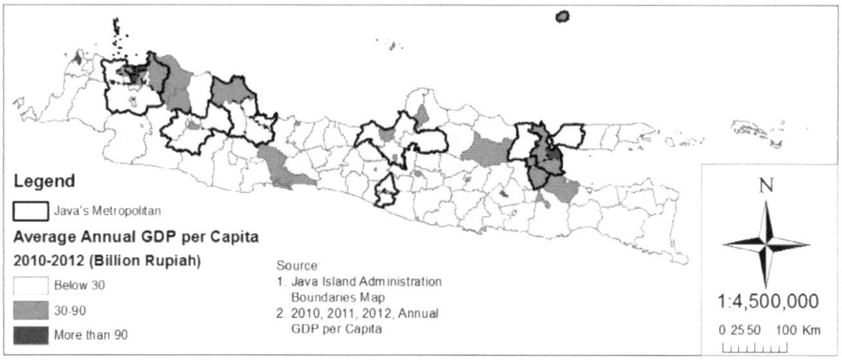

Figure 4. Average of Annual GDP per Capita Year 2010 – 2012 (Million Rupiah)

There is income inequality in Java Island. The annual GDP per capita is very contrast in the cities on Java Island and the non-non-city area. Most of cities have a very high income compare to regencies in Java Island.

4.2 Merging Metropolitans into West Java North Coast Mega Region

In deciding delineation of mega region, we use scoring analysis. The scoring analysis identifies basic character of mega region from available secondary data. The explanation of scoring method is in the table below:

Table 3. Scoring Regulation

Indicators	Score		
Average of Annual GDP per Capita Year 2010-2012 (Million Rupiah)	Below 30	30-90	More than 90
Score	*1*	*2*	*3*
Average of GDP Annual Growth Rate Year 2010-2012 (Million Rupiah)	Below 4%	4-6%	More than 6%
Score	*1*	*2*	*3*
Population Density Year 2010	Below 500 person/km^2	500-1000 person/km^2	More than 1000 person/km^2
Score	*1*	*2*	*3*
Average of 10 years Population Growth (2000-2010)	Negative Growth	0-5%	More than 6%
Score	*1*	*2*	*4*

From the scoring regulation above, then we score each city and regency in Bandung then accumulate the score. From the score accumulation, we could see the pattern that presented in the map below.

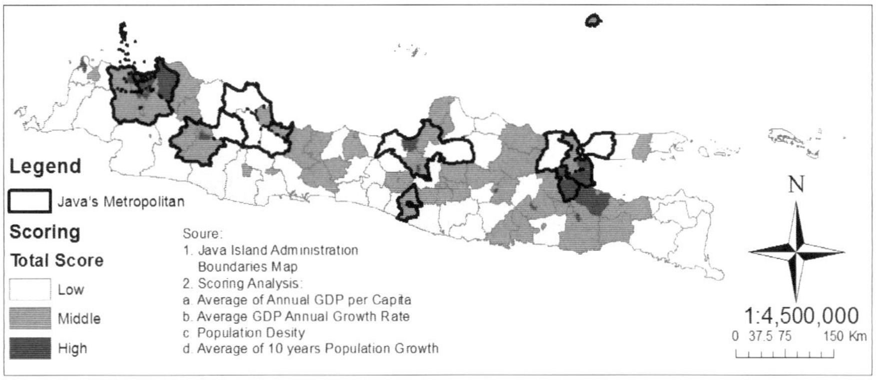

Figure 5. Mega Region Scoring Analysis

Map above show the total score that has been classified into three classes: low, middle, and high score. The higher the score, it's possible that area to be a mega region. In contrary, the lower the score, it is not possible for that area to be a mega region. In conclusion, there are two salient urban centres in Java: Jakarta and Surabaya. Jakarta Area influence sprawls to three metropolitans: Jabodetabek, Bandung Raya, and Cirebon Raya. Meanwhile, Surabaya mostly influences area Gerbangkertosusila. Later on, this paper will discuss about Western Java more intensively, where Jakarta as the prime mover of the urbanization in Java Island that influence the development of metropolitan around it, all the way across the island. The influence of mega region must have limited distance, not too far from Jakarta and also have physical continuity and tight connectivity, so Surabaya Metropolitan who is also the second most influential metropolitan is not accounted in this mega region because Surabaya is too far from Jakarta and also there are some physical discontinue between them. For those reasons, it can be concluded that the area of mega regions are encompassed these territory:

- Jakarta, Bandung, and Cirebon Metropolitans form a mega region situated in Northern Coast of West Java.
- The other areas, Banten Province, such as: Cilegon City, Serang City, and Serang Regency are not entitled as a metropolitan but share a quality to become a megaregion from the Figure 6. If we took a closer look this region has a great deal in foreign investment and also industrial sector.
- West Java North Coast province regencies, such as: Subang Regency, Karawang Regency, and Purwakarta Regency are not entirely entitled by national metropolitan system, but only in West Java metropolitan's version. However, they have a good score in the map above and also connect Metropolitan Jakarta to Bandung Metropolitan and Cirebon Metropolitan. Moreover, they are passed by Pantura Road System.

Meanwhile, southern part of West Java – the non-metropolitan area of West Java Province such as Sukabumi Regency, Cianjur Regency, and Garut Regency has a low score. It can be because they located in Southern Road System of Java, different from other three areas above that located in Pantura road network. Pantura is the most develop road network in Java Island where most economic activities take place. Moreover, Pantura road network is forming a very strategic network and have intriguing upcoming development programs from government and private institution, such as Karawang airport; and Majalengka Airport and Aero city. In conclusion that the development of mega

IOP Conf. Series: Earth and Environmental Science **79** (2017) 012011 doi:10.1088/1755-1315/79/1/012011

region in Northern Coast of Java is situated and influence by fast growing development in Pantura Road System.

In delineating mega region, sure the quantitative scoring map in Figure 6 will be accounted in this research. In addition to that, we also account non-quantitative qualities such as economic condition; physical continuity and accessibility; and also, socio political dynamics that cannot be scored in subsuming city and regency as area of mega region. In brief, Northern Coast of West Java Mega region can be understood from this tabulation:

Table 4 Mega Region Selection Background

Mega Region Area	Economy	Infrastructure	Socio Politic
Jakarta Metropolitan Area	• Economic centre in Indonesia and also gate for global economy. • Centre of global capital accumulation in Indonesia. • National government centre	• Soekarno Hatta airport (main airport in Indonesia) • Halim Perdanakusuma airport • Tanjung Priok Seaport • Pantura Highway • Railway • New Tanjung Priok New Seaport	• Headquarter of regional politic: Asean. • A megacity according to National government • Headquarter of International Bank, Finance Institution, and Companies
Bandung Metropolitan Area	Economic centre in West Java.	• Husein Sastranegara Airport • Pantura Highway • Railway • High Speed Train • Cipularang Toll Road	A metropolitan according to National government
Cirebon Metropolitan Area	Emerging economic centre in West Java	• Cirebon Seaport • Pantura Highway • Majalengka Airport and Aero City • Railway	A metropolitan according to West Java Metropolitans.
Banten (Cilegon City, Serang City, Serang Regency)	• International and National industrial hub in Indonesia. Manufacture, chemical, and metal industry. • Concentration of	• Pantura Highway • Railway • Merak Seaport • Industrial Seaports	

Mega Region Area	Economy	Infrastructure	Socio Politic
	international labour and investment. • SEZ Tanjung Lesung	• Airports: Budiarto, and Tanjung Lesung	
West Java (Subang Regency, Karawang Regency, and Purwakarta Regenc (Bodebekkarpur)	Spill over of economic growth because of Pantura Road	• Pantura Highway • Patimban Seaport, in Subang	Karawang and Purwakarta are part of Bodebekkarpur Metropolitan according to West Java Metropolitans.

From qualitative and quantitative consideration, we conclude the area of mega region as illustrate by Figure 7. The mega region in Indonesia is illustrated below, named West Java North Coast. It is situated in the area where Jakarta as the major forces in creating mega region. This mega region as illustrated below is having physical continuity; and also share infrastructure, socio, economical, and political connectivity. The main influences for connectivity inside this mega region are the Pantura transportation networks, that open relations inter mega region cities and also relation abroad.

Figure 6. West Java North Coast Megaregion

5. Conclusions

This research is intended to explore latest dynamics in urbanization form across Java Island. This paper initiates to delineate the megaregion and also characterizing megaregion in Indonesia, specifically in the Java Island. In case of Java Island, demography, economy, infrastructure, and socio – politic analysis can lead us megaproject delineation in West Java North Coast: Jakarta Metropolitan Area, Bandung Metropolitan Area, Cirebon Metropolitan Area, Banten (Cilegon City, Serang City, and Serang Regency,), West Java (Subang Regency, Karawang Regency, and Purwakarta Regency).

This area is predicted to be the most salient economic growth center in Indonesia where majority of population and economic activities took place. The advancement of Northern Coast of West Java mega region will continue on to triumph through the ambitious, appreciable, capital intensive, pervasive future development projects in this region. When we discuss about mega region, the scale of project and development is escalating from metropolitan level, we expect more of mega urban and mega project development. Notably, through: Karawang Airport; Jakarta- Bandung Fast Train; Kertajati Airport and Aerocity; and Patimban Seaport.

As the author mentioned in the introduction before, this paper is initial research to identify urbanization in the coastal area in Indonesia. Mega region in nature is has the network and importance to global social, economic, and political affairs. Realizing the important of this urbanization form in our coastal area, therefore, the further analysis can be done after we have mega region delineation, including the research identifying magnitude on social, economy, and political importance; the environmental impact from this coastal mega region development; governing mega region; and more

References

[1] McGee T G 2012 Deconstructing the mega-city: A case study of the Jakarta mega-urban region (Jabodetabek) in the first decade of the 21st century *Preliminary Draft. Kyoto University*

[2] Firman T, Dharmapatni I A I 1995 The emergence of extended metropolitan regions in Indonesia: Jabotabek and Bandung Metropolitan Area *Review of Urban & Regional Development Studies* **7(2)** 167–188.

[3] Firman T 2009 The continuity and change in mega-urbanization in Indonesia: A survey of Jakarta–Bandung Region (JBR) development *Habitat International* ***33*(4)** 327–339.

[4] Geddes S P 1915 *Cities in evolution: An introduction to the town planning movement and to the study of civics.* (London: Williams & Norgate) Retrieved from http://library.wur.nl/WebQuery/clc/461010

[5] Gottmann J 1957 Megalopolis or the Urbanization of the Northeastern Seaboard. *Economic Geography* 189–200

[6] Gottmann J 1961 Megalopolis *Twentieth Century Fund.*

[7] Hagler Y 2009 Defining US megaregions *America 2050* 1–8

[8] Florida R, Gulden T, Mellander C 2008 The rise of the mega-region. *Cambridge Journal of Regions, Economy and Society* ***1*(3)** 459–476.

[9] ESCAP 2013 Urbanization trends in Asia and the Pacific.

[10] McGee T G, Robinson I M 1995 *The mega-urban regions of Southeast Asia* (Vol. 1). UBC Press. Retrieved from https://www.google.com/books?hl=id&lr=&id=BOHsEzpJhx4C&oi=fnd&pg=PR8&dq=The+Mega-urban+Regions+of+Southeast+Asia&ots=VznBKLpyPu&sig=53aHq92GsLrEXNnik2rmFkeqr9Y

[11] West Java Province West Java Regulation No 12 year 2014 on Development Management, Metropolitan Development, and Growth Center in West Java, Pub. L. No. 12 (2014).

Planning for Resilient Settlement in Port Fairy, Australia

Sri Tuntung Pandangwati
Department of Architecture and Planning, Faculty of Engineering, Universitas Gadjah Mada, Indonesia

Abstract. This paper aims to discuss some strategies to address problems and challenges faced by a community living in Port Fairy, a coastal town located in Southwestern Victoria, Australia. It is about three hours' drive from Melbourne's CBD. Like other coastal towns, Port Fairy is also affected by climate change, severe weather events and sea level rise. The impact of these problems can be seen in East Beach, an important tourist destination in Port Fairy. Soil erosion and flood due to severe wind and wave, are main problems in East Beach. This will not only affect the housings, but also the landfill site located in this beach. Therefore, it is challenging to develop solutions to enhance the resilience of this coastal town. This study may inspire Indonesian planners to further develop the resilience of hundreds coastal towns in Indonesia.

1. Introduction

Planning for resilient settlement has been increasingly important due to climate change issue. This issue is particularly critical for coastal towns. They are vulnerable to disturbances such as sea level rise, severe winds and wave. Problems such as flood and soil erosion are faced by most coastal towns and these problems may endanger the communities living the settlement nearby in a long term period. Moreover, relocating people to another place which is safer is usually challenging and it does not solve the problem. Therefore it is critical to develop a set of strategies to deal with this kind of issue.

Port Fairy is one of important recreational destinations in Victoria, Australia. It is a small coastal settlement located in Great Ocean Road, a historical road in Australia. People visit this town for seeing beautiful scenery of white sand beach in East Beach, seeing wallabies and other animals in Griffith Island or just for fishing. Port Fairy is also the second largest port in Victoria, after Port Melbourne. However, some environmental changes which happened in this coastal town created some problems such as inundation, sand dunes erosion and bio-diversity loss that significantly affect the communities living there. People living in the area close to the beach, are vulnerable to disaster and had to spend more money on building renovation because of flood that often occur.

Hence, this study aims to analyse the problems in Port Fairy and formulate recommendations to further develop the resilience in this coastal town. The lesson learned from Port Fairy can be used for developing the resilience of other coastal towns, including the ones in Indonesia. Coastal settlements in Indonesia are also facing similar problems which occur in Port Fairy, Australia.

2. Literature Review

2.1. The Definition of Resilience

There are several definitions of resilience. In terms of engineering principle, resilience is an ability of a system to bounce back after a disturbance or a shock [1][2]. In terms of ecology, resilience is defined as not only an ability to return to a normal condition, but also an ability to absorb or accommodate a certain amount of disturbance [1]. It rejects the idea of single equilibrium and suggests that a system can create a new equilibrium. Kuei-Hsien (2012) argues that ecological resilience means ability to survive and it is not limited to just returning back to stability [2]. The last one is evolutionary resilience. It defines resilience as an ability of a system to change, adapt and transform in order to address a disturbance [1]. In this principle, there is a possibility of changing threshold. This also considers that socio-ecological system is complex, uncertain and keep changing [1].

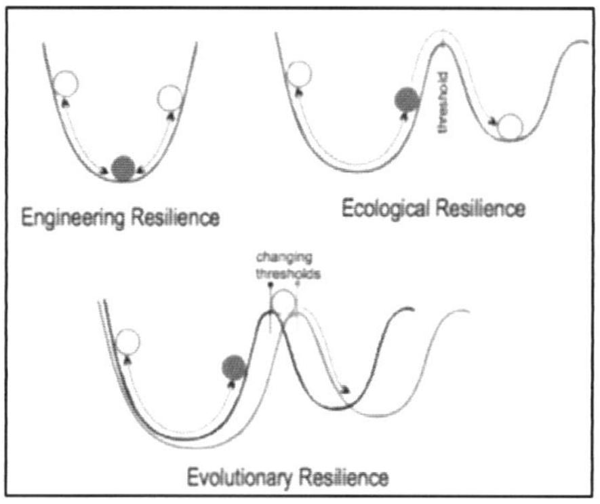

Figure 1. *Different illustrations of resilience [2]*

2.2. The Resilience components

Resilience consists of four components: diversity, modularity, redundancy and feedback [3]. Firstly, diversity means the more diverse a system, the more options it owns and the more resilient it is [3]. A system that consists of various components is considered as a resilient system because it provides many options. Secondly, modularity is owned by a system that consists of modules that each of them can be changed without seriously affecting other modules [3]. High modularity requires links among components of a system that do not make them highly dependent on each other. Thirdly, redundancy means having extra components that enable a system to survive. A loss of one component will not endanger the whole system because it still has some similar components [3]. Lastly, feedback means an ability of a system to create fast and strong feedback for other parts of the system due to change in a part of the system [3].

2.3. Types of Responses to Disturbances

According to IPCC, there are three types of responses to disturbances: retreat, accommodate and protect [4]. Retreat means preventing development in an area that is vulnerable to disaster [4]. In this case, it is possible to abandon that vulnerable area. It seems that this is the simplest and the best idea, but in practice it is very hard to be implemented. Social factor such as emotional connection to the place and economic factor such as working place are the challenges of this type of response. Another type of response is accommodating. The idea of accommodating disturbances is like an effort to live with the disaster by adapting to it. Some of the examples of this type of respond is creating houses that resilient to earthquake and flooding [4]. Lastly, protect is a type of respond that aims to create a barrier that can keep the settlement safe [4]. Both advance technology and natural-traditional elements can be used to protect the coastal settlement. Even though there are various definitions of resilience, this study will focus on ecological and evolutionary resilience. These definitions are suitable to address present day climate change issues that are unpredictable [5].

3. Method

The main inquiry of this study is what should be done to enhance the resilience of the community and physical environment in Port Fairy, Australia. This question can be broken down into following sub-questions:

- What are the problems and challenges faced to establish a more resilient settlement in the future?
- What are the existing responses to disturbances and what are their limitations?
- What are alternative solutions that can address the limitations of the existing responses?

This study focus on examining the East Beach area in Port Fairy because it is one of the most vulnerable area in this town in terms of soil erosion, flooding and landfill site. Moreover, any problems occurred in this area may affect the settlement nearby.

The secondary data regarding the existing condition in East Beach, Port Fairy was collected from the websites. Moreover, field observation was done in December 2014. This field observation includes brief non-formal interview with the local residents and community leaders. Communication with the local communities enabled researcher to better understand the condition in the study area. Therefore, the recommendations suggested in this study can be accommodated by the local communities and the municipal government to further develop the resilience of their town.

The data collected was analyzed by using SWOT analysis. The SWOT analysis was used to include challenges and opportunities in the problem analysis. This is important to see the problems in broader spatial and temporal context. The result of this analysis was used to formulate some suggested solutions to solve the problems.

4. Result and Discussion

4.1. Problem Analysis

Considering that coastal zone has a strong human-nature relationship [6], the problem analysis discusses both social and ecological aspects of East Beach site. It is critical to relates these aspects because they are interrelated and dependent on each other. The community in Port Fairy has strong social network, intense sense of place and belonging. Moreover, most part of this area is under heritage overlay [7]. These are the positive social aspect that should be taken into account in developing a resilient settlement in this community. However, ageing population is a problem faced by this community [7]. It is critical to attract young people to live in and develop this area.

In terms of ecological aspect there are several environmental problems in East Beach area (see Figure 2). It has a high risk of flooding from the river, the lough and sea level rise. There is also a significant erosion occurs because of waves and lack of sedimentation [8]. This erosion also leads to land contamination because it is affecting the landfill and night soil site located under the sand dunes [9].

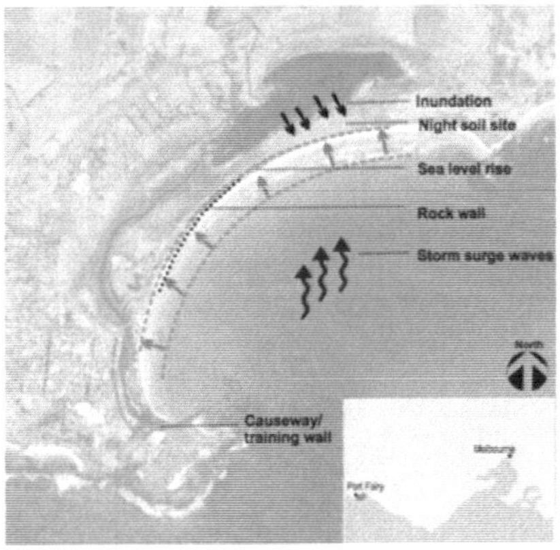

Figure 2. Environmental Problems in East Beach, Port Fairy

Flooding, coastal erosion and the issue of night soil site are major problems in East Beach. Based on the problem analysis (see Figure 2), coastal erosion is caused by both climate change and the presence of causeway [8]. Flooding is also caused by two climate change impacts, sea level rise and storm surge. Moreover, these environmental problems are affecting both human and nature. They endanger human settlement with its economic, social and cultural aspects as well as damaging the coastal ecological system. Solutions for addressing these issues should consider four aspects, minimizing the exposure of night soil site, adapting sea level rise, accommodating severe winds and waves and redesigning the causeway to bring back the sedimentation to East Beach.

In terms of opportunities, there is a local movement called 'sea change' and 'tree change' that aim to create more jobs opportunities in coastal towns [7]. This movement may attract more skilled migrants to live in Port Fairy. In terms of environmental aspect, wind and waves in this coastal area can be used as the sources of renewable energy. The environmental problems such as severe wind and waves can be an opportunity for this town to develop an approach of climate change adaptation. Furthermore, a pilot project of resilient settlement in Port Fairy may be potential for developing an educational tourist destination in Port Fairy. This destination can be a part of the Great Ocean Road tour, one of famous tourist destination in Australia. However, the uncontrolled development of tourism and gentrification in this town are challenges that may lead to the decline of the local identity, volunteering activities and environmental degradation. The summary of problem analysis is depicted in Figure 3.

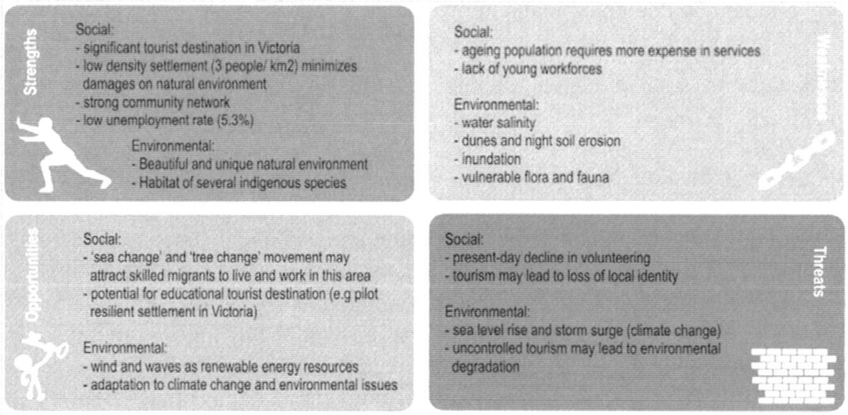

Figure 3. Problem Analysis for East Beach, Port Fairy

4.2. Suggested Solutions

According to SWOT analysis (see Figure 3), several weaknesses and threats are possible to be addressed by utilizing strengths and opportunities owned by this site. Social problem such as ageing population and diminishing young workforces can be managed by the opportunity to encourage skilled migrants to move there. Climate change challenges and environmental issues can be addressed by adaptation. This also creates a potential for developing an educational tourism and an opportunity to creates jobs attractive for young people. Educational tourism can minimize damage on coastal natural environment unlike uncontrolled commercial tourism. These opportunities are supported by some strengths such as strong community network and the beauty and uniqueness of natural environment. These positive aspects provide quality of life there. However, the community must be aware of the negative trends such as decline in volunteering and loss of local identity caused by urban gentrification.

Three strategies were developed to address the issues mentioned before. The first strategy is remediating any kind of land contamination in the coastal area. This strategy can be achieved by treating the landfill site by using phytocapping and bioremediation (see objective 1 in Figure 4). The second strategy is reducing coastal erosion and its impact on East Beach site. This strategy can be implemented by protecting and managing the sedimentation (see objective 2 and 3 in Figure 4). The last strategy is establishing resilient community and settlement. This can be done by accommodating the inundation and flooding problems in the existing settlement (see objective 4 in Figure 4).

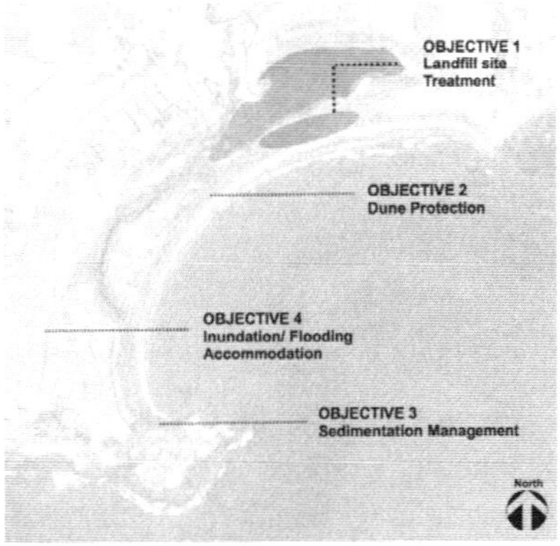

Figure 4. The mapping of the suggested solutions

Objective 1: Landfill site treatment

Considering that excavation of the night soil is expensive and takes time, phytocapping is an alternative solution to deal with this landfill issue. This is a relatively new method to reduce contamination from landfill site. This can reduce water infiltration into the landfill site by using vegetation to keep rain water. This is beneficial to minimize pollution (leachate and methane) from the landfill [10].

The use of vegetation is not only beneficial for avoiding pollution from the landfill, but also useful to stabilize the sand dunes, break storm wind and increase bio-diversity in this area. The plants' roots can protect the dunes from erosion caused by both wind and waves. The presence of vegetation in this site also has a function as wind breaker to protect the residential area from storm wind. Norfolk Island pines are one example of site specific vegetation that is resilient to storm wind. Furthermore, this trees may facilitate some species to reclaim this area and increases bio-diversity there.

However, phytocapping suggests a single type of vegetation namely, Norfolk Island Pines. This response does not create diversity and can be a weak resilient solution. There are several types of indigenous vegetation that can be use in Phytocapping such as Tea Trees and Marram Grass [11].

Moreover, phytocapping does not remediate the night soil site. Other solution such as Bioremediation can be an alternative long term solution complementing phytocapping (see Figure 5). This enables contaminated land to degrade and self remediate [12].

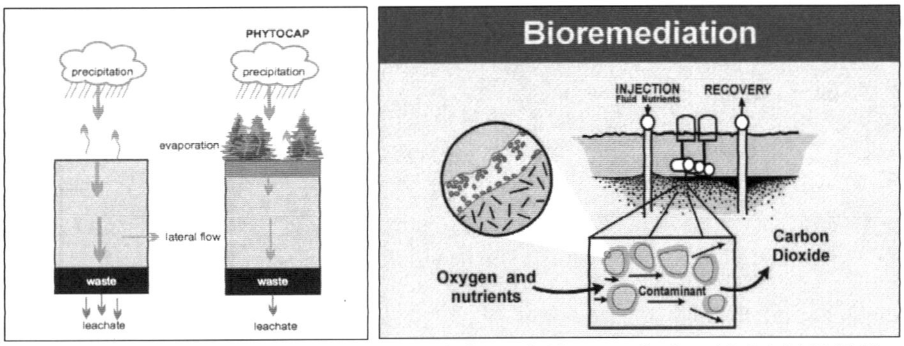

Figure 5. The concept of Phytocapping (left) and Bioremediation (right) [13] [14]

Objective 2: Dunes Protection

Coastal erosion is caused by wind and wave that affect sand dunes. East beach does not have any barriers that can protect its dunes from wind and strong wave. This area has been eroded by roughly 10 meters in 2009 - 2014. A sea wall was established in this area to address this issue. However, this existing solution is not an effective solution because it is just protection and there is no effort to accommodate the storm waves. Artificial reef is one solution that is more effective than sea wall. It does not just protect the dunes from erosion but also can create a new habitat for some fish and a new diving attraction. This offers protection as well as ecological and recreational advantages.

There are several designs of artificial reef that use prefabricated materials such as concrete. This material is relatively less expensive than natural rocks or volcanic rocks [15]. Artificial reef has been installed in South Beach to protect the dunes from strong waves. The design of this artificial reef can be varied based on the need of the site. There is a possibility for community to involve in the process of designing this reef.

Figure 6. Artificial reef can be used for reducing erosion as well as creating new habitat for fish [15]

However, artificial Reef consists of similar moulded concretes that lacks of diversity and aesthetical value. Furthermore, if it is poorly designed, it may be collapse during severe storms and become trash that pollutes marine ecosystem [16]. Its installation also may damage existing natural habitat and it enables concentration of fish unnaturally that may create imbalance ecosystem [17][18]. These problems can be addressed by utilizing natural material for wave breaker such as timber and stones. However, artificial reef is still possible to be successfully implemented as long as planned well. This offers many social and ecological benefits. Additionally, the reef design can be modified as sculptured statues. This may create new attraction in East Beach and attract local artists to contribute to the design.

Figure 7. Wooden Poles and Sculptured Artificial Reef [19][20]

Objective 3: Sedimentation Management

Port Fairy has a significant training wall, the Moyne River Training Wall, that was built in the nineteenth century in order to facilitate the navigation for vessels. However, this wall has an impact on the amount of sedimentation on East Beach. The erosion at East Beach is approximately 3.000-4.000

m3 per year in the past 150 years [9]. This problem occurs because the training wall block the transportation of the sand from the river to the bay. This has created a substantial sediment accumulation on the wall and this had an influence on the erosion on East Beach. The amount of sand brought back to the beach has been decreasing continuously in the last hundred years. Therefore, this training wall should be redesigned to improve the sedimentation on the bay. Simply removing the wall is not a solution because this training wall is listed in the Victorian Heritage Register. It is considered as a historically significant building that represents a nineteenth century regional port infrastructure in Victoria, Australia. Furthermore, removing the wall will affect the vessels' navigation.

This problem may be addressed by using sand by-pass system. This technology has been developed in Goid Coast, Queensland, Australia since 1980s. This system pumps the sand trapped in the training wall system, to the bay. This system may replace the natural movement of sand from the river to East Beach. This system usually consist of several jet pumps located in approximately 11 meters below the sea level [21]. Sand by-pass system is relatively more effective than coastal reclamation.

Objective 4: Inundation Accommodation

The inundation affecting the housings in East Beach can be addressed by constructing flood resilient home designs. Nowadays, there are various kinds of design that enable the housings to accommodate severe flood events. It is important to design a house that is not just resistant to a particular level of flood, but the design must ensure that the house can accommodate any levels of flood. Most modern flood resilient house designs use the concept of traditional elevated house. However, this type of design is limited to a certain level of flood. Practically, people cannot predict exactly the maximum level of flood. Therefore, a kind of elevated house that is able to inflate in a case of a great flood is needed. This kind of design has been developed in Netherland because more than half of its settlement is below the sea level [22] (see Figure 8). Besides, educating the community to be ready for any storm surge and flooding is also important.

Figure 8. Wooden Poles and Sculptured Artificial Reef [22]

5. Conclusion

The existing engineering approach implemented in East Beach, Port Fairy cannot effectively address the complex environmental problems that occur there. The existing solutions mostly focuses on short-medium term and does not address the root problem. The community also has not been engaged in the implementation process. Community should be actively involved in solving the root problems and enhancing diversity in both short-medium and long term scale. The creativity of the local community should be accommodated to develop eco-tourism and enhance the economic sustainability in East Beach particularly and Port Fairy in general.

This study provides recommendations that aim to overcome the limitations of the existing approach implemented in East Beach. Landfill site should be treat by using natural-long term method such as phytocapping. The dunes should be protected by installing beautiful artificial reef that designed by local artists. This can protect the dunes as well as creating a new tourist attraction in this area. Furthermore, the sedimentation problem should be tackled by adding a sand by-pass system. Lastly, the inundation problem should be addressed by implementing a kind of floating house construction especially for the dwellings located near the shore.

The lesson learned from this study can be used to improve the condition of coastal settlement in Indonesia. The problems occur in East Beach are also common in Indonesia. The strategies mentioned to address coastal erosion and inundation in East Beach can also be adopted with modification in Indonesian costal town such as Jakarta, Semarang, Makassar and Denpasar.

Acknowledgments
I would like to thank Australia Awards for providing scholarship to study in the University of Melbourne, Australia. This paper was written based on my experience involved in a field trip of a summer course in my master's degree.

References

[1] Davoudi S, Shaw K, Haider L J, Quinlan A E, Peterson G D, Wilkinson C, Fünfgeld H, McEvoy D, and Porter L 2012 Resilience: A Bridging Concept or a Dead End? "Reframing" Resilience: Challenges for Planning Theory and Practice Interacting Traps: Resilience Assessment of a Pasture Management System in Northern Afghanistan *Urban Resilience: What. Planning Theory & Practice*, **13**(2), pp. 299-333.
[2] Kuei-Hsien L 2012 A Theory on Urban Resilience to Floods--A Basis for Alternative Planning Practices. *Ecology & Society*, **17**(4), pp. 627-640
[3] Beilin R., Reichelt N T, King B J, Long A, and Cam S 2013 Transition Landscapes and Social Networks: Examining On-Gound Community Resilience and its Implications for Policy Settings in Multiscalar Systems. *Ecology & Society*, **18**(2), pp. 99-113.
[4] IPCC 1997 *The Regional Impacts of Climate Change.* Victoria. Viewed on 18 December 2014. http://www.ipcc.ch/ipccreports/sres/regional/index.php?idp=223
[5] McEvoy, D., Fünfgeld, H., & Bosomworth, K. 2013 Resilience and Climate Change Adaptation: The Importance of Framing. *Planning Practice & Research*, *28*(3), 280–293. http://doi.org/10.1080/02697459.2013.787710
[6] Wilkinson, C. 2012 Social-ecological resilience: Insights and issues for planning theory. *Planning Theory, 11*(2), pp. 148-169.
[7] Moyne Shire 2007 *Moyne Futures: Moyne Shire Community Plan.* Port Fairy. Moyne Shire Council. viewed on 6 December 2014. http://www.moyne.vic.gov.au/Files/Moyne_Futures_Community_Plan.pdf
[8] BMT 2007 *Port Fairy East beach Coastal Erosion Engineering and Feasibility Study.* Victoria. Viewed on 18 December 2014. www.moyne.vic.gov.au/Files/EastBeachReport.pdf
[9] Flocards F, Carley J, Rahman P, and Coghlan 2013 *Future Coasts: Port Fairy Coastal Hazard Assessment. UNSW Water Research Laboratory for Shire of Moyne.* Victoria. Viewed on 18 December 2014. www.moyne.vic.gov.au/Files/Port_Fairy_Coastal_Hazard_Assessment.pdf
[10] Venkatraman K and Ashwath N 2007 Phytocapping: An alternative technique to reduce leachate and methane generation from municipal landfills. *The Environmentalist*, **27**(1), pp. 155-164.
[11] Moyne Shire Council 2014 *Indigenous Plants of Moyne Shire Booklet.* Victoria. Viewed on 12 December 2014. http://www.moyne.vic.gov.au/page/page.asp?Page_Id=1015&h=0
[12] Choi, S.-C., Lee, Y.-H. and Oh, Y.-S. 2003 Treatability tests for the bioremediation of unsanitary landfill waste soils. *JOURNAL OF MICROBIOLOGY-SEOUL-*, **41**(2), pp. 169-173.
[13] Pacific Environment 2014 *Phytocapping Landfill Technology.* Viewed on 18 December 2015. https://www.pacific-environment.com/solutions/services/phytocapping-landfill-technology/
[14] Mohan, Ravi (2014) *Cleaning up the planet with bioremediation.* Viewed on 18 December 2014 http://followgreenliving.com/cleaning-planet-bioremediation/
[15] Reef Ball Foundation 2014 *Reef Ball Artificial Reefs: Shared Resources for all Reef Ball Companies.* Viewed on 18 December 2014. http://www.artificialreefs.org/
[16] Yip, M. (1998) *An Overview of Artificial Reefs.* Salzburg. Viewed on 18 December 2014. biophysics.sbg.ac.at/ar/reef.htm
[17] Aspen Point Productions Inc. 2014 *Artificial Reefs: The Good, The Bad, The Controversial.*

Viewed on 18 December 2014. http://www.discover5oceans.com/2010/01/artificial-reefs-the-good-the-bad-the-controversial/

[18] Wilding T A and Sayer M D 2002 Evaluating artificial reef performance: approaches to pre-and post-deployment research. *ICES Journal of Marine Science: Journal du Conseil*, **59**(suppl), pp. S222-S230.

[19] Taylor, Jason de Caires. 2014 *Eerily Beautiful Underwater Sculptures: Art Transformed into Artificial Reefs.* Viewed on 18 December 2014. http://www.lovethesepics.com/2013/10/eerily-beautiful-underwater-sculptures-art-transformed-into-artificial-reefs-pics/

[20] Tours and Tales.com 2014 *Beach in Zeeland.* Viewed on 18 December 2014. https://www.flickr.com/photos/58593105@N00/8138170996

[21] Gold Coast Waterways Authority 2016 *The Sand By-Pass System Fact Sheet.* Viewed on 20 September 2016. http://www.gcwa.qld.gov.au/userfiles/resources/static/GCWA_Sand_Bypass_Fact_Sheet.pdf

[22] Boo M D 2005 *Living on Water: A floating town on an unsinkable-foundation – complete with gardens, streets and squares.* Viewed on 20 September 2016. http://www.tudelft.nl/fileadmin/UD/MenC/Support/Internet/TU_Website/TU_Delft_portal/Akictueel/Magazines/Delft_Integraal/archief/2005_DI/2005-1/doc/DO051-18-22.pdf

Assessing the Sustainable Development of Coastal Reclamation: A Case of Makassar Using GIS Application

A Yurnita[1], S Trisutomo[2], M Ali[3]
[1]Post Graduate Student, Department of Architecture, Hasanuddin University, Makassar 90245, Indonesia
[2]Professor, Department of Architecture, Hasanuddin University, Makassar 90245, Indonesia
[3]Associate Professor, Department of Architecture, Hasanuddin University, Makassar 90245, Indonesia

E-mail: nita75_andi@yahoo.com

Abstract. Reclamation has been made in many areas in Indonesia including Makassar, as a response to the present needs of land as the impact of human activity in urban area. This research aims to assess the sustainable development of coastal reclamation and focus on environmental dimension of sustainable urban development. Assessment will be done by reclamation sustainability index (RSI) and analysis by GIS as the tools. RSI was built from previous research that has simplified from many researches and analysis by Structure of Analytic Hierarchy Process (AHP) and expert choice. RSI uses 9 indices from three indicators of environment factor which are coastal resources, building and infrastructure.

1. Introduction

The paper explores the latest urban dynamics across Java, the most developed and most densely Given the rapid development and growing population in the city of Makassar, the housing needs are also getting higher. The coastal zone in Makassar shows a rapid change during the past few decades, including large-scale reclamation projects which have been carried out along the coastal line [1]. However, these large-scale reclamation projects have already caused a series of negative effects to the marine ecological environment [2].

By definition, sustainable development of coastal zone does not only meet the increasing demand, but also protects ecology and environment, without prejudice to the future generations access to adequate food security [3]. In order to provide a scientific basis for decision-makers, therefore, it is necessary to assess the status of coastal reclamation development [3].

Reclamation Sustainability Index (RSI) is a tool of assessing the process of coastal reclamation areas in terms of their sustainability development. RSI can assist in assigning coastal reclamation development to sustainable environment and help to shape (or reshape) coastal development plans to create a well laid-out coastal environment. The planners need the process of appraising and grouping specific coastal areas in terms of their sustainability for defined uses [4]. The process is using geographic information system (GIS) as a tool for data input, storage, retrieval, manipulation, analysis and output of the spatial data [5].

IOP Conf. Series: Earth and Environmental Science **79** (2017) 012013 doi:10.1088/1755-1315/79/1/012013

However, because of the different research purposes, time and condition, no further study related to assessment of sustainability of the reclamation project on the coastal environment using RSI has been carried out yet. This paper aims to assess the sustainability or coastal reclamation area in Makassar city using RSI and analysis by GIS.

2. Theoretical Transition to Asian Background

2.1 Study Area

Geographically, Makassar city is located on the West coast of the southern part of South Sulawesi, at coordinates between 119 ° 18 27.97 " "up to 119 ° 32 " 31.03" East longitude and 5 ° 30 18 " "-" 14 49 5 ° South latitude." [6] The city is the capital of Makassar of South Sulawesi Province and is the fourth largest city in Indonesia whose area is 175.79 km^2 with coastal line 52.8 km consisting of main land coastline 36.1 Km, and along small island 16.7 km. This research was conducted in south coastal zone of Makassar, South Sulawesi Province [7] based on South Sulawesi Governor Regulation number 17 of 2008 concerning Detail Spatial Planning of Global Integrated Business Area.

Figure 1. Location of the Study
Source: Satellite Image, 2014

2.2 Developing Assessment for Sustainability Index

The AHP-Expert Choice evaluation method has been refined into four steps [8]; Determination of Evaluation Index System, Weighting of Indices, Arranging Matrix, Testing Consistency, and Determination of Priority. The AHP structure of the indices can be seen in next figure.

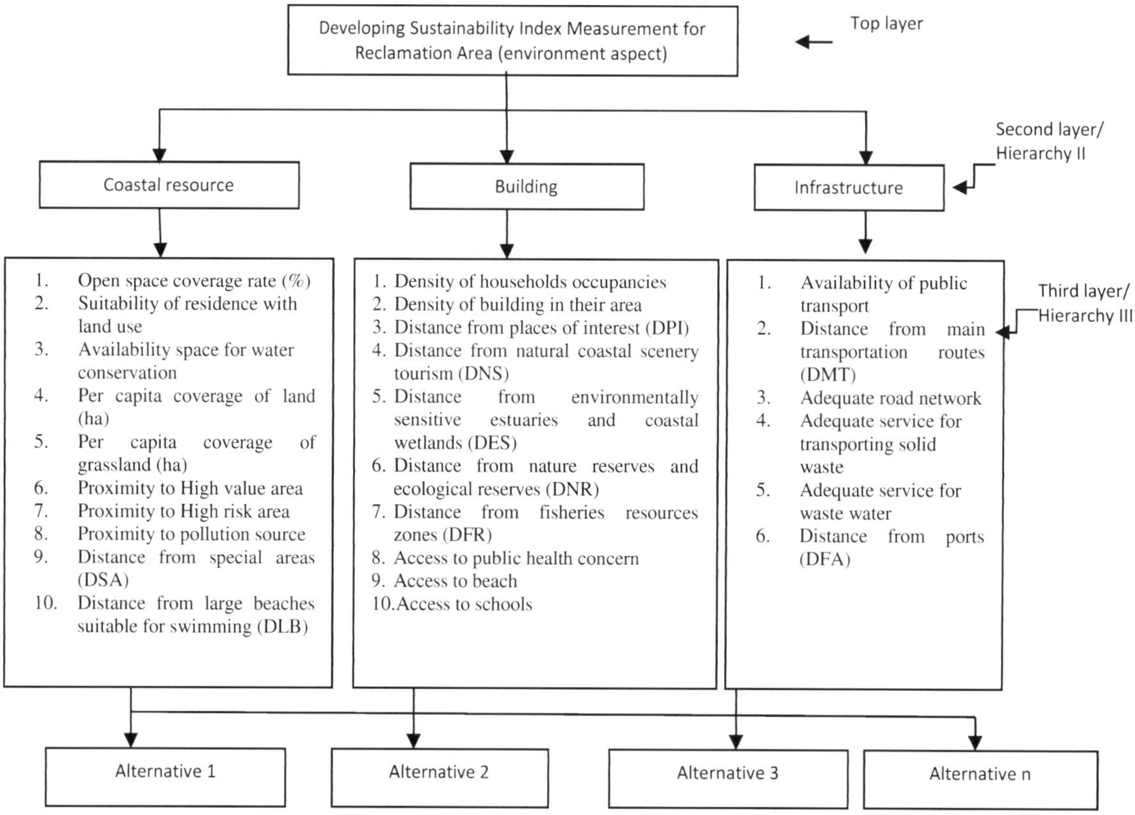

Figure 2. Structure of Analytic Hierarchy Process (AHP)
Source: Yurnita, 2016

To simplify the indices, this research is analysed through AHP and Expert choice. There were three types of experts who served as respondents whose opinions are considered and included in the matrix. They were from professional fields, government staffs and academics. They chose the most important indicators by pair wise system, using expert choice [9].

The summarized criteria obtained from experts' opinion from 72 criteria disclosed 26 applicable indicators, and from AHP, it listed 9 indicators which were most important according to the experts' choice. The experts assessed whether an indicator has more important influence than other indicators with a range of assessment 1 – 9, then ranked and graded. Top rank is environment effort rather than manmade, consistent across all interviewees, as shown in table 2.

2.3 Digital image processing

The study area was represented by Geo eye satellite images as of June 2014. The images were geometrically corrected. Rectification method (image for map) was followed. A mosaic process was elaborated to overlay the images.

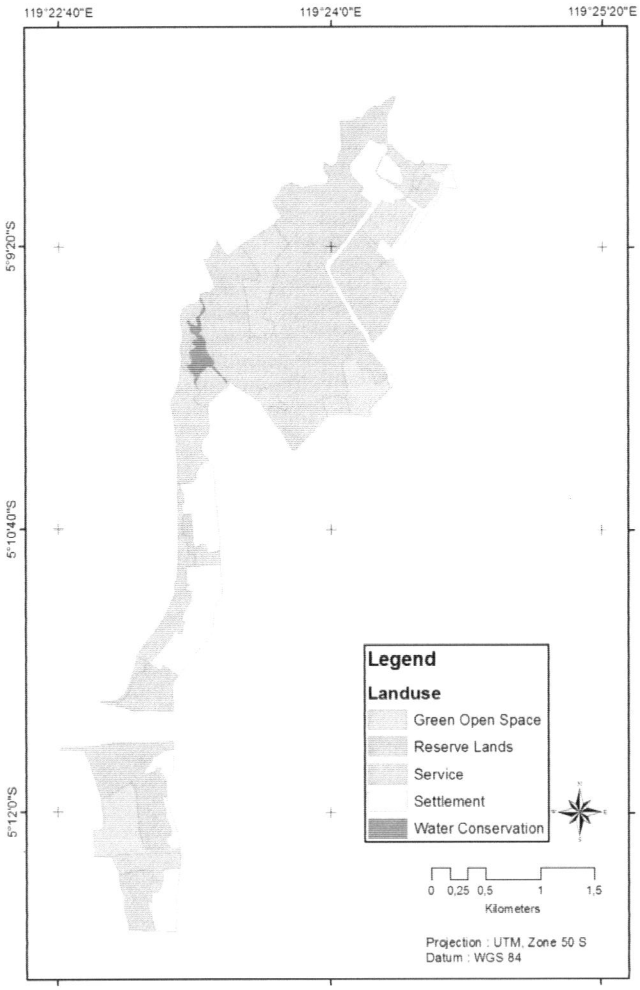

Figure 3. Land Use of Study Area.
Source: Result of GIS

2.4 Sustainability Index

Three factors which were selected for monitoring and assessing [10] the sustainability of reclamation in the study area which could be categorized as coastal resources, building and infrastructure. This factor consisted of 9 indices, that were found from previous research simplified from many indicators from many researches by Analytic Hierarchy Process (AHP) and Expert Choice.

Table 1. The Criteria of Reclamation Sustainability Index (RSI).

Category	Sub Category	Indices of sustainability
Coastal resource	Open space coverage rate (%)	3. >30 % of the area: good 2. 10–30 % of the area: poor 1. 0–10% of the area: bad
	Availability Space for water conservation	3 available: good 2. less available: poor 1. no available: bad
	Per capita coverage of land (ha)	Ha
Building	Distance from environmentally sensitive estuaries and coastal wetlands (DES)	3. >500m away from DES :good 2. 1–500m away from DES: poor 1. Within DES: bad
	Distance from nature reserves and ecological reserves (DNR)	3. >500m away from DNR: good 2. 1–500m away from DNR :poor 1. Within DNR: bad
	The Density of building in their area	3. Suitable: good 2. less Suitable: poor 1. not Suitable: bad
Infrastructure	Adequate road network	3. available: good 2. less available: poor 1. no available : bad
	Availability of public transport with weights	3. available: good 2. less available: poor 1. no available : bad
	Distance from main transportation routes (DMT)	3. 100-200 away from DMT : good 2. 200–500m away from DMT : poor 1. >500m away from DMT: bad

Source : Yurnita, 2016, (9)

2.4.1 Coastal Resources Indicator

The coastal resources indicator was calculated by three indices, open space coverage rate (%), availability space for water conservation and per capita coverage of land (ha) factor.

2.4.2 Building Indicator

Building indicator was assessed based on survey work, which identified the existence of the building in reclamation or nearby reclamation area. The building indicator considered three determining indices: distance from environmentally sensitive estuaries and coastal wetlands (DES), distance from nature reserves and ecological reserves (DNR), The density of building in their area.

2.4.3 Infrastructure Indicator

The infrastructure index was calculated using three factors: adequate road network, availability of public transport with weights, and Distance from main transportation routes (DMT).

Determination of the sustainability status was expressed by index and status of sustainability. Scale of index system of city sustainability examined in this study had three (3) categories of status of the sustainability of the city. The value of the index and status of sustainability can be shown in Table 2 [11]

Table 2. The Criteria for the Sustainability Index

Index value	Sustainability Status
3	Good (very sustainable)
2	Poor (sufficient sustainability)
1	Bad (not sustainable)

Source: Renald, 2015 and modified by author

3. Result and Discussion

Reclamation Sustainability Index (RSI) in the current study was focused on three factors: coastal resources, building and infrastructure and the was done by making use of geographic information system.

3.1 Sustainability indicators

This paper made use of geographic information system. The sustainability of the reclamation area was identified based on the Land sat ETM+ images, that has been derived from ASTER images as in Figure 2. The results obtained as shown in Figure 2, revealed that sustainability indicators in the current study were focused on nine factors; open space coverage rate (%), the availability Space for water conservation, per capita coverage of land (ha), distance from environmentally sensitive estuaries and coastal wetlands (DES), distance from nature reserves and ecological reserves (DNR), The density of building in their area, adequate road network, availability of public transport with weights and distance from main transportation routes (DMT).

3.1.1 Coastal resource index

The results showed that an area about 15% of the total investigated area is open space. From the RSI, it means that the area is classified as less sustainable, because less than 30% and the water conservation area is about 16 % it means it is available. If so, it is classified as good or sustainable. Then the last index from the coastal resource is per capita coverage of land (Ha). In Makassar, there is no certain amount to measure this indicator, because it has to count the land that people need to live, not only the house but also the land of school, work, recreation which all of them must be countable. This condition of the index is as shown in Fig. 4.

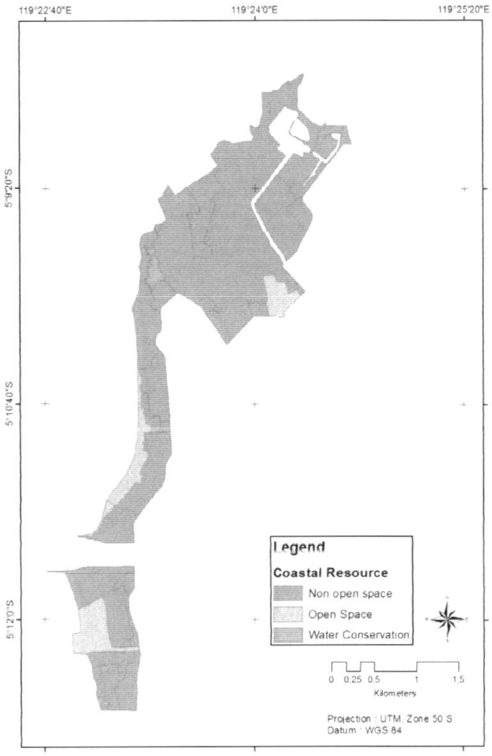

Figure 4. Coastal resource in the study area.

3.1.2 Building index

The results of the building index which consist of three indices show that the distance from environmentally sensitive estuaries and coastal wetlands is bad or not sustainable, because some part of the reclamation area is within the sensitive estuaries and coastal wetlands. The second indices show that the distance from nature reserve and ecological reserves is far from the reclamation area, which is more than 500m away. The next index is the density of building in their area which is suitable with Makassar Spatial Planning [6] it means it is sustainable as shown at figure 5(a).

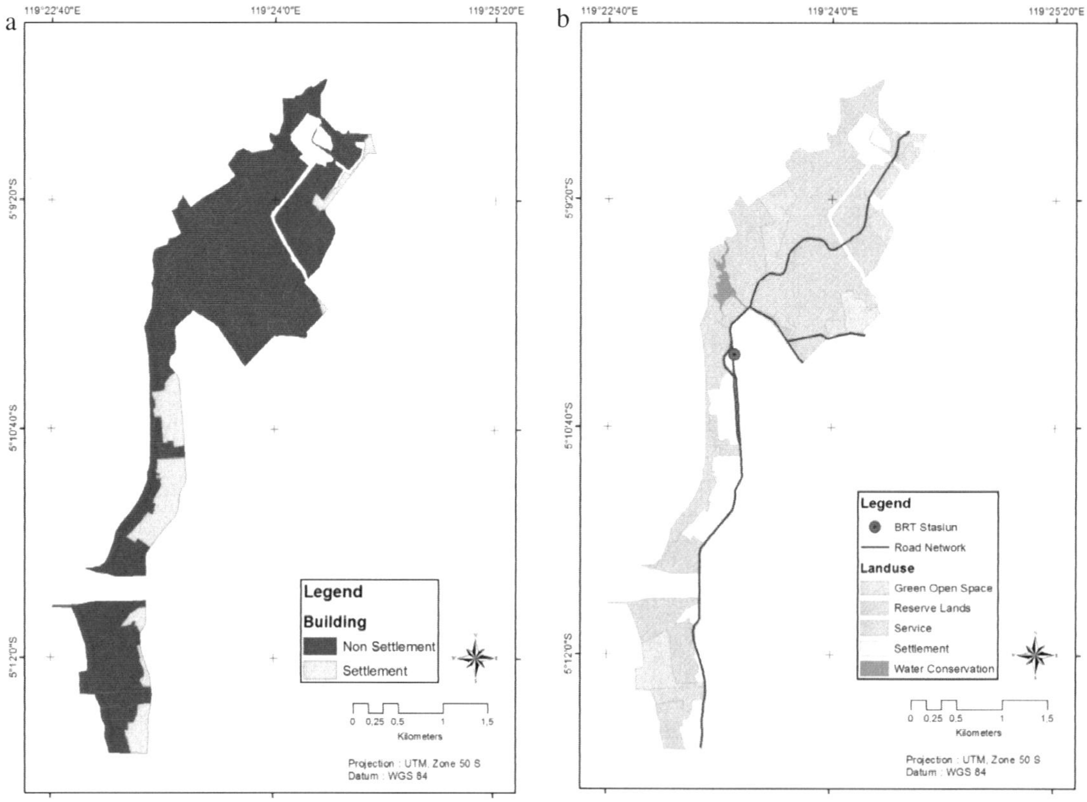

Fugure 5. a) Building index in the study area; (b) Infrastructure in the study area
Source: Result of GIS

3.1.3 Infrastructure index

The infrastructure index obtained was classified into three indices, as shown in the Fig 5(b). First of these three indices is adequate road network. From the map, it can be seen that the reclamation area has adequate road network, meaning to say it is sustainable. The availability of public transport in the reclamation area is available, so it is sustainable. The last is the distance from main transportation routes is about 420M, it means poor sustainable.

4. Conclusion

The current work assessed reclamation in terms of sustainable development in the Makassar coastal area. It depended on three factors (coastal resource, building and infrastructure) by Reclamation Sustainability Index (RSI) that was built in previous research. At coastal resource index, it can be said that the open space was classified as less sustainable, because it accounted for less than 30% and the water conservation area was about 16 % which means it is available, so it can be said it is sustainable, for index per capita coverage of land (Ha) in this paper is not counted because in Makassar, there is no certain amount to measure this indicator.

The building index from three indices show that there is one factor classified as not sustainable that is the distance from environmentally sensitive estuaries and coastal wetlands which is bad or not sustainable. Another index is classified as sustainable status.

The infrastructure index in this paper is sustainable because the from analysis found that reclamation area in Makassar has adequate infrastructure. This paper concludes that land reclamation in Makassar city is sustainable. Finally, unused index was found for research in Makassar that is index per capita coverage of land (Ha) because in Makassar, there is no certain data to measure this indicator.

Acknowledgements
The authors wish to express their gratitude to the teamwork in Lab of Urban and Regional Planning, Hasanuddin University for their valuable and helpful comments, which have improved the quality of the paper. The paper was carried out in the Program Doctoral, Department of Architecture Hasanuddin University 2016.

References

[1] Azwar SA, Suganda E, Tjiptoherijanto P, Rahmayanti H 2013 Model of Sustainable Urban Infrastructure at Coastal Reclamation of North Jakarta *Procedia Environmental Sciences* **17** 452-461
[2] Li K, Liu X, Zhao X, Guo W 2010 Effects of Reclamation Projects on Marine Ecological Environment in Tianjin Harbor Industrial Zone *Procedia Environmental Sciences* **2010(2)**
[3] Yua L, Xiyong Houa, Meng Gaoa PS 2010 Assessment of coastal zone sustainable development: A case study of Yantai, China *Ecological Indicators* **10(25)**.
[4] Feng L, Zhu X, Sun X. 2014 Assessing coastal reclamation suitability based on a fuzzy-AHP comprehensive evaluation framework: A case study of Lianyungang,China *Marine Pollution Bulletin* **89**
[5] Ismail M, Abdel Ghaffar MK, Azzam MA 2012 GIS application to identify the potential for certain irrigated agriculture uses on some soils in Western Desert, Egypt. *The Egyptian Journal of Remote Sensing and Space Science* **15** 39-51.
[6] Makassar PDK. Rencana Tata Ruang Wilayah (RTRW) Kota Makassar Tahun 2015 – 2035 In: Makassar BK, editor. Makassar: Pemerintah Daerah Kota Makassar; 2015.
[7] Suhayati, Bachri S, Patittingi F 2013 Aspek Hukum Kebijakan Pemanfaatan Tanah Haasil Reklamasi Pantai Losari di Kota Makassar.
[8] Yurnita A, Trisutomo S, Ali M 2016 Developing Sustainability Index Measurement for Reclamation Area *Fourth International Conference on Sustainable Built Environment*
[9] Mohamed ES, Schütt B, Belal A 2013 Assessment of environmental hazards in the north-western coast - Egypt using RS and GIS. *The Egyptian Journal of Remote Sensing and Space Science* **16(2)**
[10] Renald A 2015 Model Adaptasi Ketahanan Kota Rawan Bencana Banjir untuk Keberlanjutan Kota Studi Kasus: DKI Jakarta (Jakarta: Universitas Indonesia)

Visual Assessment on Coastal Cruise Tourism: A Preliminary Planning Using Importance Performance Analysis

S Trisutomo
Department of Urban and Regional Planning
Hasanuddin University
Makassar, Indonesia

E-mail: trisutomo2gmail.com

Abstract. Importance-Performance Analysis (IPA) has been widely applied in many cases. In this research, IPA was applied to measure perceive on coastal tourism objects and its possibility to be developed as coastal cruise tourism in Makassar. Three objects, i.e. Akkarena recreational site, Losari public space at waterfront, and Paotere traditional Phinisi ships port, were selected and assessed visually from water area by a group of purposive resource persons. The importance and performance of 10 attributes of each site were scored using Likert scale from 1 to 5. Data were processed by SPSS-21 than resulted Cartesian graph which the scores were divided in four quadrants: Quadrant I concentric here, Quadrant II keep up the good work, Quadrant III low priority, and Quadrant IV possible overkill. The attributes in each quadrant could be considered as the platform for preliminary planning of coastal cruise tour in Makassar

1. Introduction

Importance-Performance Analysis (IPA) was introduced and proposed by Martilla and James in 1977 [1]. Initially it was applied to measure the satisfaction level of the marketing condition and the behavioural of consumers. Later, this method was applied in e-Government, higher education service quality [2], health care service, service quality of administration in private universities [3], the performance of organisation or public policy [3], tourism accommodations [4], and tour guide evaluation [5]

By the times, the IPA was improved by many scholars. Huan and Beaman in 2006 criticized the method and contributed to develop the solitude for effective use [6]. Abalo, Varela, and Manzano also suggested to improve the method by seeking a formula for spreading out values derived from preference ranking [7]. Shortly, IPA was widely applied in any kind of management issues.

IPA model bases on importance and performance attributes. The term of importance is seen as view as a reflection of the relative value of various quality attributes. It also reflects the expectation of assessors. By meant of performance reflects the real condition of the objects based on perceive of assessors.

Indonesia as an archipelagic country which has more than 13.000 islands and has the second longest coastline in the world. Having a tropical climate, this country riches with many natural resources that possible to be developed as maritime tourism. Coastal cruise tour (CCT) is one the maritime tourism that the visitors cruise along the coastline and enjoy the lure of them. Moscardo in 1996, Morrison, et al. in 2003, Kim and Kim in 1996, describe that CCT is the new trend of global scale as a part of marine tourism [8][9][10]. According to them the CCT is to explore the world by using a very large vessel capable of thousands of tourists, sail the ocean between countries, so often called ocean cruise tour. The essence of this kind of tour is the exploration of the island and the ocean on a global scale. Although this country has potential coastal resources, CCT is being paid less attention by local government. This research was addressed to assess the potentials sites as the objects of CCT by applying the IPA model. It will be a preliminary planning effort to develop the CCT in Makassar.

2. Literature Review

Referring to Moscardo, Morrison, et al., Kim and Kim cruise tour is traveling down the coast, see the scene from the sea towards the land [9][10][17]. This kind of tour is very different from sea travel in general terms, which enjoy the beauty of the sea from the land. Conversely, coastal cruise tour is to see the sights on land from the sea. Just an example, in Sydney, Brisbane and Hong Kong, they operate opened deck vessel with a capacity of 100-150 seat, moving gently to enjoy the beautiful of the beaches. The tour creates unique sense for the tourists.

Gungor and Bozyigit urged to measure the components of the urban landscape based on the value of culture (cultural value), identity, artistic, water, historical, and heritage [11]. Sari and Darmawan suggested that the potential of a coastal tourism should have an element of interest, historical marker (historical landmark), and special attractions [12]. Butler and Waldbrook add access as one of the requirements for coastal tourism [14]. Edy (described that element of public space includes neatness, visual variety. Kher et.al suggested that the element should be assessed is the cleanliness, identity, coolness, uniqueness, beauty and convenience [15]. Furthermore, Ciftci measured the visual landscape elements beach with pollutants, landscape, natural elements (currents and waves), and the weather [16]. Based on these previous studies and related to the problem and the purpose of this research, visual objects can be classified into 3 groups: water, landscape and buildings. In the waters of the elements include sea water cleanliness, beautiful of the coast, public space. For natural elements include landscape, color harmony, panorama, freshness of atmosphere. And building elements include uniqueness, landmark, and feeling save.

Summing up, in this research the importance and performance were assessed using 10 attributes. Following are descriptions of the attributes:

(1) sea water cleanliness. The water body shows no garbage, trash, oil, and any other disposal. It also shows low water turbidity

(2) beautiful of coast. This attribute describes the nice condition of the coast

(3) public space. It indicates that availability of place that enables people to enter to come and to do any activities

(4) natural landscape means the availability of natural components such any vegetations

(5) harmony of colour means the combination of many colours

(6) panorama. Same with scenic or view

(7) atmosphere. This attribute figures out the freshness of air condition

(8) uniqueness. Unique means attractive and has differences than other, and very special

(9) landmark. This describes the availability of an attractive building

(10) feeling of save. The object provides a feeling that it was save for visitors, not endanger

All these were explained clearly to the assessors.

3. Methods

This research is a combination of quantitative and qualitative approach using IPA method. The research was focused on visual assessment of three sites i.e. Akkarena, Losari, and Paotere. Akkarena is a recreation site at the coastlines, Losari is a waterfront public space, and Paotere is a traditional port. Those three visual objects currently are the most interesting sites among existing objects at the coast of Makassar. Fourteen persons were selected purposively as assessors, 3 from local government staffs of Tourism Agency, 4 from academicians, and 3 from local government staffs who concern with the coastal development, 2 from local persons who know well about the coastal area of Makassar, and the rest were persons who work with travel agency. Before doing the assessment, the assessors were introduced and explained the attributes. A set of questionnaires consisted of 10 attributes i.e. (1) sea water cleanliness, (2) beautiful of coast, (3) public space, (4) landscape, (5) harmony of colour, (6) panorama, (7) atmosphere, (8) uniqueness, (9) landmark, and (10) feeling save. The questionnaire was supported by a set of photos of the objects which were taken from sea in many angels; this may help the assessors to recognize three objects. Using rental speedboat, the assessors were cruising along the coastlines, approaching the three sites at 25-50 m distance away, and stopped there in about 10-15 minutes to allow assessors to enjoy and to assess the objects and surrounding panorama. Than assessors were pleased to pay careful attention to the objects, then gave the score by filling the questionnaires. To understand the objects, assessors may see the available photos of the sites. With the 10 attributes, assessors should score the importance and the performance level. The importance level showed the attributes indicates had high priority level in creating the objects, while the performance level represented the existing condition based on the view of assessors. The scoring used the Likert scale, from 1 as the lowest to 5 as the highest one. In term of performance, score 1 = means the very poor and 5 = very nice view; in term importance 1 = the worst and 5 the best one. Collected data than were analysed using IPA method through SPSS 21. To create and resulted Cartesians graph, a two-dimensions grids, importance value became the X axis and perform value became the Y ordinate. The graph was broken into 4 quadrants (Q): QI concentric here, QII keep up the good work, QIII low priority, and QIV possible overkill.

4. Result and Discussion

Makassar metropolitan is the capital of South Sulawesi Province, and holds important role as a centre of national development according to the National Spatial Planning. This waterfront city has 31 km length of coastline of Makassar Straits and this city was inhabited by 1.6 million population. Having beautiful sunset, there are many potential tourism objects site at the waterfront. Those are resources that have high possibility to be developed as cruise tour objects that could be enjoyed by visitors by cruising from the bridge of Barombong in the south side to the mouth of Tallo River in the east in the future. Sitting at the strategic point as a hub in terms of sea and air transportation, this position enables to accelerate the growth of urban development.

This visual assessment selected tree tourism objects at the coast (Fig 1). They were Akkarena, a recreation site at south side, Losari public waterfront at the downtown, and Paotere a traditional Phinisi port. Fig. 2 to 4 show pictures figuring the existing condition seen from the sea water while cruising along the coastlines.

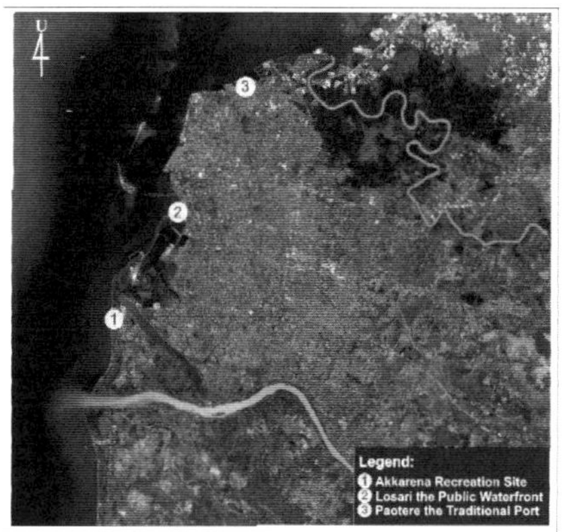

Figure 1. Map of Coastal and The Location of Research

4.1 Description of The Sites

a. Akkarena, The Coastal Recreation Site

This recreation area is owned by private real estate of Tanjung Bunga area. The site has 5 ha open space, consists of clean coastlines, green area, open landscape, restaurant, jetty, and parking area. People are able to enjoy many activities here: swimming, boating, sailing, surfing, and diving. As seen from water side, this area provides nice panorama and colorful landscape especially during the sunset.

Figure 2. Pictures of Akkarena recreation area

b. Losari, the waterfront public space

Starting in 1998, the revitalization work was beginning by doing sheet piles construction and land reclamation. Because of very limited budget, the works were done in many phases, part by part. After taking long period of coastal reclamation and construction, in 2002 the first plaza, named Pantai Losari, was completely built. Next was Bugis-Makassar plaza in 2007 and Mandar-Toraja plaza in 2008 were built. The name of the plazas represents ethnics groups which do exist in South Sulawesi. The waterfront plazas perform many uniqueness. Pantai Losari plaza, the first spacious plaza, was completely built in 2002. It was completed with benches, big planting boxes, two sculptures of Adipura Award trophies, and a jetty for boating. Bugis-Makassar plaza was the next plaza which has been completely built in 2004. It was furnished by sculptures of *becak, paraga,* Phinisi traditional boat, and *pepe-pepe* dance. In the northern side, Toraja Mandar plaza was built in 2008. It presented white buffalo (a magic animal of Toraja custom), miniature of Toraja architecture, a dancing lady with Toraja tribe dress, and a woman with traditional woven equipment. In 2012 the Amirul Mukminin mosque, a floating mosque, and 20 sculptures South Sulawesi warriors – who fought against Dutch Colonial - were built. These are objects of religious and educative arena. Many activities also available in the water area, such boating and sailing, skiing, and just walking or watching people

Figure 3. Map and pictures of Losari waterfront public space

The spacious public space has a clean and fresh atmosphere provide an openness and healthy feeling. As seen from water side, Losari waterfront is nice to look. Although the edge lines of the reclamation construction are looked hard and rigid, the harmony of colour among components create very nice panorama, particularly during the sunset times.

c. Paotere, the Phinisi traditional port

Since the colonial era, Makassar was known as a huge of maritime port. The city is not only having and operates large and modern port in Sukarno-Hatta quay, but also has Phinisi boat port at the northern side. Hundred traditional Phinisi boats, small and large, are anchored there to load and unload commodities to the eastern part of Indonesia. The architecture of Phinisi traditional ships are interesting scene seen for from the sea side. Unfortunately, less control in discharged waste from the boats influence the cleanliness of sea water area.

Figure 4. Map and pictures of Paotera traditional phinisi boat

4.2 Analysis using IPA method

Table 1, 2, and 3 show the result of assessment on 10 attributes of importance and performance. Each attribute of importance and performance were score with Likert scale from 1 to 5.

Table 1. Assessment on Akkarena and the Cartesians graph

	Attributes	Mean Importance	Mean Performance	Discrepancy
1	Seawater cleanliness	4.29*	3.97*	-0.32
2	Beautiful of coast	4.14*	4.07*	-0.07
3	Public space	3.90	4.14*	0.24
4	Landscape	4.05*	3.59	-0.46
5	Harmony of color	4.00*	3.76	-0.24
6	Panorama	3.86	3.62	-0.24
7	Atmosphere	4.19*	4.21*	0.02
8	Uniqueness	3.29	3.38	0.09
9	Landmark	3.67	4.28*	0.61
10	Feeling save	4.48*	4.38*	-0.10
	Mean	3.987	3.940	

*higher than the mean

Akkarena, a recreation area at the coast, the mean of importance score was 3.987; there were 6 attributes higher the mean i.e. (1) sea water cleanliness, (2) beautiful of coast, (4) landscape, (7) atmosphere, and (10) feeling save. In terms of performance the attributes have score higher than mean were (1) sea water cleanliness, (2) beautiful of coast, (3) public space, (7) atmosphere, (9) landmark, and (10) feeling save. Based on the mean, the real conditions of Akkarena is below the expectations of assessors.

For Losari waterfront as public space, the importance attributes which have score higher than the mean (3.488) were (1) sea water cleanliness, (4) landscape, (8) uniqueness, (9) landmark, and (10) feeling save. For the performance assessment on the attributes (7) atmosphere, (8) uniqueness, (9) landmark, and (10) feeling save, are higher than mean. (3.274). The mean of importance was higher than the importance, means assessors had higher expectations than the real conditions.

Table 2. Assessment of Losari

	Attributes	Mean Importance	Mean Performance	Discrepancy
1	Seawater cleanliness	3.95*	2.90	-1.05
2	Beautiful of coast	3.24	2.38	-0.86
3	Public space	2.71	3.14	0.43
4	Landscape	3.67*	2.31	-1.36
5	Harmony of colour	3.10	3.10	0
6	Panorama	3.10	2.76	-0.34
7	Atmosphere	3.10	3.66*	0.56
8	Uniqueness	3.71*	3.28*	-0.43
9	Landmark	4.43*	4.66*	0.23
10	Feeling save	3.90*	4.55*	0.65
	Mean	3.488	3.274	

*higher than the mean

Table 3. Assessment of Paotere

	Attributes	Mean Importance	Mean Performance	Discrepancy
1	Seawater cleanliness	4.00*	3.21	-0.79
2	Beautiful of coast	2.81	2.68	-0.13
3	Public space	4.95*	3.00	-1.95
4	Landscape	2.67	2.04	-0.63
5	Harmony of color	3.57	3.00	-0.57
6	Panorama	3.67	3.07	-0.6
7	Atmosphere	3.19	3.32*	0.13
8	Uniqueness	4.76*	3.21*	-1.55
9	Landmark	3.38	3.71*	0.33
10	Feeling save	4.33*	3.43*	-0.9
	Mean	3.733	3.067	

*higher than the mean

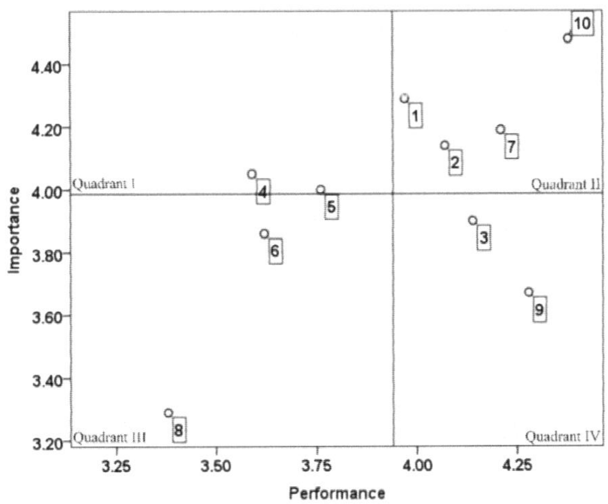

Figure 5. Cartesian graph for Akkarena

The last, Paotere the traditional Phinisi port, the attributes which have performance score than the mean (3.733) were (1) sea water cleanliness, (3) public space, (8) uniqueness, and (10) feeling save. The performance attributes have higher than the mean (3,067) were attributes (7) atmosphere, (8) uniqueness, (9) landmark, and (10) feeling save. Comparing between importance and performance, the mean of importance was higher than the performance; it means that the conditions of the Paotere was below the expectation of assessors.

This research resulted 3 Cartesian graphs (Fig 5 to 7) represented IPA of each sites. The graphs were grid into 4 quadrants as shown in Fig 5 to 7. In each graph, the 10 attributes fallen and distributed into the quadrants. Quadrant (Q) I indicate concentric here, QII keep up the good work, QIII low priority, and QIV possible overkill.

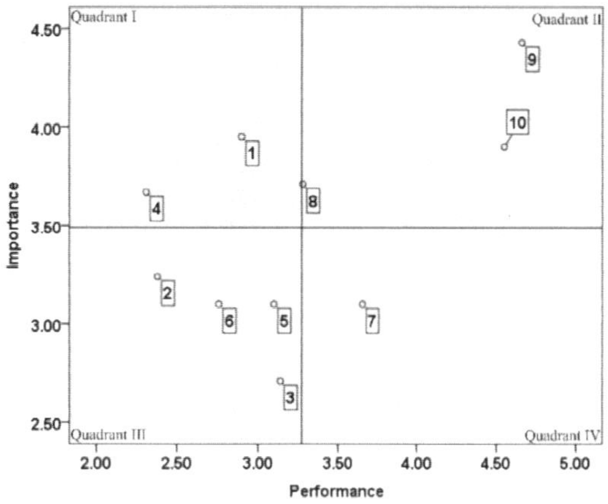

Figure 6. Cartesian graph for Losari

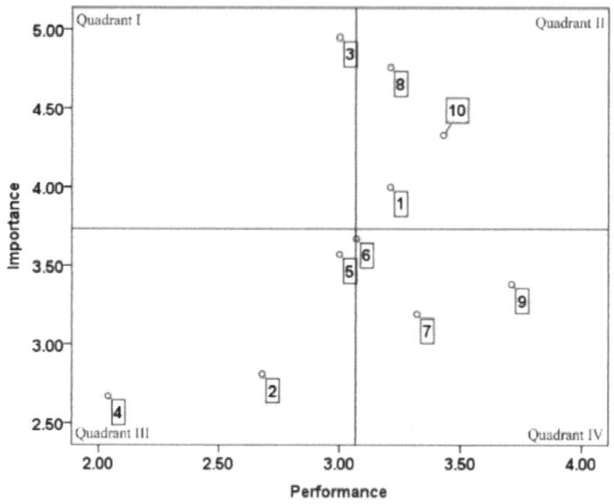

Figure 7. Cartesian graph for Paotere

Table 4. Comparative the position of attributes in quadrants

Sites	QI. indicates concentric here	QII. keep up the good work,	QIII. low priority, and	QIV. Possible overkill.
Akkarena	4	1, 2, 7, 10	5, 6, 8	3, 9
Losari	1	8, 9, 10	2, 3, 5, 6	7
Paotere	3	1, 8, 10	2, 4, 5	6, 7, 9

Note: Attributes: (1) sea water cleanliness, (2) beautiful of coast, (3) public space, (4) landscape, (5) Harmony of color, (6) panorama, (7) atmosphere, (8) uniqueness', (9) landmark, and (10) feeling save

4. Conclusions

This research analysed the importance and its performance attributes of three coastal sites that were proposed to be objects of coastal cruise tour (CCT). In the purpose of getting attributes to be improved for coastal cruise tour, the conclusions were:

(1) For Akkarena, landscape should be paid more attention. Sea water cleanliness, beautiful coast, atmosphere, and feeling save, were attributes that have high performance and to be kept sustainably. Other attributes, due to low performances, they were not to be considered.

(2) In Losari waterfront, only seawater cleanliness was the only attribute should be paid more intension; Attributes that required to be kept due to having high performance as well as importance were uniqueness, landmark, and feeling save

(3) For Paotere, the Phinisi ship port, concentration to be addressed to public space. Sea water cleanliness, uniqueness, and landmark, were the attributes that should be kept up due to having high performance and importance.

In the purpose of initiating coastal cruise tour, those attributes may use as the basic components for preliminary planning towards a comprehensive coastal tourism. It should be sustained with further research, with wider perspectives.

Acknowledgements

I would like to address my great appreciation to Hasanuddin University for providing fund through BOPTN research scheme. My appreciation also goes the resource persons who joining this research. May thanks go to undergraduate students of Laboratory of Waterfront Planning of Urban and Regional Planning Study Programs who contributed valuable data. This research will not be accomplished without any contribution of you all.

References

[1] Martilla J, James J 1977 Importance –Performance Analysis *Journal of Marketing* **41 (1)** pp 77-79
[2] Silva F, Fernandez O 2010 Using Importance-Performance Analysis in Evaluating of higher Education: *A Casucation and Management Technology. pp. 121-123*
[3] Kitcharoen K 2004 The Importance-Performance Analysis of Service Quality in Admnistrative Departments of Private Universities in Thailand *ABAC Journal* **24(3)** pp 20-46
[4] Aminian A 2012 Environmental Performance Measurement of Tourism Accomodations in pilgrimage Urban Area: The Case of Holy City of Mashhad, Iran *Social and Behaviorial Sciences* **35** pp 514-522
[5] Zhang H, Chow I 2004 Application of Importance-Analysis model in Tour Guides Performance: Evidance from Mainland Chinise Outbond Visitors in Hongkong *Tourism Management* **25 91** pp 81-91
[6] Huan T C, Beaman J 2006 *Importance Performance Analysis: the Need to Bridge Solitudes for Effective Use* (China: National Chia-ye University)
[7] Abalo J, Varela J, Manzano V 2007 Importance-Performance Analyses: A Formula for Spreading out values Derivied from Preference Ranking *Journal of Bussiness Research.* **60** pp 115-121
[8] Moscardo G 1996 Tourist Perspectives on Cruising: Multidimensional Scaling Analyses of Cruising and Other Holiday Types *Journal of Tourism Studies* **7** pp. 54-63
[9] Morrison A M 2003 Comparative Profiles of Travellers on Cruises and Land-based Resort Vacations, *Journal of Tourism Studies* **14** pp 99-111
[10] Kim S G, Kim Y J E 1996 Overview of Coastal and Marine Tourism in Korea. *the Journal of Tourism Studies* **7**
[11] Gungor S, Bozyigit R 2011 Tourism Potemntial of Gazipasa (Antalya-Turkey) *Procedia Social and Behaviorial Sciences* **19** pp 231-239
[12] Sari S R, Darwawan E 2008 Coastal Area of Ujungnegoro Batang Regency *Journal of Coastal Development.* **9(1)** pp 43-51
[13] Butler R W, Waldbrook L A 2003 A New Planning Tool the Tourism Opportunity Spectrum *Jurnal of Tourism Studies* **14(1)** pp 21-33
[14] Kher M et.al. 2008 Visual Assessment and Factors Affecting Visual Rating on Human-Made Landscape Elements of Wetland *Journal Arsitektur Landscape* **2(1)** pp 53-73
[15] Ciftci C 2010 A Visual Assessment for Land Use Analysis at the Coastal Area of Beysehir Lake. BALWOIS 2010 (Republic of Macedonia: Ohrid)
[16] Moscardo G 1998 Interpretation and Sustainable Tourism: Functions Examples and Principles *Journal of Tourism Studies* **9(1)** pp 2-13
[17] Eskildsen J K, Kristensen K 2006 Enhancing Importance-Performance Analysis. International *Journal of Productivity and performance Management* **55(1)** pp 40-60

IOP Conf. Series: Earth and Environmental Science **79** (2017) 012015 doi:10.1088/1755-1315/79/1/012015

Spatial Planning on Riverfront Urban Area in Banjir Kanal Barat Semarang

EDarmawan[1,2], T W Murtini[1,2], HWerdiningsih[1,2] and A S Enis[1,2]
[1]Architecture Department, Engineering Faculty, Diponegoro University
[2]Prof. Soedarto SH Street No. 1 Tembalang, Semarang 50241, Indonesia

Phone : 081575337300; fax: 62-24-76480840.
E-Mail : edy_darmawan@yahoo.com

Abstract. Riverfront in Banjir Kanal Barat (Western Flood Canal) develops rapidly since its western part of the east banks side is developed as a recreation area. Banjir Kanal Barat is also used for the venue for special occasions. Along the banks of rivers there are street vendors that sells second-hand goods and behind it there are densely populated residential. It is quite interesting, but it also takes high risk because the surroundings are prone to flood. The existing recreation area was built without considering flood-prone locations. Therefore, on this research, the researchers will design spatial planning in the area Banjir Kanal Barat river in order to make it more interesting for tourism. To achieve this goal, several aspects in Hamid Shirvani's theory will be applied, so the important aspects able to attract people to visit Banjir Kanal Barat will be identified. Comparative qualitative research methods used with a direct view conditions on the field and study literature. From the analysis it can be known the lack of harmony between the use space in Banjir Kanal Barat with buildings in the surrounding areas so it needs the optimization the use of space to be around it with the function of Banjir Kanal Barat.

1. Introduction

Canal is a water duct made by human for various needs and is usually a part of river flow that has been widened and deepened at some areas [1]. The functions of canal are as a part of flood controlling system and as a means of transportation. *Banjir Kanal Barat* Semarang is one of the canals controlling the longest river in Semarang. It is the downstream of *Kali* (river) *Garang* which flows to the north from Mount Ungaran to two main branch streams, *Kali Kreo* and *Kripik*. *Banjir Kanal Barat* is located at the boundary of West Semarang and South Semarang, from *Kali Garang* bridge to *Bojong Salaman*. It was built by Dutch to control flood frequently happening in Semarang. Maximum discharge of *Banjir Kanal Barat Semarang* is 740 m^3 / second. Up to now, the existence of this canal is still preserved, and it works as the main drainage system of Semarang City that moves water or liquid waste to Java Sea.

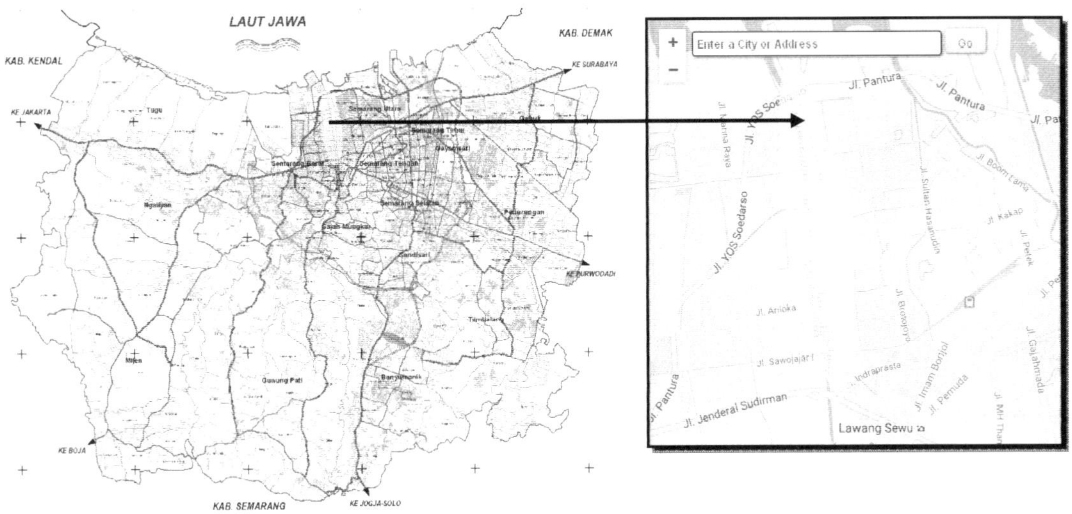

Figure 1. The Location of *Banjir Kanal Barat.*

In 2010, *Banjir Kanal Barat Semarang* is normalized in order to organize its riverside and to optimize its function. Before being normalized, there were many bushes and reeds. In its centre, there was sediment caused by erosion from *Kali Garang*. Hardening of sedimentation is also seen in the riverside, and it shrinks the river. This problem makes *Banjir Kanal Barat Semarang* cannot optimally function because it reduces the volume of water that can be accommodated which eventually makes the water overflow and causes flood around the area. After being normalized, *Banjir Kanal Barat Semarang* is wider, cleaner, and more well-maintained, and it also works normally as flood controlling canal. In fact, the river, ±5.3 km in length and 160 m in width, has become the recreation ground of Semarang. Semarang City Government has also been preparing to make this area as a tourist destination.

Development of the *Banjir Kanal Barat Semarang* as a location for recreation and tourism destinations is quite interesting, because the arrangement of the riverside area by utilizing the potential of natural and accessibility aspect can provide a maximum view and allows users to perform recreational activities or travel. However, in its development should not only focus on the area along the river, but also need to consider the surrounding area in order to obtain maximum results. On the banks of the river along the way in the *Banjir Kanal Barat Semarang* itself there are stalls of vendors selling second-hand goods, where the presences of these traders sometimes use the area which can cause congestion. This condition can certainly interfere with access to people who want to visit the *Banjir Kanal Barat Semarang*. On the back of the stalls traders are densely populated settlements that require further arrangement because the condition tends to be seedy. Based on these problems, then the purpose of this study was to design the spatial planning in the area of the *Banjir Kanal Barat* river to be more attractive for recreation. Aspects of the theory of Hamid Shirvani will be implemented in detail, both land use, building form and massing, pedestrian ways, open space, signage, support activity, parking and circulation, and conservation. Of these theories will be determined aspects in accordance with the conditions around the *Banjir Kanal Barat Semarang*. The results of this study are aspects that can attract people to visit the *Banjir Kanal Barat Semarang*.

2. Methods

The research method is basically a way to get data with a specific purpose and usefulness [2]. Into this report, methodology used is descriptive qualitative. Implying the methodology qualitative descriptive as a method used to describe and outline the findings in the field that aims to collect information in actual and detail, identify problem, making comparisons or evaluation, and determine what anybody else is doing in the face of the same problem and learn from experience them to set the decision in time to come [3]. The use of the methodology qualitative descriptive in this report used to get an overview of design spatial planning at *Banjir Kanal Barat Semarang* and outline the findings in the field in actual and detail in the form of object in a field made in the form sketches shots and photos. Types of research methods will be implemented are included in the survey were used to obtain data from a particular place in nature (not artificial), but the researchers did treatment in data collection. For example, by distributing questionnaires, structured interviews, and observation [2].

This research aims to design spatial planning at *Banjir Kanal Barat Semarang*. For this aims our research strategy using the survey method. The first step in our research is identify physical condition and activities in *Banjir Kanal Barat Semarang*. This step is taken to determine characteristics in this area. For this, researchers conducted direct observations in the field and supported by documentation to reinforce the data obtained from the observation. The second step is applying Hamid Shirvani's Theory in *Banjir Kanal Barat Semarang*. Data obtained through literature study and field observations.

The results of the data collection and analysed by qualitative descriptive analysis techniques incorporating photos or pictures and tables as supporting information. Data analysis is the process of systematically searching for and compiling data obtained from interview, fields notes, and other materials that can be easily understood and its findings can be communicated to others [4]. Activity in the data analysis is based on classifying data based on variables and types of respondents, tabulate the data based on the variables of all respondents, presenting the data of each variable studied, perform calculations after that drawing conclusions from what has been done.

3. Result and Discussion

The development of *Banjir Kanal Barat* as recreation area of Semarang should not only focus on the riverbanks, but it should also consider the surrounding to obtain the optimal result. Therefore, on this research, researchers will observe and analyse the condition around *Banjir Kanal Barat* by applying *Elemen Pembentuk Kota* (elements that form a city) theory of Hamid Shirvani, so the aspects which can attract citizens to visit *Banjir Kanal Barat* can be identified, and these are the explanations:

a. Land use

Tata Land use is regulating of land to decide the best choice of allocating certain function [5]. Land use around *Banjir Kanal Barat* is for settlement, offices, education, trades, and conservation. The existing settlement is densely-populated settlement with mixed-functions along the riverbanks of *Banjir Kanal Barat*. Those houses are not well-arranged, so it looks slummy. Besides, there are street vendors building their stalls along the riverbanks as seen in Figure 2 below. The growth of street vendors classified as informal sector in one area is a natural symptom. However, if it is uncontrollable, it will affect the land use quality around *Banjir Kanal Barat*. Therefore, the possible design is optimizing land use in the riverside for green open space that can be used social activities and creating open spaces integrated with residential areas. Optimizing room can be achieved by rearranging residential area or relocating the residential areas and then directing them into vertical residence to reduce density. Besides, land use should also be provided with smooth movement. Road planning should be done appropriately and should consider ongoing activities, location, width, material, and construction.

(a) (b) (c)

Figure 2. Street Vendors in Riverbanks.

b. Building Form and Massing

Building form and massing discuss how shapes and masses of buildings can form a city and how the relation among the masses [6]. Building form and massing are not only determined by the height or the size of the building and the appearance or the configuration of the building mass; but they are also determined by the environmental impact in order to create a building with good quality and appearance. By focusing on the previous elements, a more harmonized design of building view that blends with surrounding historical buildings can be realized. Around *Banjir Kanal Barat,* the building forms and massing are mostly square or rectangular because most of the people dwelling there live in one or two-storey houses, and there is also a line of shops located along the road. The orientation of those buildings is toward the canal. The façade of the buildings, seen in Figure.3, looks dull because the area is densely populated and not well-arranged. Based on the planning, the height and the scale of the buildings will be adjusted with human scale (except for the landmarked buildings), and the appearance of the buildings will be improved by adding a new architectural concept or style which is more attractive or dynamic and also compatible with its development planning for recreational area. The planning, furthermore, will be conducted without neglecting factors such as safety, convenience, and stability of the building and will consider its environmental topography in order to avoid damaging its natural condition.

(a) (b) (c)

Figure 3. Looks Buildings Around *Banjir Kanal Barat.*

c. Circulation and Parking

Circulation is a part of urban planning elements that is directly able to shape and control pattern of activities in a city while parking is the supporting element for urban circulation system that determines whether an area is lively or not [6]. *Banjir Kanal Barat* is located at the main line of transportation in Semarang City. This area is actually accessible and supported by sufficient number of public transportation vehicles having relatively long duration of service. The problems are that the circulation system is still not well-managed and that the parking lot is unavailable. These problems have made many people use road as the parking area which are seen in Figure.4. This 'on the road' parking system quite disturbs the circulation, and it gets worse especially when there is an event held near the area because the event suddenly creates additional temporary parking areas. These temporary parking lots are managed by people who live surrounding the areas in order to avoid conflict triggered by harsh competition for areas and incomes. Circulation and parking are the common problems faced by big cities, and therefore, the planning will include designing streets to be open spaces having good view and provided with attractive landscape elements. In addition, it is necessary to provide parking spots for the people doing activities in *Banjir Kanal Barat*, and these parking spots must be in the specific areas, so the circulation of vehicles does not overlap the circulation of pedestrians.

(a) (b) (c)

Figure 4. Conditions Parking Around *Banjir Kanal Barat*.

d. Open Spaces

An open space is defined as an area in a city which is not developed for some uses [7]. Open spaces give contribution in shaping the character of a city and has a function as rooms for social interaction, economic activities, activities related to cultural appreciation [5]. The open spaces at *Banjir Kanal Barat* consist of park, road, and field, yet their condition has not been maximally utilized because most of the areas are occupied by street vendors. These street vendors, who are not well-managed, have made the area looks slummy. In order to return the real function of the open space at *Banjir Kanal Barat*, the planning will include developments of creative spaces and a space for communities that harmoniously blend with the environment surround them. The developments will consist of adding more rooms for attractions or shows and adding areas for children to play, so people can do their outdoor activities and have rooms for discussion or social gathering. In the area *Banjir Kanal Barat* actually there had been park provided to accommodate activities the citizens, only because of a lack of vegetation as seen in Figure. 5. This area feels hot during the day, so people are refusing to come to this place.

(a) (b)

Figure5. Park in *Banjir Kanal Barat*.

e. Pedestrian Ways

Pedestrian ways are one of open spaces becoming a part of the street and having dynamic character because it is next to the road [8]. According to Nuzuluddin (2015), a pedestrian way has function of a circulation room for pedestrians which is separated from the vehicle circulation [9]. Pedestrian ways are considered as the boundary between roads and buildings, and therefore it is important to consider their safety and convenience. Unavailability of walkways causes pedestrians to not maximally feel safe and convenient at *Banjir Kanal Barat*. These pedestrians usually use some part of the road since there is no real boundary between road for vehicles and pedestrians. A pedestrian way is only available along Jl. Jend Soedirman and Jl. Soegijopranato where both streets are classified as main arteries. There is actually a pedestrian way along Jl. Kokrosono, but this pedestrian ways, seen in Figure. 6, is too narrow and occupied by street vendors. Therefore, the planning will include building a wide pavement covered with paving blocks and provided with shade trees and chairs for pedestrians to ease tiredness, and for some places, there will be bus stops giving easy access for pedestrians to take the bus.

(a) (b) (c)

Figure 6. Conditions Pedestrian Ways Around *Banjir Kanal Barat*.

f. Activity Support
Activity support are all functions of buildings and activities that support a city area [10]. Activity support around *Banjir Kanal Barat*, trades done by street vendors where permanent and semi-permanent stalls, are blended with dwellers' activities which can be seen in Figure. 7. In addition, considering that this site is built for recreational area there are also sport and recreational activities. The commercial activities around this area have specific characters and have taken place since long time ago, and for that reason, the planning for the commercial activities at this place will include developing 'people's market' that will support the recreational activities. This development, also parking area development, will certainly not disturb the previous business activities and the circulation at *Banjir Kanal Barat*.

(a) (b) (c)

Figure7. Activity Support Around *Banjir Kanal Barat*.

g. Signage
Signage can be defined as directions, road signs, advertisements, and other signs [6]. The existence of signage will influence visual appearance of a city. Well-designed signage can show the character of buildings' façade and make streets lively. Several things needing to consider in setting signage are the velocity of vehicles, the distance of reactions, numbers of listed words, and the dimension of writings on boards. Based on the study, the location of signage is not well-arranged, and therefore it is still unable to give information clearly. Referring to the problems, the

use of signage will be regulated based on its distance, size and also its existence that reflects the characters of the area. Additionally, the use of signage must also be harmonious with the buildings' architecture around *Banjir Kanal Barat*.

h. Conservation
Conservation must protect the existence of a city's environment and spaces namely building sites or historical areas [6]. *Banjir Kanal Barat* that was built by Dutch in colonial era should be preserved, and this effort can be a way to earn back people's trust [6]. The conservation of *Banjir Kanal Barat* is connected to activities of people who grow and dwell around the area. Moreover, there is also a weir called *Bendungan Simongan* which is included as one of the conserved buildings, can be seen in Figure. 8. This weir was built by the Dutch in the beginning of twentieth century to control water flow from *Kali* (River) *Garang* to *Kali Semarang* and *Banjir Kanal Barat*. The existence of the weir is still preserved until now, but there are some problems related to treatments, maintenance, and cleanliness causing the accumulation of trash around the weir's gate. If *Bendungan Simongan* were well-preserved, it would bring a beautiful panorama of water. Thus, the regional planning will also contain development of *Bendungan Simongan* to be recreational area offering a journey in which visitors explore the rivers and enjoy views along riverbanks.

(a) (b) (c)

Figure 8. *Bendungan Simongan*.

4. Conclusions

Based on the result and discussion can be known that the lack of harmony between the use of space in *Banjir Kanal Barat* with buildings in the surrounding areas that can be seen from:

1. Land use around *Banjir Kanal Barat* has mixed function such as for settlement, for trade, and for education.
2. Building form and massing is in square or rectangular, and height of the building ranges from 1 until 2 storeys facing toward *Banjir Kanal Barat*.
3. Circulation around *Banjir Kanal Barat* is still overlapped each other, and the pedestrian ways is unavailable, but it's not accessible clearly. The parking area still occupies some part of the street and is not well-arranged.
4. The existing open spaces have not been utilized because they are occupied by street vendors, open space available at the edge of the river.
5. The needs of pedestrians for convenience and safety have not been fulfilled because there is no pedestrian ways and there is no a clear boundary separating vehicles and pedestrians.
6. Activity support like trades are not well-managed, so the area looks slummy and untidy.
7. The existing signage is still not arranged well and less informative.
8. *Bendungan Simongan* is one of conserved buildings that has not got maximum treatment and maintenance even though this weir has vital role for Semarang City.

Acknowledgements

The authors say thanks to all those who have helped in collecting data used in this article as a part of the objective of our research for spatial planning on riverfront urban area.

References

[1] Jacksi, E. 2011. *Mengenal Kanal dan Fungsinya. www.elisabethjacksi.blogspot.com*
[2] Sugiyono. 2008. *Metode Penelitian Kuantitatif, Kualitatif dan R & D.* Bandung: CV. Alfabeta.
[3] Arikunto. 2006. *Procedure Research: A Practical Approach.* Jakarta: PT Rineka Cipta.
[4] Sugiyono. 2014. *Memahami Penelitian Kualitatif.* Bandung: Penerbit Alfabeta.
[5] Darmawan, E. 2003. *Teori dan Implementasi Perancangan Kota.* Semarang: Badan Penerbit Universitas Diponegoro.
[6] Shirvani, H. 1985. *The Urban Design Process.* New York: Van Nostrand Reinhold.
[7] Rapuano, Met. al. 1964. *Open Space in Urban Design.* Ohio: The Cleveland Development Foundation.
[8] Mulyandari, H. 2011. *Arsitektur Kota.* Yogyakarta: Penerbit ANDI.
[9] Nuzuluddin, T R. 2015. *Sirkulasi dan Parkir, Activity Support di Kawasan Peterongan Semarang (Penggal Jalan MT. Haryono Mulai Perempatan Lamper Sari Sampai Pertigaan Sompok). Jurnal Universitas Pandanaran Semarang.*
[10] Darmawan, E. 2003. *Teori dan Kajian Ruang Publik Kota.* Semarang: Badan Penerbit Universitas Diponegoro.

IOP Conf. Series: Earth and Environmental Science **79** (2017) 012016 doi:10.1088/1755-1315/79/1/012016

Urban Principle of Water Sensitive Design in Kampung Kamboja at Pontianak City

N Hasriyanti[1] and **E Ryanti** [2]
[1]Department of Architecture State Polytechnic of Pontianak
[2]Department of Civil Engineering State Polytechnic of Pontianak
A. Yani Street, Pontianak City 78124, Indonesia

E-mail: niexnoe@gmail.com

Abstract. This study will define the design principles of settlement area banks of the Kapuas Pontianak to approach the concept of water sensitive urban design (WSUD) in densely populated residential areas. Using a case study of a region densely located on the banks of the river with engineering literature to formulate the aspects taken into consideration and the components are arranged in the design, analysis descriptive paradigm rationalistic to identify the characteristics of residential areas riverbank with consideration of elements WSUD and formulate design principles residential area that is sensitive to water. This research is important to do because of problems related to the water management system in the settlement bank of the river in the city of Pontianak do not maximize. So that the primacy of this study contains several objectives to be achieved is to identify the characteristics of the settlement area riverbanks under consideration aspects areas design that is sensitive to water and principle areas design that will formulate the structure of the existing problems related to the needs of the community infrastructure facilities infrastructure neighborhoods and formulate and create guidelines for appropriate technology for integrated water management systems in the residential area of the riverbank and engineering design for the settlements are sensitive to water (WSUD). The final aim of the study is expected to achieve water management systems in residential areas by utilizing the abundant rainwater availability by using LID (Low Impact Development) through the concept of urban design that sensitive water

1. Introduction

The settlement is a complex problem associated with and related to the social, economic, cultural, ecological, and so on. Settlement banks of the river have the potential and the problems are diverse. The potential is directly related to the river banks to provide benefits for the surrounding community to use the river as a source of water for the fulfillment of activities and daily necessities. Along with population growth and increased economic activity, utilization and management of water in life face problems such as excessive water exploitation, reduce the function of the watershed as a water catchment areas and improving woke leads to the reduction of green open space. The result of all it is the occurrence of flooding in the settlements and the decline in water quality and quantity. The problem of water in residential areas is very important because it requires the integrated water management through the design of the city known as the Water Sensitive Urban Design (WSUD).

However, there are still things that have not implemented the concept WSUD, among others, yet it uses vegetation to absorb water and improve water quality, rain water is not used to irrigate the vegetation in the area, not the presence of the trenches to minimize run-off, and has not been used open space design that can withstand the flow of water in the area and others.

The purpose of this study is to formulate a design principle which is a guideline to achieve the implementation of the principle WSUD which basically aims to resist and absorb the water contained in the surface into the ground. Besides the application of this concept can also reduce runoff water so as to reduce flooding and flooding. Design principles to be formulated is the main purpose to establish a residential area that is sensitive to water.

WSUD is one part of the concept of green infrastructure approach. The goal of the approach is to approach the concept of urban planning and design related to water resources and environmental management and to minimize the impact caused by the presence of water in urban surfaces. WSUD concept focuses on sustainability management of the water cycle in the city where the management is the source of this concept approach. The forms of sustainable management of the water cycle, namely the management of drinking water, the water flow caused by rainwater quality, wastewater management and the water cycle (water conservation). (Yunus, 2007)

2. Methods

The method used is to make a field trip to obtain a picture of the condition, as well as the characteristics of the formulation of the problems related to water issues in the settlement area of the Kapuas river banks. Secondary data acquisition is done through literature studies and surveys related agencies. The analytical method used is qualitative descriptive analysis with a rationalistic paradigm to describe the approach to the concept of settlement WSUD on the riverbank, so that later can be formulated principles of settlement areas design.

3. Result and Discussion

LID is a way of designing region with a view to maintaining and imitating the hydrological characteristics of an area before development. Function hydrology, such as storage, infiltration and recharging groundwater or also the volume and frequency of surface flow will be maintained during use of handling storm water runoff on a small scale comprehensive and integrated, both retention and detention areas, reduction of impermeable surface and extending groove and the flow of time concentration (Coffman in the United States Environmental Protection Agency, 2000). The basic concept that describes the core of LID is a regional plan must process thoroughly to achieve success and a plan that could hold. The basic concept is, among others (Department of Environmental Resources Maryland, 1999):

a. The use of hydrology as a framework work. Hydrology approach in the process Regional planning begins with identification and keep sensitive area effect on regional hydrology, including flow and buffer, floodplains, wetlands, slope, soil high permeability and protected areas. Schematic of the area to be built then evaluated to reduce, minimize the total impervious area of the region. Subsequent analysis is making watertight areas that cannot be avoided to minimize surface watertight interconnected directly. Biodefense area, extending the flow stream, local infiltration, drainage, area retention, and other practices can be used to control and change the impermeable area water. The end result of the use of hydrology in Regional planning is to maintain the conditions of the development hydrological characteristics other than it can add aesthetics and provide recreational areas.
b. The concept Micromanaging. The key to making LID is to concept micromanaging. It requires change our perspective in the approach area to be controlled (Micro-subsheds), by controlling (Micro-techniques), location and size and frequency control of the water flow rain-controlled. management techniques small scale is done on a small sub-watershed or in residential areas also on a public area, which provides the division controlling the flow of rainwater thoroughly the observation area. This gives the opportunity to maintain hydrological function regions such as infiltration, shallow bin, and the interception of rainwater as well as good in reducing the concentration time (Tc).

c. Control the flow of rain in the watershed. Key to the repair function hydrology is to minimize and reduced as a result of land-use change land in the area of the river. This matter based on the distribution strategy control in the watershed and implemented using the techniques of micromanaging on the whole region.

WSUD is one part of the concept of green infrastructure approach. The aim of this concept approach to urban planning and design approach that related to water resources and environmental management and to minimize impact posed by the presence of water in urban surfaces. Initially, this concept appears motivated by the role of water in the life of the city, which requires setting up a harmonized between city development and the need for water. The concept focuses on the WSUD sustainable management of the water cycle in the city where the management is the source of approach to this concept. The forms of sustainable water cycle management are the management of water drinking, water flow caused by rain, water quality, wastewater management and the water cycle (Water conservation). (Yunus, 2007)

In this concept for the Kapuas River Settlements of integration between the management and the sustainability of the water cycle designing cities to adopt urban design techniques that are sensitive to water. Could concluded that the keyword of this concept is the integration. It is mentioned more clearly the principles of the concept WSUD. Those principles mentioned in Urban Stormwater (2009), namely: protecting the natural systems one of which is a natural water systems through the development of the city; integrating prevention methods rainwater into urban design; protect water quality and improve the quality of irrigation water system; reduce the flow of water by flowing water runoff either rain or flooding and to integrate landscape into the environment; utilizing water runoff for various needs; give the value of land by minimizing the cost of infrastructure one of them is drainage infrastructure.

Such principles can be applied in a region at Kapuas River Settlements actually in Kampung Kamboja with analysis first. Required are three kinds of analysis, ie the analysis of the site, and the land use planning capability land. Footprint analysis is useful to identify and explain the state of the (existing) which occurred at the site. What is needed is a data topography, drainage networks, soil type, land cover data on the size, the sensitive regions, and the water flow corridors. By doing all three analyses They can be found where the development concept needs to be applied with WSUD the impact to a minimum.

This concept has several important elements, namely the use of water (water reuse) and water treatment (water treatment). Rainwater and wastewater is one of the water note on this concept. In practice, this concept sees rain water management and water effluent as an opportunity to design a city and not as waste. Element The lowered to urban design elements to consider in the application of WSUD are: land that can absorb water (precinct wide infiltration basins); swales or ditches; road design; ponds; gutter; water storage tanks; similarity roof; type forming material is porous pavement (porous pavements); type of drainage cover material; sand filter; retaining walls of water flow; types of vegetation including vegetation covering the waterfront; city Forest. Based on this, the engineering LID eligible to apply is as follows.

a. Biodefense, Biodefense form in the bin aquifer combined with vegetation built around the environment buildings were still empty and got enough sunlight
b. Channel impregnation, Impregnating channel built in the neighbourhood around the building which is empty but not getting enough sunlight.
c. Rainwater tank, or Bak, was built to accommodate the flowing rainwater from the roof of the building and later can be used for raw water sources, especially for flushing toilets.
d. These new tools will LID implemented in the future adapt to the development settlement infrastructure and River Laboratories.

4. Conclusions

The discussion in advance can conclude matters as follows.

a. LID engineering that could be applied to a model of this study is to implement a drainage system is slow, and other engineering is like the making bio-retention pools and water tanks rain which will be held on next time.
b. By using a simulation program EPA-SWMM, estimated channel drainage and an existing storage of Environment Settlement at the River Centre treated as a drainage system slow for the research model can lowering the peak discharge flow surface becomes 1/16.
c. Models and the combination means LID slow drainage can lower peak discharge into a 0,018 m3 / s compared to the channel conditions flowed freely, drain water with discharge of 0.34 m3 / s.
d. The effectiveness of the channel is slow to lowering the peak discharge flow surface can be enhanced by expense occurs puddle the surface of the land, until the high limit and a certain duration.

With the above conclusion suggested several cases as follows.

a. The need for efforts to change the paradigm long associated with drainage environment, namely from wasting water as quickly and as much as possible be hold water as long as possible along does not cause an environmental disturbance.
b. The need for the implementation of drainage systems slows in a residential neighbourhood as realization paradigm change at the top.
c. It should be observed that the model is intensive in order to obtain parameters for designing dimensional drainage is slow and LID means more.

References

[1] Department of Environmental Resources (DERPGC) *Low Impact Development Modeling Programs In Prince George's County, Maryland*
[2] Department of Environmental Resources (DERPGC) 1999 *Low Impact Development Design Strategies* Integrated Design Approach (Maryland: Prince George's County)
[3] Coffman L 2000 *Low-Impact Development Design Strategies* An Integrated Design Approach (Maryland: Prince George's County)
[4] Committee on Reducing Stormwater Discharge Contributions to Water Pollution National Research Council 2009 *Urban Stormwater Management in the United States.* (United States: National Academics Press)
[5] Darsono S 2007 Sistem Pengelolaan Air Hujan Lokal Yang Ramah Lingkungan *Berkala Ilmiah Teknik Keairan* Vol. 13, No. 4-Desember 2007
[6] EPA 2000.*Low Impact Development (LID) A Literature Review* (Washington DC: Office of Water)

Understanding Urban Communication in Information Era: Analyzing Development Progress of Coastal Territories in the Context of West Java's Metropolitan Regions

Ridwan Sutriadi[1] and Meta Indriyani Kurniasari[2]

[1] *Urban and Regional Planning Department, School of Architecture, Planning and Policy Development, Institut Teknologi Bandung, Labtek IX-A Jalan Ganesha 10 Bandung 40132*
[2] *Alumni of Urban and Regional Planning Department, School of Architecture, Planning, and Policy Development, Institut Teknologi Bandung, Labtek IX-A Jalan Ganesha 10 Bandung 40132*

Abstract. This paper explores a consequence of metropolitan and development centers policy to the development progress of coastal territories by analyzing municipal website base on urban communication functions of communicative city concept. In terms of coastal territories as a part of development center, efforts have to be made in enhancing the role and function of municipal website to show their development progress. Perceptual analysis is taken as a method to measure their position, especially kabupaten/kota as coastal territories in regional context (West Java Province). The results indicate that the availability of public information in coastal territories cities lower than other cities in metropolitan area. Innovation in specifying coastal features has to be promoted in illustrating development progress of coastal territories as a part of development centers in West Java Province.

1.Introduction

The insistence of metropolitan development in West Java for driving the modern development and increasing regional competitiveness to face the dynamics of rapid global change lead to the cities innovations development. Technology utilization become another way for cities to be able to compete in this information age. However, the majority of the development direction of the city is still not considering the role of the technology in city development. The virtual of regional competitiveness in provisioning digital information reflected in municipal website of each city which can be viewed as an extension of the city [1], [2], [3]. The utilization of municipal website in informing public service aspects are still low, whereas it actually is the benchmark for basic and the easiest way to see the cities capability in the provision of public information digitally. Under Law 14/2008 on *Public Information Disclosure,* said that government bodies are required to provide, deliver, and/or regulate public information that is accurate, accountable, true and not mislead. Public information dissemination is organized through electronic media, one of them is website.

The municipal website comes as one of the government's approach to accommodate the public aspiration through digital communication. The use of ICT in accommodating public participation and spread development information also performed on development planning through *e-tools*. The differences of information content in municipal website depend on the characteristic of each city. This led to the information disparity between the cities which already developing ICT and cities which has not.

One of the urban development issue in West Java metropolitan area is the perception of city development in coastal territories in order to be in line with the development on the national level in ICT era and the perception of urban coastal development progress. Along with the competition among cities, the perception of the representative of coastal areas is necessary, particularly through their municipal website. There are several benefits of the proposed research, ranging from theoretical benefits and in response to the issues of the coastal areas development (the latest issue of the urban development in West Java Province). This research attempt to exercise the implementation of the theory of communicative city to the context of metropolitan cities, especially coastal metropolis in West Java. The practical benefits can give you a practical extent of the attractiveness of an urban coastal area represented by official information from the municipal as a comparison for the development of social media that can give positive or negative impacts towards urban development in existing and in the future.

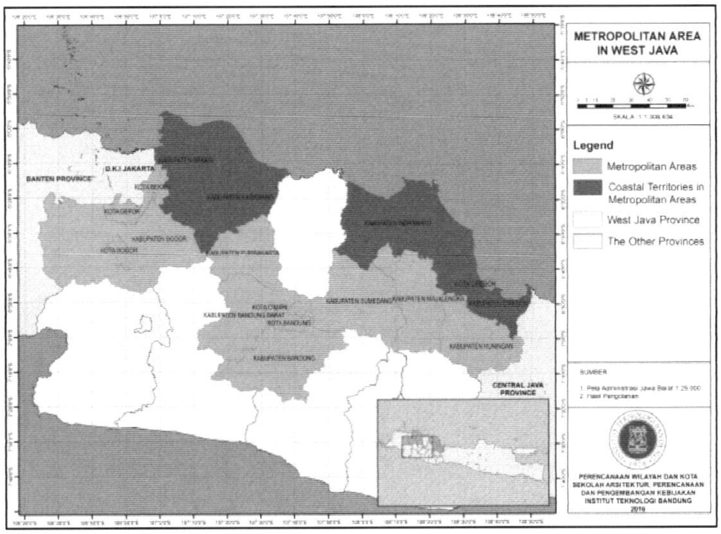

Figure 1. Coastal Area in Metropolitan West Java

2. Methods

This research used a mixed-method approach or qualitative and quantitative approach. The descriptive qualitative method used to find out more about municipal website which potentially able to encourage public participation in the development in terms of the availability of public information. The descriptive quantitative method used to explain the results of the input calculation of the availability of information from each municipal website.

2.1 Data Collection Method

This research uses secondary data with literature studies and content website. The data used in this research is the result of the information input from municipal website of *Kota Bogor, Kabupaten Bogor, Kota Depok, Kabupaten Bekasi, Kota Bekasi, Kabupaten Karawang, Kabupaten Purwakarta, Kabupaten Bandung, Kota Bandung, Kabupaten Bandung Barat, Kota Cimahi, Kabupaten Sumedang, Kota Cirebon, Kabupaten Cirebon, Kabupaten Kuningan, Kabupaten Majalengka* and *Kabupaten Indramayu.*

The focus of this research is to compare the public information services provided on municipal website in coastal areas (*Kabupaten Bekasi, Kabupaten Karawang, Kabupaten Indramayu, Kabupaten Cirebon* and *Kota Cirebon*) to another city in metropolitan areas.

2.2 Data Analysis Method

This research uses a perceptual analysis approach in the era of the advancement of ICT which developed based on *e-tools* (picture below) as an inseparable part of urban development in the digital era. The representation of this analysis is the municipal website content analysis which is related to metropolitan areas in West Java. In addition, this research uses communicative city approach to the context of the metropolitan areas in the ICT era by focusing on ten municipal functions such as surveillance function, coordination function, entertainment function, mass communication function,

IOP Conf. Series: Earth and Environmental Science **79** (2017) 012017 doi:10.1088/1755-1315/79/1/012017

local government function, public service provider functions, economic developers function, community function identifier, and community mobilizer function [4].

This research uses qualitative and quantitative analysis techniques. Qualitative analysis method used is the analysis of website's content. Quantitative analysis method uses is schallogram analysis that aims to determine the hierarchy rank of public information services in metropolitan areas and to determine the potential of availability of information on municipal website and accommodate public participation in development through website.
The formula used in schallogram analysis:

- To determine range
  Maximum value (MXV) and minimum value (MNV), the upper limit (UL), the lower limit(LL) and A = *the number or range.*
  High : UL <= X <= MXV
  Middle : LL <= X < UL
  Low : MNV <= X < LL
- Determine "R" (*Reproducibility Coefficient)* and Error

$$R = \frac{\sum \text{Nodes in Schallogram Ladder}}{\sum \text{Total Nodes}} \times 100\% \qquad (1)$$

$$\text{Error} = 100\% - R \qquad (2)$$

3. Schallogram

3.1 Schallogram analysis of surveillance function
Surveillance Function measures the extent to which the municipal website informs residents of city activities, provides news about current programs, and alerts residents to potential problems under the city's purview by inputting the information about this variables:

Table 1. Surveillance Variables

No	Surveillance Variables
1	Up to date news or stories about city
2	Information about city social service program
3	Information about schools/education program, services
4	Information about city housing program, services
5	Information about city transportation program, services
6	Information about city garbage collection, environment program, services
7	Information about city recreation, sports programs, services
8	Information about city health program, services

According to the result, *Kabupaten Karawang*, *Kabupaten Cirebon*, *Kota Cirebon*, *Kabupaten Indramayu*, and *Kabupaten Bekasi* are on the "lower" category in the provision of information surveillance function. Based on the "nodes" of the ladder, *Kabupaten Karawang, Kabupaten Cirebon and Kabupaten Indramayu* should improve the availability of variables point 1-5,7,8 on Table 1. While *Kabupaten Bekasi and Kota Cirebon* should improve the availability of variables point 1-8 on Table 1.

Table 2. Schallogram Ladder: Surveillance Function

NO	CITIES	HIGH (3)								MIDDLE(2)								LOW(1)								TOTAL	RANGE
		1	2	3	4	5	6	7	8	1	2	3	4	5	6	7	8	1	2	3	4	5	6	7	8		
1	Kota Bogor	X	X					X						X						X	X		X		X	24	H
2	Kota Bandung		X	X		X	X											X			X			X	X	23	H
3	Kabupaten Kuningan							X	X		X							X		X	X	X	X			18	M
4	Kabupaten Karawang														X			X	X	X	X	X		X	X	11	L
5	Kabupaten Cirebon														X			X	X	X	X	X		X	X	11	L
6	Kabupaten Indramayu														X			X	X	X	X	X		X	X	11	L
7	Kota Cimahi				X			X										X	X	X		X	X		X	11	L
8	Kabupaten Bandung													X				X	X	X	X		X	X	X	8	L
9	Kabupaten Bekasi																	X	X	X	X	X	X	X	X	8	L
10	Kabupaten Majalengka										X							X		X	X	X	X	X	X	8	L
11	Kabupaten Sumedang										X							X		X	X	X	X	X	X	8	L
12	Kabupaten Bandung Ba										X							X		X	X	X	X	X	X	8	L
13	Kota Bekasi																	X	X	X	X	X	X	X	X	8	L
14	Kota Cirebon																	X	X	X	X	X	X	X	X	8	L
15	Kota Depok																	X	X	X	X	X	X	X	X	8	L
16	Kabupaten Purwakarta																	X	X	X	X	X	X	X	X	8	L
17	Kota Depok																	X	X	X	X	X	X	X	X	8	L

3.2 Schallogram analysis of coordination function

Coordination Function measures the extent to which the website provides audiences with information or opportunities to contact city departments or officials or to participate in official business by attending public meetings by inputting the information about the variables:

Table 3. Coordination Variables

No	Coordination Variables
1	Provide information for contacting city departments
2	Solicits citizen initiatives to report problems online
3	Provides information for contacting mayor
4	Provides information for contacting city council reps
5	Provides information on where, when how to attend, speak at council's meetings

According to the result, *Kabupaten Cirebon* and *Kabupaten Indramayu* are on the "High" category in the provision of coordination function, but they need to improve the availability of variables point 2,3 and 5 on the Table 3. *Kota Cirebon* is on the "Middle" Category, but this city need to improve the provision of variables point 2-4 on the Table 3. Kabupaten Beaksi and *Kabupaten Karawang* are on the "Lower" Category, they need to improve the provision of variables 2-5 (Kab.Bekasi) and variables 1-5 (Kab.Karawang) variables on Table 3.

Table 4. Schallogram Ladder: Coordination Function

NO	CITIES	HIGH (3)					MIDDLE(2)					LOW(1)					TOTAL	RANGE
		5	4	3	2	1	5	4	3	2	1	5	4	3	2	1		
1	Kota Bogor	X	X		X	X								X			15	H
2	Kabupaten Bogor	X	X			X				X				X			15	H
3	Kota Depok		X	X		X						X			X		14	H
4	Kabupaten Cirebon		X			X						X		X	X		14	H
5	Kabupaten Indramayu		X			X						X		X	X		14	H
6	Kota Bandung			X		X						X	X		X		13	H
7	Kota Cirebon						X				X		X	X	X		10	M
8	Kota Cimahi									X	X	X	X	X			7	L
9	Kabupaten Bekasi										X	X	X	X	X		6	L
10	Kabupaten Kuningan		X									X		X	X	X	5	L
11	Kabupaten Majalengka					X						X	X	X	X		5	L
12	Kabupaten Bandung			X								X	X		X	X	5	L
13	Kota Bekasi					X						X	X	X	X		5	L
14	Kabupaten Purwakarta							X				X		X	X	X	5	L
15	Kabupaten Sumedang											X	X	X	X	X	5	L
16	Kabupaten Bandung Barat											X	X	X	X	X	5	L
17	Kabupaten Karawang											X	X	X	X	X	5	L

3.3 Schallogram analysis of socialization function

Provided residents, potential residents, and visitors with historical information about the city and its culture by inputting variables:

Table 5. Socialization Function

No	Socialization Variables
1	Statistics about city
2	Information about regional development
3	Information about spatial urban planning
4	Link to spatial and development planning department
5	Website's social media account
6	Website's link share

According to the result, *Kabupaten Cirebon* is on the "High" category in the provision of coordination function, but this city needs to improve the availability of variables 3-5 on the Table 5. *Kabupaten Bekasi*, *Kabupaten Indramayu*, *Kota Cirebon*, and *Kabupaten Karawang* are on the "Lower" category. They need to improve the provision of variable 1,3,4,5,6 (*Kab.Bekasi),* variables 3-5 (*Kab.Indamayu),* variables 1-5 *(Kota Cirebon)* and variables 1,2,3, and 5 *(Kab.Karawang*) on the Table 5.

Table 6. Schallogram Ladder: Socialization Function

NO	CITIES	LOW(1)						MIDDLE (2)						HIGH (3)						TOTAL	RANGE
		1	2	3	4	5	6	1	2	3	4	5	6	1	2	3	4	5	6		
1	Kabupaten Bandung Bar	X	X	X	X	X							X							6	L
2	Kota Cimahi	X	X	X	X	X													X	6	L
3	Kota Cirebon	X	X	X	X	X							X							6	L
4	Kabupaten Karawang	X	X	X		X											X		X	6	L
5	Kabupaten Bandung		X	X	X		X							X				X		7	L
6	Kabupaten Majalengka		X	X	X	X	X	X												7	L
7	Kabupaten Bekasi	X		X	X	X	X		X											8	L
8	Kabupaten Indramayu			X	X	X			X				X	X						8	L
9	Kabupaten Kuningan			X		X	X		X					X			X			8	L
10	Kota Depok			X	X	X			X					X					X	8	L
11	Kabupaten Purwakarta			X	X	X		X	X				X							8	L
12	Kota Bekasi			X	X	X			X					X					X	8	L
13	Kabupaten Sumedang			X	X	X		X	X										X	8	L
14	Kabupaten Cirebon			X	X	X			X				X	X						13	M
15	Kabupaten Bogor				X	X							X	X	X	X				17	H
16	Kota Bandung					X			X		X			X		X			X	18	H
17	Kota Bogor		X		X	X								X		X			X	18	H

3.4 Schallogram analysis of Ertertainment Function

Entertainment Function reflects whether the municipal websites provided residents with entertainment via such venues as fun photos of community highlights or provided information about community entertainment/activities that would fulfill this function externally by inputting variables:

Table 7. Entertainment Variables

No	Entertainment Variables
1	Provides gallery/set of city photos and videos
2	Information about city recreation
3	Provides information about lodging in city for visitors

According to the result, *Kota Cirebon* is on "Middle" category but this city needs to improve the provision of variable 1-2 on the Table 7. *Kabupaten Bekasi*, *Kabupaten Karawang*, *Kabupaten Cirebon* and *Kabupaten Indramayu* are on the "Lower" category. They need to improve the provision of variable 2-3 (Kab.Bekasi), variables 1 and 3 (Kab.Karawang), variables 1-3 (Kab.Cirebon and Kab.Indramayu) on the Table 6.

3.5 Schallogram analysis of mass communication function

Mass Communication Function aims to know whether the municipal websites provided link to another mass communication channel by inputting variables:

Table 8. Mass Communication Variables

No	Mass Communication Variables
1	Links to local metro newspapers
2	Links to local television stations
3	Links to local radio stations
4	Links to local cable television systems
5	Links to community/non-metro newspapers

According to the result, *Kabupaten Cirebon* is on "Middle" category but this city needs to improve the provision of variables 1,3,5 on the Table 8. *Kabupaten Bekasi*, *Kabupaten Karawang*, *Kabupaten Cirebon* and *Kabupaten Indramayu* are on the "Lower" category. They need to improve the provision of variables 2,3,5 (*Kota Cirebon*), variables 1,2,3,5 (Kab.Indramayu, Kab.Karawang and Kab.Bekasi) on the Table 9.

Table 9. Schallogram Ladder: Mass Communication

NO	CITIES	HIGH (3)					MIDDLE (2)					LOW (1)					TOTAL	RANGE
		1	2	3	4	5	1	2	3	4	5	1	2	3	4	5		
1	Kota Cimahi	X			X								X	X		X	15	H
2	Kota Bogor			X	X	X		X				X					13	H
3	Kota Bandung				X	X						X	X	X			12	H
4	Kabupaten Bandung						X		X				X		X	X	10	M
5	Kabupaten Cirebon							X		X		X		X		X	9	M
6	Kabupaten Majalengka								X			X	X		X	X	8	L
7	Kabupaten Bogor					X				X		X	X	X			7	L
8	Kota Cirebon	X								X			X	X		X	7	L
9	Kabupaten Purwakarta		X							X		X		X		X	7	L
10	Kabupaten Bandung Barat									X		X	X	X		X	7	L
11	Kabupaten Indramayu									X		X	X	X		X	7	L
12	Kabupaten Sumedang				X							X	X	X		X	5	L
13	Kota Depok				X							X	X	X		X	5	L
14	Kabupaten Karawang				X							X	X	X		X	5	L
15	Kota Bekasi				X							X	X	X		X	5	L

3.6 Schallogram analysis of local government function

Local Government describes the extent to which municipal websites provided information about elected officials, representation and election processes, government organization, formal lines of authority and offices, and links between officials and constituent by inputting variables:

Table 10. Local Governent Variables

No	Local Government Variables
1	Provides information for contacting city mayor
2	Provides map and/or list of wards and city council reps
3	Provides information for contacting city council reps
4	Provides information on where, when, how to attend, speak at city council meetings
5	Provides information about Act and Law
6	Provides city council agenda
7	Provides information about city budget

According to the result, *Kabupaten Cirebon* is on "Middle" category but this city needs to improve the provision of variable 1,3,5 on the Table 10. *Kabupaten Bekasi*, *Kabupaten Karawang*, *Kabupaten Cirebon* and *Kabupaten Indramayu* are on the "Lower" category. They need to improve the provision of variable 1-3 (Kab.Cirebon and Kab.Indramayu), variables 1,2,3,5 (Kab.Karawang), and variables 1-5 (Kab.Bekasi) on the Table 11.

Table 11. Schallogram Ladder: Local Government

NO	CITIES	LOW (1)							MIDDLE (2)							HIGH (3)							TOTAL	RANGE
		1	2	3	4	5	6	7	1	2	3	4	5	6	7	1	2	3	4	5	6	7		
17	Kabupaten Bekasi	X	X	X	X	X	X	X															7	L
16	Kota Cirebon	X	X		X	X	X	X			X												7	L
15	Kabupaten Indramayu	X	X		X	X	X	X			X												7	L
14	Kabupaten Bandung Barat	X	X		X	X	X				X				X								7	L
13	Kabupaten Cirebon	X	X		X		X	X			X									X			7	L
12	Kabupaten Sumedang	X	X		X		X	X					X					X					7	L
11	Kota Bekasi	X	X		X	X	X	X										X					7	L
10	Kota Cimahi	X	X		X	X	X											X				X	7	L
9	Kabupaten Karawang	X	X		X	X	X											X				X	7	L
8	Kabupaten Bandung		X	X	X	X	X	X	X														8	L
7	Kabupaten Majalengka		X	X	X		X	X	X											X			8	L
6	Kota Depok		X		X	X	X	X								X		X					8	L
5	Kabupaten Kuningan	X		X	X	X	X	X		X													8	L
4	Kabupaten Purwakarta	X			X	X	X	X			X						X						9	L
3	Kabupaten Bogor	X						X			X		X				X		X		X		16	M
2	Kota Bandung	X										X	X	X			X	X				X	17	H
1	Kota Bogor		X			X		X								X		X	X		X		21	H

3.7 Schallogram analysis of public service provider function

Public Service Provider indicates the extent to which municipal websites provided information about the traditional city services—from police and transportation to garbage collection, schools, and health—and solicited feedback about the delivery of such services to the public by inputting variables:

Table 12. Public Service Provider Variables

No	Public Service Provider Variables
1	Solicits citizen initiatives to report problems
2	Provides information about city garbage collection/environment, programs, services
3	Provides information about city transportation programs, services
4	Provides government's Apps
5	Provides information about city health programs, services
6	Provides information for contacting city council reps
7	Provides information about city housing programs, services
8	Provides information about school, education programs
9	Provides information about city social service programs
10	Provides information about city recreation/sports programs
11	Provides information about taxes

According to the result, all the coastal cities are on "Lower" category. They need to improve the provision of variables 1,2,3,4,6,7 (*Kabupaten Karawang*), variables 1,2,3,5,6,7,8,9 (*Kabupaten Bekasi*), varables 1,3,5,6,7,8,9 (*Kabupaten Cirebon*), variables 1,2-10 (*Kabupaten Indramayu*) and variables 1-10 (*Kota Cirebon*) on the Table 13.

Table 13. Schallogram Ladder: Public Service Provider

NO	CITIES	HIGH (3)											MIDDLE (2)											LOW (1)											TOTAL	RANGE
		1	2	3	4	5	6	7	8	9	10	11	1	2	3	4	5	6	7	8	9	10	11	1	2	3	4	5	6	7	8	9	10	11		
1	Kota Bandung	X	X	X	X					X		X										X						X	X	X	X				33	H
2	Kota Bekasi				X						X	X			X									X	X			X	X	X	X	X			30	H
3	Kota Bogor								X	X		X			X	X			X			X		X	X			X	X						26	H
4	Kabupaten Bogor										X	X							X					X	X	X	X	X	X		X	X			16	L
5	Kota Depok							X			X	X									X			X	X	X	X	X	X		X				14	L
6	Kabupaten Karawang					X			X		X										X			X	X	X	X		X	X				X	14	L
7	Kabupaten Bekasi															X						X	X	X	X	X		X	X	X	X	X			13	L
8	Kabupaten Cirebon				X							X		X								X		X		X		X	X	X	X	X			13	L
9	Kabupaten Indramayu											X		X								X		X		X	X	X	X	X	X	X			13	L
10	Kota Cimahi													X									X	X		X	X	X	X	X	X	X	X		12	L
11	Kota Cirebon																						X	X	X	X	X	X	X	X	X	X	X		12	L
12	Kabupaten Kuningan						X		X															X	X	X	X	X		X		X	X	X	11	L
13	Kabupaten Majalengka											X												X	X	X	X	X	X	X	X	X	X		11	L
14	Kabupaten Sumedang															X								X	X	X		X	X	X	X	X	X	X	11	L
15	Kabupaten Bandung Bar																							X	X	X	X	X	X	X	X	X	X	X	11	L
16	Kabupaten Purwakarta																				X			X	X	X	X	X	X	X	X		X	X	11	L
17	Kabupaten Bandung																							X	X	X	X	X	X	X	X	X	X	X	11	L

3.8 Schallogram analysis of economic developer function

Economic Developer profiles whether municipal websites performed their economic roles by soliciting businesses and acting as an entrepreneur forming partnerships among and with private entities in addition to appealing to tourists and potential residents by inputting variables:

Table 14. Economic Developer Variables

No	Economic Developer Variables
1	Provides information about doing business with city
2	Provides online services/forms for business permits, licenses
3	Provides information/services for potential businesses, developers
4	Solicits investments in city, promotes city for development
5	Provides statistics about city
6	Provides information about lodging in city for visitors
7	Provides information about activities for tourists

According to the result, *Kabupaten Cirebon* is on "High" category but this city needs to improve the provision of variable 1,5,6,7 on the Table 14. *Kabupaten Bekasi*, *Kabupaten Karawang*, *Kota Cirebon* and *Kabupaten Indramayu* are on the "Lower" category. They need to improve the provision of variable 1-5 (*Kota Cirebon*), variables 1-4,6 (*Kab.Karawang*), and variables 1-7 (*Kab.Bekasi and Kab.Indramayu)* on the Table 15.

Table 15. Schallogram Ladder: Economic Developer

NO	CITIES	HIGH (3)							MIDDLE (2)							LOW (1)							TOTAL	RANGE
		1	2	3	4	5	6	7	1	2	3	4	5	6	7	1	2	3	4	5	6	7		
1	Kota Bandung	X	X				X						X					X	X			X	21	H
2	Kabupaten Cirebon		X	X	X											X				X	X	X	19	H
3	Kota Bogor							X				X				X	X	X		X	X		15	M
4	Kabupaten Kuningan					X		X						X		X	X	X	X				9	L
5	Kota Cirebon													X		X	X	X	X	X		X	9	L
6	Kabupaten Karawang					X		X								X	X	X	X		X		7	L
7	Kabupaten Bandung			X												X	X		X	X	X	X	7	L
8	Kota Depok									X						X		X	X	X	X	X	7	L
9	Kabupaten Bekasi															X	X	X	X	X	X	X	7	L
10	Kabupaten Bogor															X	X	X	X	X	X	X	7	L
11	Kabupaten Indramayu															X	X	X	X	X	X	X	7	L
12	Kabupaten Majalengka															X	X	X	X	X	X	X	7	L
13	Kabupaten Sumedang															X	X	X	X	X	X	X	7	L
14	Kabupaten Bandung Barat															X	X	X	X	X	X	X	7	L
15	Kota Bekasi															X	X	X	X	X	X	X	7	L
16	Kota Cimahi															X	X	X	X	X	X	X	7	L
17	Kabupaten Purwakarta															X	X	X	X	X	X	X	7	L

3.9 Schallogram Analysis of Community Identifier

Community Identifier pinpoints the extent to which municipal websites provided their residents information that would facilitate cultivation of community identity by inputting the web visitor number from *www.statshow.com.*

According to the result, all the coastal cities are on "Lower" category. *Kab.Cirebon* is on 6th place with 215,715 visitors, *Kab.Indramayu* is on 10th place with 136,145 visitors, *Kab.Bekasi* is on 12th place with 93,440 visitors, *Kab.Kawarang* is on 13th place with 93,440 visitors and *Kota Cirebon* is on 15th place with 70,650 visitors. The website visitor's data collected for this research was the data during 2015.

Table 16. Municipal Website Visitor

NO	CITIES	WEB VISITOR	RANGE	NO	CITIES	WEB VISITOR	RANGE
1	Kota Bandung	1,533,730	T	10	Kabupaten Indramayu	136,145	R
2	Kota Depok	688,390	S	11	Kabupaten Sumedang	135,050	R
3	Kota Bogor	342,735	R	12	Kabupaten Bekasi	109,865	R
4	Kota Bekasi	316,820	R	13	Kabupaten Karawang	93,440	R
5	Kabupaten Bogor	275,210	R	14	Kabupaten Bandung Bara	73,365	R
6	Kabupaten Cirebon	215,715	R	15	Kota Cirebon	70,650	R
7	Kota Cimahi	204,765	R	16	Kabupaten Majalengka	70,445	R
8	Kabupaten Bandung	172,280	R	17	Kabupaten Kuningan	54,020	R
9	Kabupaten Purwakarta	155,490	R				

3.10 Schallogram Analysis of Community Mobilizer Function

Community Mobilizer depicts whether municipal websites exhorted residents to get involved and to assert an impact on city government and development, in addition to whether they provided information about neighborhood organizations and how to activate citizen participation by inputting variables:

IOP Conf. Series: Earth and Environmental Science **79** (2017) 012017 doi:10.1088/1755-1315/79/1/012017

Table 17. Community Mobilizer Variables

No	Community Mobilizer Variables
1	Provides information on where, when, how to attend, speak at city council meetings
2	Provides information about community planning programs
3	Provides list of individual neighborhood organizations
4	Solicits citizen initiatives to report problems
5	Number of neighborhood websites
6	Website's social media account
7	Provides information about getting a job with city/career center

According to the result, all the coastal cities are on "Lower" category. They need to improve the provision of variables 1-5 (*Kab.Cirebon, Kab.Inramayu, Kota Cirebon, and Kab.Bekasi)*, and variables 1,2,3,5,7,8 (*Kab.Karawang*) on the Table 18.

Table 18. Schallogram Ladder: Community Mobilizer

NO	CITIES	HIGH (3)								MIDDLE (2)								LOW (1)								TOTAL	RANGE
		1	2	3	4	5	6	7	8	1	2	3	4	5	6	7	8	1	2	3	4	5	6	7	8		
1	Kota Bogor	X			X	X	X		X										X	X				X		24	H
2	Kota Bandung		X				X	X	X				X					X		X		X				23	H
3	Kota Bekasi						X	X	X									X	X	X	X	X				19	H
4	Kabupaten Bogor	X						X	X					X	X				X	X	X					16	M
5	Kota Cimahi						X	X						X			X	X	X	X	X					12	L
6	Kabupaten Cirebon							X	X						X			X	X	X	X	X				11	L
7	Kabupaten Indramayu							X	X						X			X	X	X	X	X				11	L
8	Kota Cirebon							X							X		X	X	X	X	X	X				11	L
9	Kabupaten Bekasi																X	X	X	X	X	X	X	X		9	L
10	Kabupaten Bandung																	X	X	X	X	X	X	X	X	8	L
11	Kota Depok						X		X									X	X	X	X	X		X		8	L
12	Kabupaten Kuningan			X														X	X		X	X	X	X	X	8	L
13	Kabupaten Majalengka								X									X	X	X	X	X	X	X		8	L
14	Kabupaten Sumedang						X											X	X	X	X	X		X	X	8	L
15	Kabupaten Karawang						X						X					X	X	X		X		X	X	8	L
16	Kabupaten Purwakarta														X			X	X	X	X	X		X	X	8	L
17	Kabupaten Bandung Barat														X			X	X	X	X	X		X	X	8	L

3.11 Hierarchy Analysis on Public Information Services per Website Function

Hierarchy analysis aims to determine the hierarchy of public information services among cities in metropolitan areas. It also emphasizes to compare the coastal cities rank to another city. The data input for this analysis derive from all "range" resulted in schallogram ladder per website function. This "range" will be analyzed further on the combined schallogram analysis.

Table 19. The result of "range" from all schallogram ladder per website function

NO	CITIES	Website Function									
		1	2	3	4	5	6	7	8	9	10
1	Kabupaten Bandung	L	L	L	L	M	L	L	L	L	L
2	Kabupaten Bekasi	L	L	L	L	L	L	L	L	L	L
3	Kabupaten Bogor	L	L	M	L	L	M	L	L	H	H
4	Kabupaten Cirebon	L	L	L	M	L	L	L	L	H	M
5	Kabupaten Indramayu	L	L	L	L	L	L	L	L	H	L
6	Kabupaten Kuningan	L	L	L	H	L	L	L	L	L	L
7	Kabupaten Majalengka	L	L	L	L	L	L	L	L	L	L
8	Kabupaten Sumedang	L	L	L	L	L	L	L	L	L	L
9	Kabupaten Bandung Bar	L	L	L	L	L	L	L	L	L	L
10	Kota Bandung	H	H	H	H	H	H	H	H	H	H
11	Kota Bekasi	L	L	H	L	L	L	H	M	L	L
12	Kota Bogor	H	L	H	L	H	H	H	L	H	H
13	Kota Cimahi	L	L	L	L	H	L	L	L	L	L
14	Kota Cirebon	L	L	L	L	L	L	L	H	M	L
15	Kota Depok	L	M	L	L	L	L	L	L	H	L
16	Kabupaten Karawang	M	L	L	L	L	L	L	L	L	L
17	Kabupaten Purwakarta	L	L	L	L	L	L	L	L	L	L

These all variables of website function will be analyzed further by schallogram analysis. The result of the analysis will give us information about the rank among cities in the provision of public information, determine how coastal areas rank influence on their development and provide information about what variables need to be improved. So that it can be used as inputs, advice and recommendation to the government in order to improve publication of public information services on the municipal website of each city.

Table 20. All Website Function Variables

No	Website Function Variables
1	Surveillance function
2	Entertainment function
3	Mass Communication function
4	Local Government function
5	Public Service Provider function
6	Economic Developer function
7	Community Identifier function
8	Community Mobilizer function
9	Coordination function
10	Socialization function

Hierarchy of public information service is important to know so that we can know how the distribution and the availability of public information at each city. It also supports the implementation of good governance, especially in terms of public disclosure, society should be given the easiest to consume public information.

According to the analysis, cities that has the highest public information service are Bandung City followed by Bogor City. Even it has a "high" category on providing public information, Bogor City still has shortcomings which need to be improved such as entertainment function, local government function, and community mobilizer function.

According to the result, *Kabupaten Cirebon* is on 4th rank "Hierarchy 2" on providing public information. But although it's on 4th rank, this city need to improve the information availability of all website function except local government function, coordination function and socialization function which already has a good performance. Among all cities located in Cirebon Metropolitan Area, *Kabupaten Cirebon* is the best in the provision of public information. According to West Java Spatial Plan 2009-2029 *(RTRW Provinsi Jawa Barat), Kabupaten Cirebon* is directed to be a part of National Activity Center *(PKN/Pusat Kegiatan Nasional)* with integrated facilities and infrastructure, and directs the main activity in industrial sector, *marine* business and agriculture and mineral mining activities. In terms of public information availability, *Kabupaten Cirebon* needs support its role in supporting the PKN region in Cirebon Metropolitan Area and development of West Java.

Tabel 21. Hierarchy Analysis on Public Information Services per Website Function

NO	CITIES	LOW (1)										MIDDLE (2)										HIGH (3)										TOTAL	RANGE	RANK
		10	9	8	7	6	5	4	3	2	1	10	9	8	7	6	5	4	3	2	1	10	9	8	7	6	5	4	3	2	1			
17	Kabupaten Bekasi	X	X	X	X	X	X	X	X	X	X																					10	L	
16	Kabupaten Purwakarta	X	X	X	X	X	X	X	X	X	X																					10	L	
15	Kabupaten Majalengka	X	X	X	X	X	X	X	X	X	X																					10	L	
14	Kabupaten Sumedang	X	X	X	X	X	X	X	X	X	X																					10	L	
13	Kabupaten Bandung Barat	X	X	X	X	X	X	X	X	X	X																					10	L	10
12	Kabupaten Karawang	X	X	X	X	X	X	X	X	X											X											10	L	
11	Kabupaten Bandung	X	X	X	X	X		X	X	X	X						X															10	L	9
10	Kota Cimahi	X	X	X	X	X		X	X	X	X																X					10	L	
9	Kabupaten Indramayu	X		X	X	X	X	X	X	X	X												X									10	L	8
8	Kabupaten Kuningan	X	X	X	X	X	X		X	X	X																	X				10	L	
7	Kota Depok	X		X	X	X	X	X	X		X									X			X									10	L	7
6	Kota Cirebon	X			X	X	X	X	X	X	X		X											X								11	L	6
5	Kota Bekasi	X	X			X	X	X		X	X			X											X				X			12	L	5
4	Kabupaten Cirebon			X	X	X	X		X	X	X	X						X					X									17	M	4
3	Kabupaten Bogor			X	X		X	X		X	X					X			X			X	X									20	M	3
2	Kota Bogor			X				X		X												X	X		X	X	X		X		X	30	H	2
1	Kota Bandung																					X	X	X	X	X	X	X	X	X	X	30	H	1

Kota Cirebon is on 6th rank or "Lower" category than *Kabupaten Cirebon*. This city needs to improve the provision of information on all website function except community mobilizer function and coordination function which already has a good performance. According to West Java Spatial Plan 2009-2029 *(RTRW Provinsi Jawa Barat), Kota Cirebon* is directed as the core city of National Activity Center *(PKN/Pusat Kegiatan Nasional)* in West Java with facilities and infrastructure that is integrated with the sphere of influence (hinterland), as well as being the main node services and trade and industry area to the east, as well as for cultural and religious tourism activities. With all the role embedded on Cirebon City, then public information service should be improved in order to support the spreading of information related with its roles.

Kabupaten Indramayu is on 9th rank or "Lower" category in providing public information on municipal website. This city needs to improve the information availability of all website function except coordination function which already has good performance. According to West Java Spatial Plan 2009-2029 *(RTRW Provinsi Jawa Barat), Kabupaten Indramayu* directed as the *center of region activity* (PKW/Pusat Kegiatan Wilayah) in West Java with integrated facilities and infrastructure and the main activities of sustainable wetland agriculture, fisheries and *marine business*, industry, mining (oil, gas, agribusiness) and agro-industries. With all the role embedded on *Kabupaten Indramayu*, then public information service should be improved in order to support the spreading of information related with these roles.

Kabupaten Karawang is on 12th rank or "Lower" category in providing public information on municipal website. This city needs to improve all the information availability of all website function except surveillance function which already has a good performance. According to West Java Spatial Plan 2009-2029 *(RTRW Provinsi Jawa Barat), Kabupaten Karawang* is directed as a supported node of PKN (National Activity Center) development of Bodebek (Bogor, Depok and Bekasi) urban area for sustainable agricultural wetland activities, *marine business*, non-polluting industry and non-extractive which does not interfere with irrigation and water supply as well as agro-industrial activities. In terms of public information availability, *Kabupaten Karawang* needs support its role in supporting the node PKN development of Bodebek and its own role in developing West Java.

Kabupaten Bekasi is on the last rank, 13th on "lower" category in providing public information on municipal website. This city needs to improve all website function because it shows a worse performance than other cities in Metropolitan West Java. According to West Java Spatial Plan 2009-2029 *(RTRW Provinsi Jawa Barat), Kabupaten Bekasi* is directed to be a buffer zone in the National Activity Center (*PKN/Pusat Kegiatan Nasional)* center system of Jabodetabek (Jakarta, Bogor, Depok, Tanggerang, Bekasi) urban areas. With all the role embedded on *Kabupaten Bekasi*, then public information service should be improved in order to support the spreading of information related with its roles.

4. Conclusion

The importance of delivering coastal information on the municipal website to the society needs to be highlighted. The provision of public information on municipal website for citizen by the coastal city government is very important. It can create a critical, educated, well informed society which lead to the active participation of society in development of coastal area. In addition, information service via municipal website shows the competitiveness of the city in terms of the utilization of the ICT in stimulating regional development in information age.

Considering as maritime country, Indonesia needs to highlight its marine role in order to plan coastal cities. According to the West Java Spatial Plan 2009-2029 the coastal territories play an important role in the development of West Java. The north coastal territories which metropolitan cities are located, has several development directions, for instance the development of tourism in coastal regions to develop the tourism area of the north coast with development priority of Cirebon coastal cultural tourism zone, development of residential area of north coastal area that include basic facilities and infrastructure which located outside the coastal damage and disaster-prone coastal area, the development of marine business district coastal area, marine and small islands, and provincial strategic area PANTURA *(KSP/Kawasan Strategis Provinsi Coastal PANTURA).*

The availability of public information on the municipal websites metropolitan West Java shows the openness of the government in disseminating public information to the society. In delivering the public information on municipal website, each coasl city needs to adjust the communication channel with the characterictics of its own society. Based on hierarchy analysis of public information services, the information availability of coastal cities shows a poor performance and needs improvement in all website function aspects. It means, the information availability has not been able to support the roles that embedded in each city yet and has not been able to stimulating and encouraging the development of West Java in terms of utilizations of ICT in developing West Java through public information on municipal website.

Acknowledgements

We would like to thank those who support this research, especially Bappeda Jabar (West Java Planning Board) and Pusdalisbang Jabar (Pusat Data, dan Analisis Pembangunan Daerah Provinsi Jawa Barat).

References

[1] Jeffres L W 2006 Metropolitan Websites as Urban Communication. *Journal of Computer-Mediated Communication*, 957-980.
[2] Jeffres L W 2010 The Communicative City: Conceptualizing, Operationalizing, and Policy Making. *Journal of Planning Literatur*, 99 - 110.
[3] Jeffres L W, Neuendorf K, Jian G and Cooper K S 2010 *Auditing Communication Systems to Help Urban Policy Makers.* National Communication Association Conference (San Francisco: Urban Communication Reader III). Pp. 1-35.

[4] Downey E 2010 *Handbook of Public Information Systems, Third Edition.* In E. Downey (New York State: Value of Country Web Sites) pp. 188 - 206
[5] Firmino R J 2005 Planning the Unplannable: How Local Authorities Integrate Urban and ICT Policy Making. *Journal of Urban Technology, **12***, 49 - 69.
[6] INFOASAID 2010 *Characteristics of different communication channel.*
[7] Jeffres L W 2010 The Communicative City: Conceptualizing, Operationalizing, and Policy Making. *Journal of Planning Literatur*, 99 - 110.
[8] Jeffres L W, Neuendorf K, Jian G and Cooper K S 2010 *Auditing Communication Systems to Help Urban Policy Makers.* National Communication Association Conference (San Francisco: Urban Communication Reader III). Pp. 1-35.
[9] BAPPEDA 2014 *Grand Design Pembangunan Metropolitan dan Pusat Pertumbuhan di Jawa Barat.* Pemerintah Provinsi Jawa Barat, BAPPEDA Provinsi Jawa Barat.
[10] Kitchener C O 2010 *Online Communications Strategy.*
[11] Klosterman R E 1992 Evolving Views of Computer-Aided Planning. *Journal of Planning Literatur*, 249 - 260.
[11] Larsen T K 2003 *ICT in Urban Planning.*
[12] Mandarano L, Meenar M and Steins C 2010 Building Social Capital in the Digital Age of Civic Engagement. *Journal of Planning Literatur, 25*, 123.
[13] Mangini A, Perlman S, Fredrick J and Arkenberg C. (n.d.). *Strategies for Effective Communication in the New Digital Media Landscape.* City Council Ad Hoc Committee on Technology.
[14] Pearson. (n.d.). *Communication Channels.*
[15] Peraturan Daerah No 22 Tahun 2010 tentang RTRW Provinsi Jawa Barat 2009 – 2029
[16] Peraturan Daerah 12 Tahun 2014 Pengelolaan Pengembangan Metropolitan dan Pusat Pertumbuhan Jawa Barat
[17] Peraturan Gubernur No 7 Tahun 2009 tentang Pendayagunaan Website di Lingkungan Pemerintah Provinsi Jawa Barat
[18] Pradana Alvian Chris 2013 *Analisis Pemasaran Citra Kota Menggunakan Situs Web Pemerintah Studi Kasus Kota Inti Metropolitan di Indonesia.* Tugas Akhir Program Studi Sarjana Perencanaan Wilayah dan Kota.
[19] Rinaldi H and Yuardani A M 2015. Analisis Peluang Transformasi "E-Government" menjadi "E-Democracy" melalui *Website* Resmi Pemerintah Daerah. *MIMBAR, 31*, 103-112.
[20] Sari Ivanie Destila 2016 *Pemanfaatan Media Sosial Twitter dalam Perwujudan Open Government di Kota Bandung.* Tugas Akhir Program Studi Sarjana Perencanaan Wilayah dan Kota.
[21] Sugiyono 2011 *Metode Peneltian Kuantitatif, Kualitatif dan R&D.(* Bandung: ALFABETA)
[22] Sutriadi R 2015 *SMART CITY Inovasi, kota komunikasi dan kota berkeadilan.* (Bandung)
[23] Undang-Undang 14 Tahun 2008 tentang Keterbukaan Informasi Publik
[24] *Utah Government Website* 2016. http://www.utah.gov/
[25] Wulandari Agustiah 2012 *Pemanfaatan Teknologi Informasi dan Komunikasi dalam Meningkatkan Koordinasi Penyusunan Rencana Tata Ruang di Metropolitan Bandung.* Tesis Program Studi Magister Perencanaan Wilayah dan Kota Institut Teknologi Bandung.

IOP Conf. Series: Earth and Environmental Science **79** (2017) 012018 doi:10.1088/1755-1315/79/1/012018

Legislation Amendment Impact on Coastal Management Pattern: *An Edge of Chaos towards Punctuated Equilibrium*

Dian Rahmawati
Urban and Regional Planning Department,
Institut Teknologi Sepuluh Nopember,
Surabaya, Indonesia

Abstract. Management of coastal areas and small islands in Indonesia refers to a set of coordination consists of planning, utilizing, monitoring, and controlling coastal resources that are carried out by every level of the government and many related sectors. An alteration has occurred in National Act about Local Government, Act No. 23/2014, which has the implications in the coastal management territorial zone. The issues started by the shifting of the seaward delineation authorities, it is stated that from 0 – 12 Nautical Miles (nmi) are under provincial government which previously 0 – 4 nmi were under city government, and 4 – 12 nmi were under provincial government. That said, there are no territorial management that is handled by the city/local government, including permit regulation. In hierarchy, provincial government are in upper level than city government. Chaos are happening especially in the context of authorities' management level, but to reach the main purpose of the National Act there should be an equilibrium point to normalize the situation and get the stakeholders understand the principles and keep engaged in the new form of management. This article aims to assess the impact of the National Act alteration to the sustainability of coastal management. A case study to simplify the model is in the East Java province. Theoretical framework to assess the alteration impact were related to Integrated Coastal Zone Management principles and regulatory review of the coastal zones. The analytical methods used in this article is descriptive comparative to observe the alteration in the pattern of coastal zone management variables. The comparative study was then mapped in a flow diagram to be compared on different management situation. As a result, significant changes were revealed in the pattern of the coastal management factors. (i) Authorities delineation; (ii) Planning documents; (iii) Permit system; (iv) Control; and (iii) Program Accreditation.

1. Introduction

Coastal zone defined as an area of transition between terrestrial and marine ecosystems that are affected by the biophysical and oceanography dynamic; while small island defined as an island with an area of less than or equal to 2,000 km2 and its unity ecosystem (Diposaptono, 2016). Management of coastal areas and small islands according to Act No. 27/2007 jo Act No. 1/2014 on the Management of Coastal Zone and Small Islands is a coordination set consists of planning, utilizing, monitoring, and controlling of coastal and small islands resources are carried out by the Government and Local Government, sectors, between terrestrial and marine ecosystems, as well as between science and management to improve the welfare of the people. Coastal zone and small islands has large varieties of natural resource potential and very important for the development of social, economic, cultural, environmental, and sovereignty of the nation, so it needs to be managed sustainably and considering global issues, taking into account the aspirations and community participation.

Changes in regulation about Local Government that was stated in Act No. 32/2004 is now updated in Act No. 23/2014 has ample implications for the authority in the management of coastal zone and small islands. Previous policy established that marine area 1-4 miles or one-third of the provincial management of the authority of the local government that latter in this article will be defined as *regional government*. Act No. 23/2014 mandated the provincial government, the upper level of the regional government, the authority to manage the natural resources from 0 nautical mile to 12 nautical miles; from the coastline to the open sea and / or toward the archipelagic waters (Article 27 of Act No. 23/2014). This has led some chaos to the regional government which no longer has the authority to manage their coastal zone and small islands. Whereas in planning, regional government has already prepared their twenty years planning document and permit system for coastal use. The plan covers the mainland from the *kecamatan pesisir* (coastal sub-districts) to the most distant waters maximally 4 nautical miles from the coastline of the highest tides.

East Java province covers 2,128 km coastline. Marine and coastal area of East Java has the area nearly twice the land area or reach 75,700 km^2 on calculation with the 12 nautical miles limit of the province (Centre for Research and Development of Marine Geology, 2012). East Java Province not only broad in terms of area, but also rich in natural resources that are really potential to become the supporting capacity development of the region. Management of resource utilization through spatial allocation in coastal zone and small islands in East Java Province are stated in Local Regulation No. 6/2012 on the Management and Zoning Plan for Coastal Areas and Small Islands (RZWP-3-K) for the period 2012-2032. Substantial changes to the legislation on local government also has implications for the relevance of the stated local regulation, including its set of management: planning, utilizing, monitoring and controlling aspects of the whole resource management. Post changes in Local Government Act enacted that the provincial government has greater authority for the management of fisheries and marine sector.

Enactment of the new local government regulation has many impacts. Some regional government in East Java province, with previously formulated spatial planning on their coastal area has specific zoning as marine protected areas, reclamation areas, and fisheries. However, due to the change of authority in coastal zone management, what had been initiated by the regional government are no longer enacted, at least waiting to be adopted in the provincial level document. Observing the conditions of the problems described above, this article is based on a fundamental question: "In what extend does the legislation amendment impact the pattern of sustainable coastal management in East Java province?"

2. Methods

Based on the type of problems that were explored, this study is using qualitative approach. Qualitative approach in research involve activities that focused on meanings and interpretations of an event through describing, writing, and reporting. The population in this study were the resource people of the parties who give and get influenced and have interests in sustainable coastal management in East Java province. The samples in this study were taken using a non-probability sampling method, which is a sampling method that is not grounded by chance because the unknown population, instead a criterion has been specified before. Terms of selection of respondents in this study consists of two groups involved, as follows:

1. Government Group:
   a. Regional Development Planning Agency of East Java Province
   b. Department of Fisheries and Marine Resources of East Java Province
   c. Legal Bureau Regional Secretariat of East Java Province

2. Civil Society Group:
   a. Association of Indonesian Coastal Management Specialist (HAPPI) East Java Branch
   b. Expert Coastal and Marine Policy and Regional Planning Lecturer ITS Surabaya

Aspects of coastal management is based on the Law on coastal management is divided into four that consists of planning, utilizing, monitoring, and controlling. Each of these aspects are then translated into variables that were observed in the study (shown in Table 1).

Table 1. Coastal Management Aspect Observed

	Variable Observed
Planning	Area Delineation, Zoning
Utilizing	Permit System
Monitoring	Mechanism of Evaluation
Controlling	Incentives and Disincentives

In order to measure the variables in the research context, both primary and secondary data are collected and analysed. The primary data collection is specifically important to describe the effect of changes in Local Government Act on the pattern of sustainable coastal management in East Java province. The method of primary data collection in this study were using in depth interviews and discussions. Secondary data collection was conducted by reviewing the contents of the documents concerned with this study, including books, research, spatial planning documents, legislation, as well as articles in internet and mass media.

3. Result and Discussions

The coastal area, as transitional area between land and sea, is a concentration of attention and interests, where human activities are interrelated with the environment and the mainland [1]. Management of integrated coastal is the management of the utilization of natural resources and environmental services that exist in coastal areas, done by conducting a full assessment of coastal areas and their natural resources and environmental services contained therein, determines the goals and objectives of utilization, and plan the management of all activities in order to achieve an optimal and sustainable development [2].

Sustainability is the use capital in many forms i.e. natural resources, human resources, man-made resources, and socio-cultural resources in a controlled and prudent way, to ensure that the current generation and future generation can achieve good security in economy and keeping the ecological integrity as an attitude. Sustainable development depends on three basic elements, namely:

1. Natural environment, where ecological integrity is to be achieved when the public, either individually or in groups, live in harmony with the natural environment;
2. Economic development, to achieve the level of economic security, which can control the community, both individuals and groups, can live with their own socio-economic conditions;
3. Community-based development or governance-centered development, which can create democracy by the involvement of public participation.

Sustainability in the coastal area has their own specification. The so-called Integrated Coastal Management (ICM) or Integrated Coastal Zone Management (ICZM) is now become the widely-adopted term in a way countries develop their coastal area. ICZM is a process of achieving the goals and objectives of sustainable development in the coastal areas, within the constraints of physical, social, and economic, as well as within the limits of law, institutions, as well as financial and administrative systems and it is focuses on the relationship between sectoral activities to achieve more comprehensive management (UNEP, 1995). Some principles are as follows:

1. Holistic Approach: all sorts of elements relating to hydrological, geomorphological, climatic, ecological, social, economic, and cultural systems must be considered in an integrated and holistic. Sea and land area in the coastal area should be well managed jointly.
2. Ecosystem approach: conserving ecosystem as priority approach to coastal planning and management should be implemented to ensure the sustainability of coastal development.
3. Good Governance: good governance of coastal areas requires cross-sectoral coordination of the various administrative services coastal region, both horizontally and vertically.
4. Inter and Intra-Generations Solidarity: ICZM should be implemented to ensure the distribution of coastal resources is better between this generation and the future generation.
5. Maintain originality: activities that occur in the coastal area is very complex to be considered properly, and prioritized to maintain the distinctiveness of local identity.

6. The principle of prevention: deterrence is necessary for controlling the activity that may exceed the carrying capacity of coastal and to minimize the negative impacts of tourism and development.

Ecological boundary concept in is an initial part that is very important to start the sustainable coastal management. To achieve the welfare of society, through economic interests and sustainability of ecosystems through spatial planning, integrated planning needs a boundary that is not only accommodate its administrative needs as it mostly done in planning, but more to a boundary that accommodate characteristics in the coastal area, both ecology and physical (Pratikto, 2006). The approach of this boundary could adopt the principles of ecoregion planning that consists of principles such as natural domain as the main boundary, coastal area as the headwaters with specific handling, maritime integrated approach, and proportionally land use management.

In Indonesia, the law that controlling issues about Local Government was managed in Act No. 32/2004, and it has been updated on Act No. 23/2014. According to normative and empirical jurisdiction perspective, there are at least four important changes that laid in the updated law. These changes are as follow:
First, classifications of government affair. According to Article 9 Paragraph 1, Government Affairs consists of absolute concurrent, and public categories. In the previous Act it contained two categories; obligatory and optional.
Second, the divisions of government affair. According to Article 13, the division of concurrent government affairs are divided under the authority of the Central Government, Provincial Government, and Regional Government. The management is based on the principle of accountability, efficiency, and externalities as well as prioritizing on national strategic interests. The significant changes related to the divisions of government affair in the amendment laid on the specific issues for forestry sector, marine sector, and energy and mineral resources sector. As stated in Article 14 Paragraph 1, the implementation of government affairs in forestry, marine, and energy and mineral resources sectors are shared under the authority between Central Government and Provincial Government. That said, implementation for those specific sectors are released from the Regional Government authority.
Third, of the effectuation of authority. Based on the authority of the government affairs, each level of government has the governance guidelines on authority. In more detail, the authority of government management are presented in Table 2 below.

Table 2. Governance on Authority

	What to Govern
Central	• Alone by the Central Government; • Delegate to the Governor as a representative of the Central Government or the vertical institutions in the region based on the principle of deconcentration; or • Assigning to the Regional Government based on the principle of Assisting Tasks (*Tugas Pembantuan*).
Provincial	• Alone by the Provincial Government; or • Assigning to the Regional Government based on the principle of Assisting Tasks; • Assigning to *Kepala Desa* (Headman/ Village Head)
Regional	• Alone by the Regional Government; or • Delegated partly on implementation to the *Kepala Desa*

Fourth, the management on the coastal zone and small islands. In this issue, there are several important changes about the authority of management, benefit share on resources, delineation authority, and archipelagic provincial. These changes are discussed below:

a. The authority of management;
According to Article 27 Paragraph 1 of Act No. 23/2014, the Provincial Government is given the authority to manage marine resources in the area of 0-12 nmi. That said, this article stipulates that only the provincial who has the right to manage marine resources. In contrast with the previous act that was dividing the authority of Regional Government with marine areas are given the authority to manage the resources in the sea areas from 0 nmi to 4 nmi, and Provincial Government were managing the areas from 4-12 nmi (Article 18 Paragraph 1 of Act No. 32/2004). That said, Article 27 Paragraph 1 of Act No. 23/2014 has released the authority of the Regional Government in the management on the coastal zone and small islands. The authority of management is explained into several categories as presented in Table 3.

Table 3. Important Points in Authority of Coastal Management

	Explanation
Permit System	In order to optimize the management of coastal area and small islands, the State used to implement HP-3, a form of right and authority permit. It is then updated on the Act No. 1/2014 which provide a new form of licensing divided into two stages: izin prinsip (principle permit) and izin lokasi (location permit) which are more deconcentrated in the activities systems. These new licensing form are then making HP-3 no longer applicable. Licensor for HP-3 were given to the regional government, the updated form of licensing are now under the authority of the governor as the authorities are shifted to the provincial level.
Recommendation	The rules of authority of the Recommendation by the Regent / Mayor were disrupted. As stated in Article 11 paragraph (1) PP. 62 of 2010, that the utilization of PPKT as meant in Article 5 paragraph (1) letter c shall obtain permission from the Minister after receiving recommendation from the governor and / or regent / mayor in accordance with their authority.
Polsus (Special Agent for Controlling Coastal Area) Management	Proposals for the appointment and dismissal of the *Polsus* from Civil Servants at the Regional Government shall be the authority of the Province; but for coaching process are able to be an assisting task of the regional government.
Reclamation	Article 5 paragraph (2) minister regulation No 17 / Permen-Kp / 2013 on Reclamation Licenses in Coastal Areas and Small Islands stated that the Reclamation Site Permit and the Reclamation Implementation Permit as meant in paragraph (1) for Certain National Strategic Area, Coastal areas within the National Strategic Area, and cross-province reclamation

	Explanation
	activities are published after consideration and recommendation given by the bupati / mayor and governor to the minister of sea and fisheries.
Activities	The authority of the Provincial Region to manage resources in marine areas as set forth in Article 27 paragraph (2) of Law Number 23 Year 2014 includes: (a) exploration, exploitation, conservation, and management of marine resources excluding oil and natural gas; (B) administrative arrangements; (C) spatial arrangements; (D) help maintain security at sea; And (e) help maintain the sovereignty of the State. This provision are specifically given for non oil and gas resources.

b. Benefit share on resources;
Although regional management authority is repealed, the resources producing and non-producing regional having the right to receive revenue sharing from the administration of Government affairs. Determination of regency / municipality producing region for marine sharecropping is a marine product within the 4 (four) mile mileage boundary measured from the baseline to the high seas and / or toward the archipelagic waters. In the case of the Regency / City borders referred to in paragraph (5) less than 4 (four) miles, the boundaries of the territory shall be divided equally or measured in accordance with the centerline principle of the adjacent territory. That is, the district / city government still gets the their right in different form for the resource sharing of 4 nautical miles.

c. Delineation authority;
According to Article 27 paragraph (3) of Law No. 23 of 2014, the Provincial Authority is to manage natural resources at sea is at most 12 (twelve) nautical miles measured from the coastline towards the high seas and / or towards archipelagic waters. If the territorial sea between two (2) Provincial Areas is less than 24 (twenty four) miles, the authority is to manage natural resources at sea whereas divided equally or measured in accordance with the centerline principle of the territory between the two Provinces (Article 27 Paragraph (4) This means that there is a change of authority on the management of marine resources that is only for the Provincial Region, and also the change of distance sharing of marine resource management area only for Provincial Region, so that the division of marine resource management area for regional is removed.

d. Archipelagic provincial.
According to Article 28 paragraph (2), in addition to have the authority managing natural resources at sea, for the island-characterized Provincial Region, the Central Government shall implement its authority in the field of maritime based on the Task of Assistance principle. A new assignment may be undertaken if the Provincial Government of the island characterizes the norms, standards, procedures and criteria established by the Central Government (Article 28 paragraph 3 of Law No. 23 of 2014).

Based on the aspects of coastal management (Planning, Utilization, Supervision, and control). Identifiable changes affected by changes in existing coastal management-related legislation are presented in Table 4.

Table 4. Pattern on Coastal Management due to the Legislation Amendment Impact

	Management Pattern	
	Before the enactment of Act 23/2014 and Act 1/2014	After the enactment of Act 23/2014 and Act 1/2014
Planning; Area Delineation	Authority to manage resources in marine areas furthest 12 (twelve) nautical miles is measured from the coastline to the open sea and / or towards the archipelagic waters for the province and 1/3 (one third) of the provincial authority area for the district / Law No. 32 of 2014 on Regional Government).	The authority of the Provincial Region to manage natural resources at sea is at most 12 (twelve) nautical miles measured from the coastline towards the high seas and / or towards the archipelagic waters. What is meant by "coastline" is the boundary of the meeting Between the sea and the mainland at the time of the highest tidal sea.
Planning Documents	District / Municipal Government is authorized to prepare Zoning Plan, Management Plan, and Action Plan of Management with coastal area consisting of the mainland, the outer coastal subdistrict and the waters area, which is 4 miles from the coastline.	The authority to draft a Strategic Plan, Zoning Plan, Management Plan, and Action Plan for Management with a coastal sphere of 12 miles from the coastline is fully available in the Provincial Government.
Utilization; Permit	The Bupati / mayor is authorized to grant and revoke Location Permit And Management Permits in the region Coastal waters and small islands according to their authority. (Law No. 1 of 2014 on the Management of Coasts and Small Islands)	Since the Regency / City Government is no longer authorized in water management, granting and revoking the location permit and water management permit 0-12 miles becomes the authority of the Governor. On the other hand, a coastal area permit on the mainland is granted and revoked by the Regent / Mayor in accordance with its authority
Supervision; Patrol	Supervision over planning and implementation of management of Coastal Zone and Small Islands shall be coordinated by related institutions in accordance with their authority. This means that the waters 0-4 miles coordinated by the Government District / City. (Law No. 27 of 2007 in conjunction with Law No. 1 of 2014 on the Management of Coasts and Small	Supervision over planning and implementation of management of Coastal Zone and Small Islands shall be carried out in a coordinated manner by the relevant agencies of the Provincial Government.

	Management Pattern	
	Before the enactment of Act 23/2014 and Act 1/2014	After the enactment of Act 23/2014 and Act 1/2014
	Islands)	
Control; Program Accreditation	The Bupati / Mayor has the authority to draft and / or propose the accreditation proposal for the Management of Coastal Areas and Small Islands which is the authority of the Governor and / or Government in accordance with the standards and guidelines (Law No. 27 of 2007 in conjunction with Law No. 1 of 2014 on the Management of Coasts and Small Islands)	Regents / mayors are no longer authorized to draft and / or propose accreditation programs for the Management of Coastal Areas and Small Islands, but the authority becomes the authority of the governor.

4. Conclusions

The enactment of Act 23/2014 provides significant implications for the pattern of coastal management, as shown in the study case in East Java Province. Significant changes were revealed in the pattern of the coastal management factors: (i) Authorities delineation; (ii) Planning documents; (iii) Permit system; (iv) Control; and (iii) Program Accreditation. In order to answer the main question about the extent of how these changes leads to a sustainable pattern in coastal management, it should be seen from the main activities in the whole management process, i.e. planning, utilizing, monitoring, and controlling of coastal resources based on spatial allocation and authorities. Planning aspect revealed that on the delineation of its territory, the authority of the provincial government for coastal areas is as far as 12 nautical miles. The authority is to draw up a Strategic Plan, Zoning Plan, Management Plan, and Action Plan for Management is entirely the responsibility of the provincial government. The existence of community involvement in the planning process is needed still. In terms of licensing, changes in authority have implications for the granting and revocation of location permits and a 0-12 nautical miles water management permit that eventually becomes the authority of the Governor. On the other hand, a coastal area license on the mainland is granted and revoked by the regional government in accordance with its authority. In utilizing aspect, the elimination of the mechanism for granting the Right of Coastal Waters (HP-3) is also marked the shifting of delegated mechanism to the non-governmental parties, especially private sector. Monitoring over the planning and implementation of the management of Coastal Zone and Small Islands is fully coordinated by related agencies of the Provincial Government. Regional government are no longer authorized to draft and/or propose accreditation programs for the management of coastal areas and small islands, and the authority becomes the authority of the governor.

The important thing is also to describe how does it stand as a part that establish a punctuated equilibrium situation. In the process, coastal zone management processes must establish a system to achieve sustainable use of coastal resources. Sustainable in this case is that the utilization of natural resources at the present time does not reduce the opportunity of future generations to utilize the natural resources in the same portion in a better quality. Changes in the pattern of management must be addressed wisely in the process, there is no impact that violates the intended sustainable indicators economically, socially, and environmentally.

Acknowledgments
Authors wishing to acknowledge assistance or encouragement from Department of Urban and Regional Planning of Institut Teknologi Sepuluh Nopember for the research funding and the opportunity to cooperation with important stakeholders in coastal zone management.

References
[1] Scura, L., Chua, T.E., et al., 1992. Lessons for Integrated Coastal Zone Management: The ASEAN Experience. Integrative framework and methods for coastal area management. ICLARM Conf. Proc. 37. T

[2] Sorensen, J. and McCreary, S., 1990. Institutional Arrangements for Managing Coastal Resources and Environments, National Park Service, US Department of the Interior & USAID.

[3] Dahuri, Rokhmin dkk. 1996, Pengelolaan Sumber Daya Wilayah Pesisir dan Lautan Secara Terpadu, P.T., Pradnya Paramitha, Jakarta;

[4] Peraturan Menteri Kelautan dan Perikanan No 12/Permen-Kp/2013 tentang Pengawasan Pengelolaan Wilayah Pesisir Dan Pulau-Pulau Kecil

[5] Peraturan Menteri Kelautan dan Perikanan No 17/Permen-Kp/2013 tentang Perizinan Reklamasi di Wilayah Pesisir dan Pulau-Pulau Kecil Peraturan Pemerintah Nomor 62 Tahun 2010 tentang Pemanfaatan Pulau-Pulau Kecil Terluar

[6] Undang-Undang No. 1 Tahun 2014 tentang Pengelolaan Wilayah Pesisir dan Pulau-pulau Kecil

[7] Undang-Undang Nomor 22 Tahun 1999 juncto Undang-Undang Nomor 32 Tahun 2004 tentang Pemerintahan Daerah

[8] Undang-Undang Nomor 23 Tahun 2014 tentang Pemerintahan Daerah

Social Capital and Public Participation on Planning in Coastal Area Development

Gunawan Prayitno* and M. Syaifurridzal

Jurusan Perencanaan Wilayah dan Kota, Fakultas Teknik, Universitas Brawijaya, Jl. Mayjen Haryono No. 167 Malang 65145

Abstract. Indonesia is one country to the world that rich with natural resources, especially on marine and coastal resources. But, the condition of rural inhabitants in coastal area still low in economic condition and public facilities and others not suitable for good living environment. The goal of this paper is to integrate the concept of social capital and public participation in the community activities. Social capital, which is interpreted with the term of the trust, networks and norm as governing human behavior is significant to motivate and coordinate collective action towards collaboration. Collective action or collaboration among people in the communities could solve the problem together. In the Grootaert research, with the title "Social Capital, Household Welfare and Poverty in Indonesia" (1999) found that active participation in decision making and memberships in heterogeneous organizations further reduce the likelihood to be poor. In this research, we found the same from Grootaert finding, that social capital (trust) has positive impact to community activities (path point 0.56) in this research location.

1.Introduction

Indonesia is one country in the world that rich with natural resources, thus putting the country in third place as a country that has high bio diversity (mega bio diversity). It's potency can be viewed from a wider variety of marine life, including a variety of fish and various types of mammals and marine animals. Indonesia also has potency in coral reefs. It is consist of various types, and dominated by corals. Coral reefs approximately 42,000 Km2 or 17% of the world's coral reefs and ranked as the second largest coral reef in the world after Australia. It is generally served as a spawning (spawning ground), foraging (feeding ground), care (nursery grounds) fish and other marine life, retaining abrasion, and can be made object / nautical tourism.

Likewise, coastal area has many potencies too. It's known as the marine ecosystem enormous resource potential and an opportunity for an increase in productive economic activity. One of the examples of the coastal region is Banyuanyar village, Sampang Regency, East Jawa province.

As a coastal region, the Banyuanyar Village has a variety of potential natural and human resources. Banyuanyar village is one of the largest fish producer in Sampang regency. The most of the inhabitants work as fishermen and processing the fish. The residents of a fishing village able to process fish into products that market oriented and has high resale value. The natural resource potential is also supported by the human resource, where the population has had a high work ethic and the spirit of togetherness (high level of social capital).

High activity inhabittants in the public or community activities, actually shows a high level of social capital. High levels of social capital will facilitate the community to solve problems together. In accordance with the concept of social capital are:

"Social capital, which is interpreted with the term of the trust, norms, networks and identity as governing human behavior is significant to motivate and coordinate collective action towards collaboration. Social capital such as trust, norms, and networks that can improve the efficiency of society by facilitating coordination action (Bourdieu, 1983; Coleman, 1994) in Kobayashi et.al., (2012).

Grootaert in his research, with the title "Social Capital, Household Welfare and Poverty in Indonesia" (1999) found that active participation in decision making and memberships in heterogeneous organizations further reduce the likelihood to be poor. As the literature on social capital has often argued, for local associations to be effective, members must participate actively. Grotaert results suggest that this is not achieved per se by attending meetings (which in Indonesia is often obligatory) but by participating actively in the decision making process. Households that active in community meeting are presumably better able to reap the benefits from the associations.

Based on the above understanding, the aim of this paper is to empirically prove the effect of "trust" as social capital construct on activity's respondent in the community activities. So, we develop a hypothesis: higher level of social capital could improve inhabitants activity in the community activities. This research tries to measure the relation between social capital and community participation in the community activities using the concept of 'network'. It is a concept explained by different factors, some of which are observable, while others are unobservable to the researcher. Observable factors are data from the survey related to the demographic characteristics (income, and education); while for unobservable factors we create data from questions in the questionnaire survey. We measured this relation with the answers of questions in questionnaire survey about the respondent network to their living environment/village, their neighbor's and how their norms.

2.Methods

2.1 The study area

The study area lies in Kampung Nelayan, Banyuanyar village administration, in the District Sampang, of Sampang Regency. Size of Banyuanyar Village (Kampung Nelayan) is ± 34.39 hectares (Figure 1).

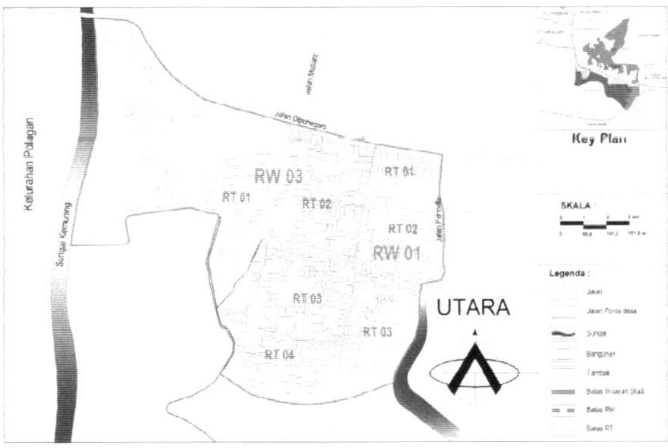

Figure 1. Administrative Boundaries

2.2 Sample selection

2.2.1 Sample selection. Sampling technique used in this research is purposive sampling (Hamidi, 2008).

2.2.2 The numbers of the sample. We choose total amount of the samples are 89 inhabitants. The criteria for the sample are representative of the people who work as fishermen, fish merchants and processors of fishery products. This sample used as a representative of the number of fisging communities in the Kampung Nelayan.

2.3 Method

We used descriptive and evaluative analysis to answer the problem identification, based on the data in the study area to assess the physical and non-physical condition. Besides, we also used PRA (participatory rural appraisal), a participatory methods to assess the potential and problems of public participation in the terms of the study area.

To measure the level of social capital and the relation with community activities in this research area we use structural equation model that we run on Mplus software.

3.Social Capital Concept

Social capital, which is interpreted with the term of the trust, norms, networks and identity as governing human behavior is significant to motivate and coordinate collective action towards collaboration. Social capital such as trust, norms, and networks that can improve the efficiency of society by facilitating coordination action (Bourdieu, 1983; Coleman, 1994) in Kobayashi et.al., (2012). Social capital improves participants' monitoring, reduces free-riders, thus mutual bonds of trust.

Communities with high levels of social capital are more effective at exercising social control over deviant and uncivil behaviors (Subramanian et al. 2002; Sampson et al. 1997). Similarly, weaker social ties directly increase vulnerability to crime by decreasing the likelihood of receiving deterrence, help or information from neighbors and limiting connections to and thus help from police and other institutions (Bellair 1997; Kennedy et al. 1998), although this effect may vary by type of community or, in the present study, cross-culturally between countries (Mihaylov and Perkins, in press).

At the individual level, Perkins and Long (2002) proposed a two-by-two social capital framework: one dimension distinguishes intrapsychic (cognitive/affective) from behavioral responses; the other contrasts informal/spontaneous versus formally organized responses. This yields four components of psycho-behavioral, social capital: the informal, affective component is social bonding (or sense of community); the informal behavior is neighboring; the organized, cognitive component is collective efficacy (or empowerment); the organized behavior is citizen participation.

People gain access to social capital through membership in interpersonal networks and social institutions and then convert it into other forms of capital to improve or maintain their position in society (Coleman 1988). As example: Migrant networks are sets of interpersonal ties that connect migrants, former migrants, and non-migrants to one another through relations of kinship, friendship, and shared community origin. Network connections increase the likelihood of international migration because they lower the costs and risks of movement and increase the expected net returns to migration. Having a tie to someone who has migrated yields social capital that people can draw upon to gain access to an important kind of financial capital, that is, high foreign wages, which offer the possibility of accumulating savings abroad and sending remittances home (Pauline et al., 2001).

Based on Grootaert research, in his study on "Social Capital, Household Welfare and Poverty in Indonesia" (1999) found that active participation in decision making and memberships in heterogeneous organizations further reduce the likelihood to be poor. The memberships in associations that bring together people from different neighborhoods and kin groups also reduce the probability to be poor.

As the literature on social capital has often argued, for local associations to be effective, members must participate actively. Our results suggest that this is not achieved per se by attending meetings (which in Indonesia is often obligatory) but by participating actively in the decision making process. Households that active in community meeting are presumably better able to reap the benefits from the associations.

4. Characteristic of the Research Area

4.1 The physical characteristic

4.1.1 General physical characteristic. The topography of Kampung Nelayan Village Banyuanyar is a lowland area has a slope of 0-6% (flat), at the height of the space between 0-2 meters above sea level. Kampung Nelayan Village Banyuanyar traversed by the river Kemuning used as transport infrastructure and a source of irrigation ponds.

4.1.2 Land use. The land use in Kampung Nelayan Village Banyuanyar can be seen in Table 1.

Table 1. Land Use

No	Land Use	Size (Ha)	Percentage (%)
1	Settlement	21.60	62.81%
2	Fishpond	6.70	19.48%
3	Port	1.45	4.22%
4	Trade and services	0.63	1.83%
5	Education building	0.60	1.74%
6	Office buildings	0.15	0.44%
7	Industrial / warehousing	1.29	3.75%
8	Open Space/Fields	1.97	5.73%
	Jumlah	**34,39**	**100**

The most dominating functions in Kampung Nelayan are a comprehensive settlement in 21,6 ha or by 62.81% of the total area.

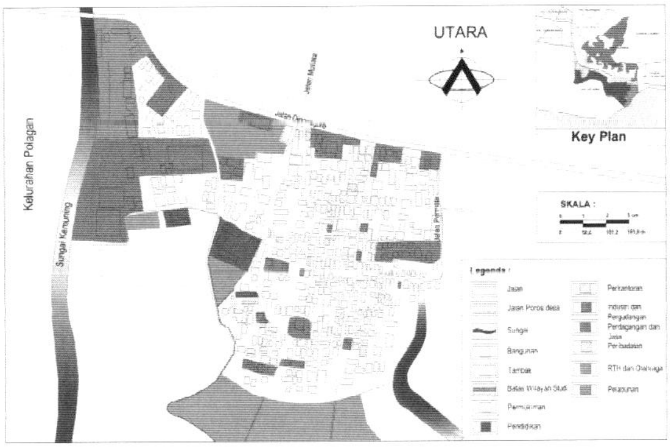

Figure 2. Land Use Map

4.2 Nonphysical characteristic

4.2.1 Population. Banyuanyar village size is 34.39 hectares, with the total population of 1,344 inhabitants divided in 448 households. Composition of population by sex are 696 male and 648 females. According to the Profile of Banyuanyar village from the total inhabitants 1,344 people (448 families), 1,241 inhabitants (413 families) are poor. Poverty is due to the majority of residents livelihood in Banyuanyar village are fishermen who do not have a steady income. The largest number of inhabitants is working as a fisherman (821 inhabitants) or 91.43%. The second type is the greatest work of civil servants by 21 inhabitants or 2.34%. The lowest is the number of private teachers (4 people) or 0.45%. The result from fishing in Kampung Nelayan Banyuanyar are fluctuations, unpredictable, and decline each year. Here is a marine fish production (Table 2).

Table 2. Marine Fish Production

NO	Fish Species	Production (days)	Price/Kg (RP)
1	Peperek	1-2 Ton	1.000-2.500
2	Ekor Kuning	1 Ton	2.000
3	Selar	1 Ton	3.000
4	Lemuru	2-6 Ton	1.000-1.400
5	Kembung	1-2 Ton	4.000
6	Tongkol	2-3 Ton	3.000-5.000
7	Udang dan lainnya	1 ton	15.000

There is also a group home industry fish processing in the fishing village, the majority industry is boiled fish (34), bloater (18) drying fish (5), fish paste (6), shrimp paste (3) fish refrigeration (7), with the number of production (ton):

Table 3. The Numbers of Production

No	Type of Fish Product	Total (Ton/production)
1	Boiled fish	49,8
2	Bloater	17,8
3	Drying Fish	12,2
4	Fish Paste	0,4
5	Shripm Paste	0,2
6	*Cool box*	35,2

The data show that the home processing industry is able to absorb the existing workforce in the Kampung Nelayan village. The number of workers absorbed by each business unit depends on its business scale, which is the majority of workers in the fishing village home industry are women.

5. Community Activities and Social Capital

There are a number of dimensions to social capital and to measure its level. Standardization in measuring social capital is still far away (Lin, 2001). There has been an abundance of ad-hoc measures, often derived from data not specifically designed to measure it but that happened to be available readily for analysis. This has made a thorough and specific testing of social capital theory difficult for structural comparison.

Latent variables were used to define the concept social capital. We have designed some questions in the questionnaire survey to measure the use of three components of social capital. They are: (1) norms; (2) networks; (3) and trust. We sought response on trust, network and trus (10 questions) and participation in community activities using 5 questions.

5.1. Community activities

The participation on community activities could measure from the technical stage of settlement planning and poverty reduction. The level of community participation in the five community activities are:

- The readiness of the community meeting (ACT1),
- Focus Group Discussion (FGD) as (ACT2),
- Participatory Rural Appraisal (PRA) as (ACT 3),
- Participatory Planning (metaplan) as (ACT4) and
- Preparation and submission of the proposed activities (ACT 5).

5.1.1 The readiness of the community meeting

Indicators Meeting Community Preparedness have four benchmarks, namely (1) the public present at the meeting (min 75% attendance); (2) the community actively discussing (the suggestions / advice) in a meeting, (3) people involved in the discussion of preparedness or unpreparedness of society to carry out participation planning; and (4) community involved in the selection of the public cadres. Here are the results of the calculation of each of its benchmarks.

Table 4. Community Preparedness Meeting

Indicators		Attend (person)	Total Sample	Percentage	Total
The community attend at the meeting (min attends 75%)	≥75%	19	89	21,3%	100%
	< 75%	70		78,7%	
Active in discussion	Active	11	89	12,4%	100%
	Not Active	78		87,6%	
Involve in every decision making regarding to the planning that will implement	Involve	19	89	21,3%	100%
	Not involve	70		78,7%	
Involved in the selection of the public cadres	Yes	19	89	21,3%	100%
	No	70		78,7%	

There are 19 participants or 21.3% of 89 attend the community preparedness of the Meeting and around 70 other people generally do not attend. This is caused by the activities did in working hour/labor time and the rest not attend because they have not got information about the activity.

In the discussion stage, participants who actively provide suggestions or advice are 11 people or 12.4%. They like to give a suggestion regarding to economic, physical and social form, whereas suggestions from the public is that the aid target.

Furthermore, in the dealing for readiness to implement participatory rural planning activities there are 19 people or 21.3% expressed their readiness. The next activities is to select community cadres. Participants involved in the selection of cadres are 19 people or 21.3% from the total samples.

5.1.2 Focus group discussion (FGD)

Focus Group Discussion (FGD) in this research study held to coincide with the implementation of the Village Council Development Plan (Musrembangkel). This activity attended by 22 participants comprising representatives of each block.

We use focus group discussion (FGD) to collect data regarding to (i) public meeting, (ii) active in the discussion and (iii) community involve in the decision of FGD result about poverty and environmental management. Here are the results of the calculation of each of its indicators.

Table 5. Indicators of Focus Group Discussion (FGD)

Indicators		Attend (person)	Total Sample	Percentage	Total
The community attend at the meeting (min attends 75%)	≥75%	22	89	32,8%	100%
	< 75%	67		67,2%	
Active in the discussion	Active	18	89	20,2%	100%
	Not Active	71		79,8%	
Involve in the decission of FGD result about poverty and environmenttal management	Involve	22	89	32,8%	100%
	Not involve	67		67,2%	

Focus Group Discussion (FGD) attended by 22 participants or 32.8% who want to involve in the decision of FGD result about poverty and environmental management. There are many inhabitants did not attend because the discussion in the working time and some people did not receive information about the activity.

In the discussion stage, participants who actively provide suggestions or advice as many as 18 people or 20.2%. Proposed discussion given which are the causes of poverty and the potential of a fishing village. Participants also provide many physical development activities proposed docks at the supporting port because it is considered very important for the advancement of the economy of the fishing village.

Furthermore, in the FGD decision, there are 22 people or 32.8% agree. The total inhabitants who involved in the selection of cadres total 19 respondents or 21.3% of the total samples taken.

5.1.3 Participatory rural appraisal (PRA)

We measure Participatory Rural Appraisal (PRA) with 5 indicators, namely: (i) attend at the meeting (min 75%), (ii) Active in the PRA activities, (iii) Involve to coincide map and the potency of social and economic, (iv) the society engaged to coincide map and problem identification of ecological, environment than physical; and (v) involve to coincide PRA. Here are the results of the calculation of each of its indicators.

Table 6. Indicators for Participatory Rural Appraisal (PRA)

Indicators		Attend (person)	Total Sample	Percentage	Total
Attend at the meeting (min 75%)	≥75%	26	89	29,2%	100%
	< 75%	63		70,8%	
Active in the PRA activities,	Active	20	89	22,5%	100%
	Not Active	69		77,5%	
Involve to coincide map and the potency of social and economic	Involve	26	89	29,2%	100%
	Not involve	63		70,8%	
The society engaged to coincide map and problem, identification of ecological, environmental and physics	Involve	26	89	29,2%	100%
	Not involve	63		70,8%	
Involve to PRA agreement	Involve	26	89	29,2%	100%
	Not involve	63		70,8%	

The implementation of the PRA attended by 26 participants or 29.2% of the samples. In the discussion stage, participants who actively provide suggestions or advice are 20 people or 22.5%. Information given them the range of economic conditions, physical and social and cultural fishing village.

The map for the project location should be discuss first to get an information regarding to the issues. The issues are economic potention, social, and cultural.

5.1.4 Metaplan

Metaplan has four indicators, namely (i) the public present at the meeting held (min 75% attendance), (ii) the community actively discussing (the suggestions / advice) in a meeting, (iii) people involved in agreement about potential and problems of the region, (iv) the people involved in the agreement about priority programs and projects.

Table 7. Metaplan Indicators

Indicators		Attend (person)	Total Sample	Percentage	Total
The community attends at the meeting (min attends 75%)	≥75%	19	89	21,3%	100%
	< 75%	70		78,7%	
Active in discussion	Active	11	89	12,4%	100%
	Not Active	78		87,6%	
Involve in agreement about potential and problems of the region	Involve	16	89	18%	100%
	Not involve	73		82%	
The people involved in the agreement about priority programs and projects.	Involve	20	89	22,5%	100%
	Not involve	69		77,5%	

The first indicator attending to community meeting there are 19 respondents or 21.3% respondent involved. All indicators for metaplan, the majority of the respondents answered not active nor not involved.

5.1.5 The preparation and submission of proposed activities

Preparation and Submission of Proposed Activities carried out concurrently with the metaplan, because this activity is a follow up to the selection of priority programs and projects that will be proposed.

There are two indicators to formulation and preparation the proposed activities in the community. (1). The community decided by themselves physical project that will implement; (2) The community makes the budged for the project by themselves. Here are the results of the calculation of each of its indicators.

Table 8. Indicators The Preparation and Submission of Proposed Activities

Indicators		Attend (person)	Total Sample	Percentage	Total
The community decided by themselves physical project	Yes	26	89	29,2%	100%
	No	63		70,8%	
The community makes the budged for the project by themselves	Yes	6	89	6,7%	100%
	No	83		93,3%	

The table shows that in the preparation and submission of proposed activities, communities involved in the small numbers (only 29,2 %). It means that the participation of the respondent is very low. Look the same for the second activities (the community makes the budged for the project by themselves), it's more difficult so less people involved.

5.2. Social Capital

We design latent variable to measure social capital level by 10 (ten) questioner related to:

Table 9 Variable to measure social capital

	Variable	Sub-variable
Social Capital	Trust	Honesty
		Help each others
		Trust to relatives
		Trust to neighbors
		Trust to government
	Network	Network each members
		Network individu with group
		Network individu with communities
	Norms	Attitudes towards norms
		Attitudes towards sanctions

Sources: Various references and ideas of the author

6. The Model

This study proposes more than one latent variable (i.e., norm, network and trust) that explains the causal relationship among observed variables based on structural equation and applies structural equation model (SEM). The level of social capital interprets from the relationship between norm, network and trust.

6.1. Estimation of the model

The next stage of the analysis involves running structural equation model (SEM) with Mplus analysis software. Structural equation modeling (sometimes called path analysis) employed to gain additional insight into causal models and explore the interaction effects and pathways between variables. This analysis (SEM) gives us more rigorously test whether our data supports our hypotheses.

Models were tested using SEM goodness of fit tests to determine if the pattern of variances and covariances in the data is consistent with structural (path) models theoretically specified. We conducted an SEM analysis on the correlation between independent variables to understand the indirect effects. SEM is multivariate regression in which the response variable in the regression equation may become predictor in another equation (Schumacker and Lomax, 2010). This allows us to account for correlations and to distinguish direct and indirect effects of our exogenous and latent variables social capital formation. For estimation where the dependent variable is a dichotomous outcome (binary discrete choice model), we use the robust (mean- and variance adjusted) method of Weighted Least Square (WLS) also known as WLSMV (Muthén and Muthén, 2012). In general this method is preferable to Maximum Likelihood (ML) estimation when the data are severely non-normal distributed (Olsson et al, 2000).

We test several different model specifications, and we chose the model with goodness of fit higher than 2 (RMSEA, CFI and TLI). For the individual and regional attributes and community activities variables, we designed a path based on the goodness of fit of each model. We finally selected variables and model structure with the highest estimation accuracy.

Our final model with the best model fit are CMIN/DF = 1.332, CFI = 0.989, TLI = 0.986 and RMSE 0.041). In general CFI>0.95, TLI>0.95, RMSEA<0.05 and WRMR<1 can be indications of good models (Yu, 2002). CFI = Comparative Fit Index; TLI = Tucker-Lewis Index; RMSEA = Root-Mean-Square Error of Approximation; WRMR = Weighted Root Mean square Residual (WRMR is suitable to evaluate models with non-normally distributed outcomes). The structure of this model is further illustrated in Figure 3.

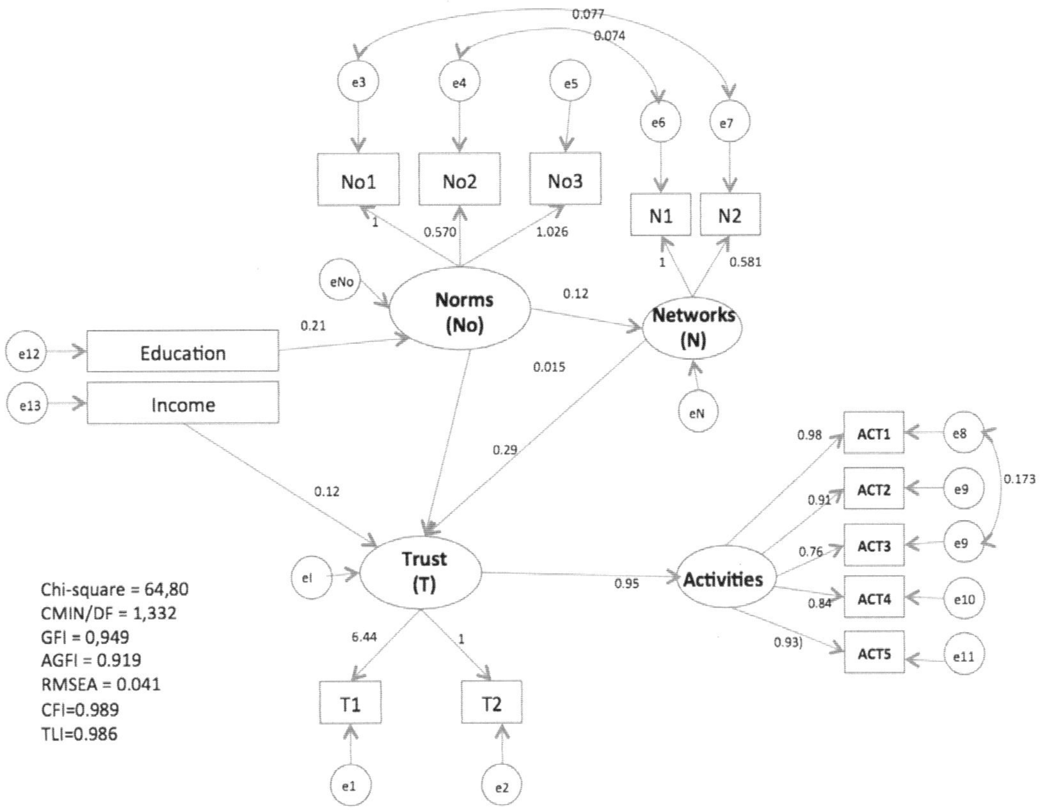

Figure 3. The Model

Furthermore, there are significant paths for activities in the communities from the latent variable "trust". Norms and networks has significant value to "trust". Trust are constructed by the two variables (T1 and T2) suggested by the PCA which all are found significant though the importance of the exogenous variables varies. Income has direct path to trust, and education giving impact on norm. The relation among social capital construct shows that norm has direct paths to network and trust, and network has direct path to t. This result confirms our chosen construct name, i.e. the trust is the central theme for this construct.

Figure 3 demonstrate that the respondents who have higher trust each other have significant impact to active in community activities. The estimated value from trust to community activities is 0.95 and the t value (C.R) is more than 1.96 (5.292). It means that if the trust increase, then it will have a significant effect to involve/active in community activities.

IOP Conf. Series: Earth and Environmental Science **79** (2017) 012019 doi:10.1088/1755-1315/79/1/012019

7. Conclusion

*7.1 Physical characteristic*The land use is dominated by settlements. It means that all conditions are fulfilled, but there is still a shortage of public health facilities. The infrastructure for housing like water infrastructure services is still not evenly distributed, and they still need for toilet and TPA / TPS.

7.2 Social capital and community activities

In this paper, we measure the relationship between social capital and community activities for coastal area development. We use the concepts of norm, network and trust to measure the relationship of social capital and community activities. The aim of this study is to investigate the relationship between social capital and community activities. The analysis is based on a survey of community activities in Banyuanyar village which is typical of coastal area in Indonesia. Section 3 first describes the concept of social capital and analyses it based on previous studies on norm, network and trust. Here we find the possibility of forming social capital from the concept of norm, network and trus, and how the participation on community activities. In Section 4, we present the characteristics of research area. Section 5 explained the respondent participation on the community activities; and the component of social capital. Last in the Section 6 we present the structural equation model to measured the relationship between latent variables and observe variables. From the structure of the model, we could estimate the relation between social capital and community activities.

Our study shows that trust positively has a significant impact to active in community activities. We have confirmed our hypothesis that households with higher social capital is will active in the community activities. Our current model so far could explain the relation between social capital and community activities, but we need to do the next the research to calculate the policy implication on it's.

References

[1] Bourdieu P 1986 *The Forms of Capital in Handbook of Theory and Research for the Sociology of Education,* edited by John G. Richardson. (New York: Greenwood Press) Pp. 241–58.

[2] Coleman J S 1988 Social Capital in the Creation of Human Capital. *American Journal of Sociology*. Vol. 94. University of Chicago.

[3] Grootaert 1999 *Social Capital, Household Welfare and Poverty in Indonesia.*

[4] Putnam R D 2000 *Bowling Alone: The Collapse and Revival of American Community* (New York: Simon and Schuster)

[5] Lin, N. Social Capital: a Theory of Social Structure and Action. Cambridge: Cambridge University Press. 2001.

[6] Schumacker, RE. and Lomax, R.G. A Beginner's Guide to Structural Equation Modelling. Third Edition. New York: Routledge, Taylor and Francis Group. 2010.

[7] Lu, Yao. "Rural–urban Migration and Health: Evidence from Longitudinal Data in Indonesia." Social Science and Medicine 70(3). 2010. p. 412-419

[8] Muthén, L.K. and Muthén, B.O. (2012). Mplus User's Guide. Seventh Edition.Los Angeles, CA: Muthén & Muthén.

[9] Olsson, U.H.. Foss, T., Troye, S.V., and Howell, R.D. 2000. The performance of ML, GLS, and WLS Estimation in Structural Equation Modeling under conditions of misspecification and nonnormality. Structural Equation Modeling7(4): 557-595.

IOP Conf. Series: Earth and Environmental Science **79** (2017) 012020 doi:10.1088/1755-1315/79/1/012020

Spatial Modeling in The Coastal Area of East Java Province

Ummi Fadlilah Kurniawati
Department of Urban and Regional Planning, Institut Teknologi Sepuluh Nopember.

E-mail: fa.ummi99@gmail.com

Abstract. The existence of gaps that occur between regions, shows that it is a reasonable process considering that each region has different initial endowment factors. The first step that can be done to controll disparity is know what is the benchmark of the gap. The revenue growth indicator is one of benchmark for measuring regional disparities. The regional output is represented by the gross domestic regional income per capita. Concerning the phenomenon of regional disparity, East Java Province is concentrated in the north-south part, especially in coastal areas is an early indication of the gap. This is what prompted the analysis of predictor factors affecting the disparity in East Java Coastal Areas through a spatial modeling approach. Spatial modeling is done on the consideration that there are different local characteristics or potentials in each regency / city. Factors Economic growth, social factors, and physical development factors are the main factors in this study will be described in derived variables to obtain a clear picture of the influence of each factor to the disparity that occurred in the Coastal Region of East Java Province.

1. Introduction

Regional disparities occur due to differences in potential resources in an area. These resources include geographical conditions, diversity of natural resources, quality of human resources, and ethnic or political ties [1]. Reducing regional disparities is one of the central themes in regional development [2]. The gap in here is the inequality of inter-regional development progress that occurs due to differences in the rate of growth. Responding to the gap that occurs between regions, it can be said that it is a reasonable process considering that each region has a factor of initial endowment which is different from each other [3].

On a regional scale, disparities can also lead to conflict between regions. In addition to the threat of disintegration and conflict between regions, the gap between regions also ultimately leads to an increase in the number of poor people [4]. Responding to these negative impacts, it is clear that regional disparity is a crucial issue in the context of regional development and it is well worth the attention. To be able to overcome the problem of regional gaps, then first must be known benchmark of the gap. In the context of regional development, the gap size is determined based on indicators of regional development success. Benchmarks for measuring disparities are measured from income inequality and GDP growth [5]. Meanwhile, the income growth indicator for measuring regional disparities is the concept of regional output using the approach area. Regional output is represented by an indicator of per capita gross regional domestic income [6]

Based on the above description, this research will analyze the level of disparity in East Java with spatial modeling approach. Consideration of spatial approach due to the different characteristics of local potential that exist in each region. The scope of this study takes Coastal Area in East Java Province. Coastal Area in East Java Province is the main area of zoning of coastal area and small islands covering regencies / cities which have sea area that is bordered by sea such as Tuban Regency, Lamongan Regency, Gresik Regency, Surabaya City, Sidoarjo Regency, Pasuruan Regency, Pasuruan

City, Probolinggo City, Probolinggo Regency, Situbondo Regency, Banyuwangi Regency, Bangkalan Regency, Sampang Regency, Pamekasan Regency, Sumenep Regency, Lumajang Regency, Malang Regency, Blitar Regency, Tulungagung Regency, Trenggalek Regency, Pacitan Regency and Jember Regency [7]. Social, physical, and economic equity issues in East Java Coastal Region become one of the main issues that become the program of East Java Province to reduce regional disparity [8].

2. Genral Description About Disparities Issues in The Coastal Area of East Java Province

Equity and efforts to open access in the underserved islands of coastal and small island areas in the East Java Province is a major issue in the East Java Coastal Area [8]. The issue is divided into social, economic, and physical issues.

a. Social issues are reducing the number of poverty, fulfilling the basic rights of coastal communities in obtaining proper education and health services..
b. Economic issues are optimizing the potential utilization of the marine economy, optimizing sources of funding and development, unstable economic growth on the north and south coasts, and increasing commercial activity in ports.
c. Physical issues concern about infrastructure aspect such as the improvement of accessibility, the fulfillment of disaster infrastructure because the coastal area is a region that has various potential of natural disasters such as earthquakes, landslides (especially in coastal areas that have a type of cliffs like in The Southern Coast of East Java) tidal waves, tsunami and floods.

3. Disparity Analysis in The Coastal Area of East Java Province

Disparity analysis is divided into economic, social and physical disparities. Adapaun to get the value of economic disparitas obtained from the average value per capita GDP ratio during 2010-2014. Social disparity is derived from the average Human Development Index during 2010-2014. Physical disparities were obtained from the total road density ratio, the ratio of clean water use, the ratio of energy sold per capita from 2010-2014. The value of economic, social, and physical disparities is illustrated in **Figure 1** to **Figure 3**.

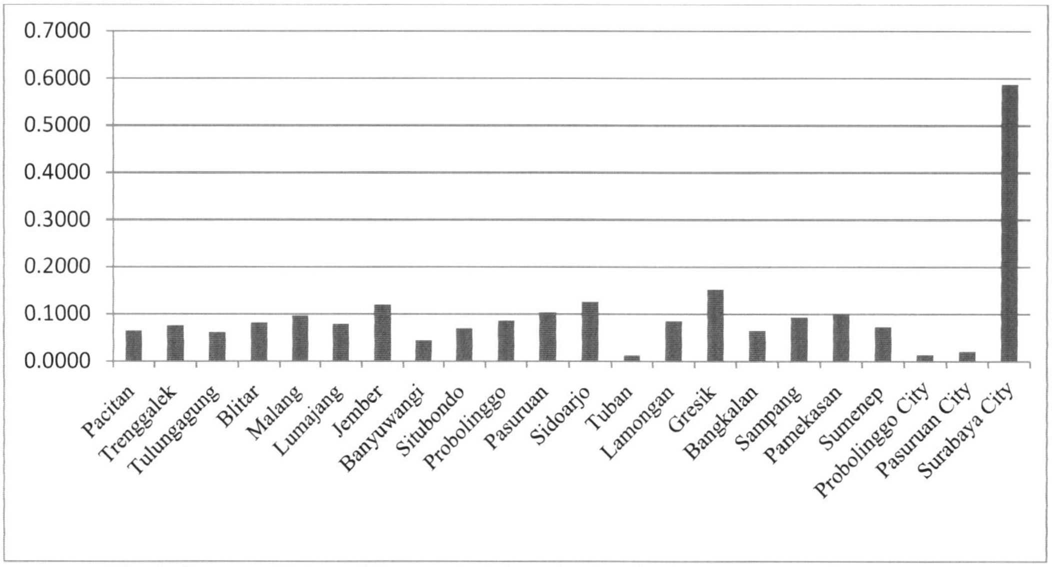

Fig 1. Average Value of Economic Disparity in The Coastal Area of East Java

Based on the Economic Disparity Chart, the average disparity value in The Coastal area of East Java Province is in the range of 0.0050 - 0.0989 during 2010-2014. The regencies / cities in the range are Trenggalek, Tulungagung, Blitar, Malang, Lumajang, Jember, Banyuwangi, Situbondo,

Probolinggo, Pasuruan, Sidoarjo, Tuban, Lamongan, Kabupaten Gresik, Bangkalan, Sampang, Pamekasan, Sumenep, Probolinggo City, Pasuruan City and Surabaya City.

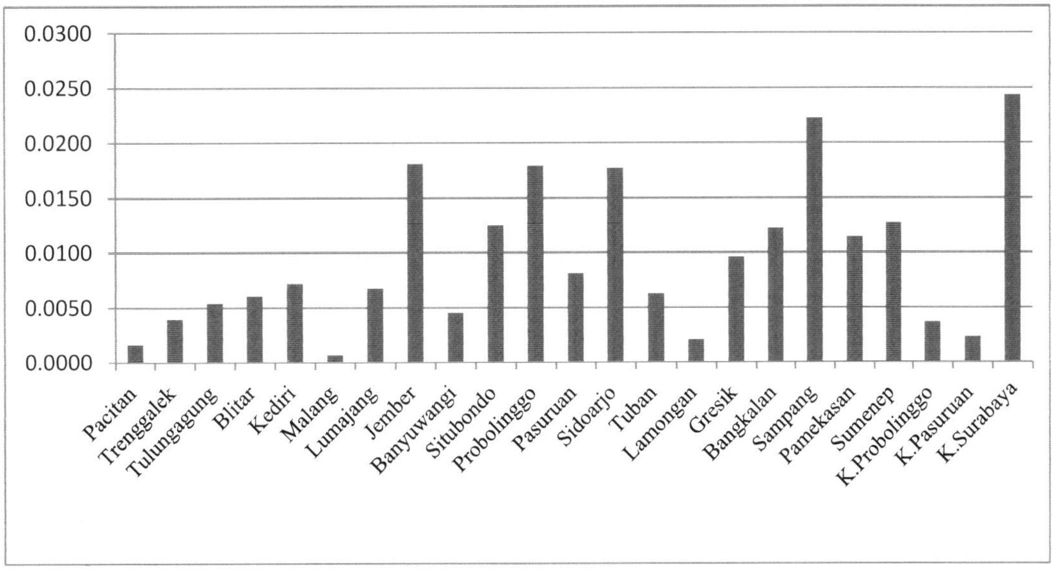

Fig 2. Average Value of Social Disparity in The Coastal Area of East Java

Based on the Social Disparity Chart, the average disparity value in The Coastal area of East Java Province is in the range 0,0003-0.0080 during 2010-2014. The regencies / cities in the range are Pacitan, Trenggalek, Tulungagung, Blitar, Kediri, Malang, Lumajang, Jember, Banyuwangi, Situbondo, Probolinggo, Pasuruan, Sidoarjo, Tuban, Lamongan, Gresik, Bangkalan, Sampang, Pamekasan, Sumenep, Probolinggo City, Pasuruan City and Surabaya City.

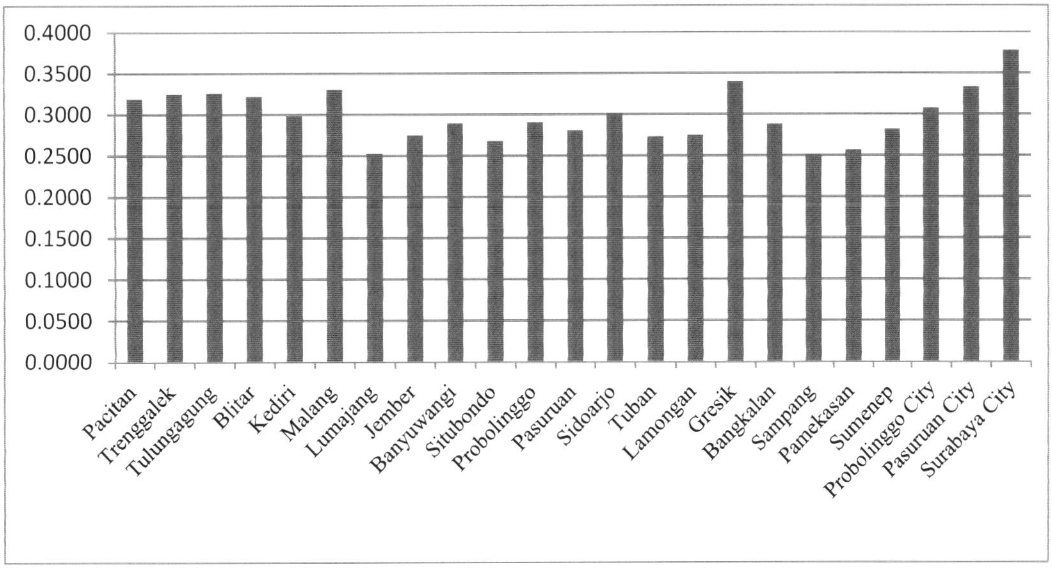

Fig 3. Average Value of Physical Disparity in The Coastal Area of East Java

Based on the Physical Disparity Chart, the average disparity value in The Coastal area of East Java Province is in the range 0.2737 - 0.3188 during 2010-2014. The regencies / cities in the range are

Jember, Bnayuwangi, Probolinggo, Pasuruan, Lamongan, Bangkalan, Sumenep, Pacitan, Ponorogo, Kediri, Sidoarjo, and Probolinggo City.

4. Spatial Modeling in The Coastal Area of East Java Province

Spatial modeling analysis to find out what factors affect the regional disparity in The Coastal Area of East Java Province through spatial regression by incorporating geographical elements in searching for influence models. However, before stepping on to the analyst required regression analysis Ordinary Least Square (OLS) which aims to determine whether this analysis requires geographical elements in it. The first step in the OLS regression analysis is to determine the variable of Regional Disparity Value as the dependent variable (Y) per regency / city in East Java Province and the GDP per capita variable, School Average School Values (RRLS), Literacy Rate (AMH), Life Expectancy (AHH), Poverty Percentage (K), Road Density Ratio (RJ), Per Capita Water (Water) Consumption and Electricity Per Capita (E) as independent (X). Before proceeding to the GWR analysis, a Classic Assumption Model Test must first be performed to determine the validity of the resulting model [9]. After the classical assumption test done, the next step do partial test with t-test.

The result from the test variable shows that variables affecting to disparities in coastal areas of East Java are life expectancy (AHH), electricity per capita (E), and percentage of poverty (K). The spatial modeling in each region shown in the **Table 1**.

Table 1. Spatial Modeling of Disparities in The Coastal Area of East Java Province

No	Regency/City	Spatial Modeling
1	Pacitan	Y= 5,872 - 0,116 AHH +0,000001 E
2	Ponorogo	Y= 5,831- 0,116AHH +0,000001 E
3	Trenggalek	Y= 5,959 -0,118AHH +0,000001 E
4	Tulungagung	Y= 5,931 -0,117AHH +0,000001 E
5	Blitar	Y= 6,009 -0,119AHH +0,000001 E
6	Kediri	Y= 5,946 -0,118AHH +0,000001 E
7	Malang	Y= 6,011 -0,119AHH +0,000001 E
8	Lumajang	Y= 5,986 -0,119AHH +0,000001 E
9	Jember	Y= 5,975 -0,119AHH +0,000001 E
10	Banyuwangi	Y= 5,976 -0,118AHH +0,000001 E
11	Bondowoso	Y= 5,842 -0,117AHH +0,000001 E
12	Situbondo	Y= 5,778 -0,116AHH +0,000001 E
13	Probolinggo	Y= 5,805 -0,116AHH +0,000001 E
14	Pasuruan	Y= 5,9105 -0,117AHH +0,000001 E
15	Sidoarjo	Y= 5,855 -0,116AHH +0,000001 E
16	Mojokerto	Y= 5,891 -0,117AHH +0,000001 E
17	Jombang	Y= 5,897 -0,117AHH +0,000001 E
18	Nganjuk	Y= 5,902 -0,117AHH +0,000001 E
19	Madiun	Y= 5,885 -0,116AHH +0,000001 E
20	Magetan	Y= 5,853 -0,116AHH +0,000001 E

No	Regency/City	Spatial Modeling
21	Ngawi	Y= 5,828 -0,115AHH +0,000001 E
22	Bojonegoro	Y= 5,849 -0,116AHH +0,000001 E
23	Tuban	Y= 5,812 -0,115AHH +0,000001 E
24	Lamongan	Y= 5,824 -0,116AHH +0,000001 E
25	Gresik	Y= 5,765 -0,115AHH +0,000001 E
26	Bangkalan	Y= 5,738 -0,115AHH +0,000001 E
27	Sampang	Y= 5,683 -0,114AHH +0,000001 E
28	Pamekasan	Y= 5,638 -0,113AHH +0,000001 E
29	Sumenep	Y= 5,421 -0,111AHH +0,000001 E + 0,601P
30	Kediri	Y= 5,943 -0,118AHH +0,000001 E
31	Blitar	Y= 5,976 -0,118AHH +0,000001 E
32	Malang	Y= 5,882 -0,117AHH +0,000001 E
33	Probolinggo	Y= 5,881 -0,117AHH +0,000001 E
34	Pasuruan	Y= 5,878 -0,116AHH +0,000001 E
35	Mojokerto	Y= 5,873 -0,116AHH +0,000001 E
36	Madiun	Y= 5,816 -0,116AHH +0,000001 E
37	Surabaya	Y= 5,948 -0,117AHH +0,000001 E

Source: Analysis, 2016

5. Conclusion

Predictor factors affecting Disparity in The Coastal Area of East Java Province based on the analysis with spatial modeling indicate that the majority of regencys / cities are influenced by Life Expectancy Numbers (AHH) and Per Capita Electricity Consumption (E). Meanwhile, only Sumenep Regency has different characteristic, that is influenced by Varibael Life Expectancy (AHH), Per Capita Electricity Consumption (E), and Poverty Percentage (P). This is because, Sumenep Regency tend to fluctuatif percentage of poverty and the average percentage of poverty during the last five years is greater than the average East Java Province.

References

[1] Ministry of National Development Planning Final Report. 2013. *Analisis Kesenjangan Antarwilayah*

[2] Firman, T,. 2000. Regional Gaps (Kesenjangan Wilayah). Department of Settlement and Regional Development, Directorate General of Spatial Planning and Regional Development.

[3] Dumairy. 1999. Indonesia's Economics (Perekonomian Indonesia). Yogyakarta : Publishing Section.

[4] Nugroho, Dahuri. 2004. Regional Development Economic, Social, and Environmental Perspectives (Pembangunan Wilayah Perspektif Ekonomi, sosial, dan Lingkungan). Pustaka LP3ES Indonesia: Jakarta.

[5] Rochana, Siti Herni. 2013. Journal of Intergovernmental Economic Gap in the Era of Regional Autonomy in Indonesia. Author of Regional Autonomy 2013 organized by APKASI (Association of Regency Government of Indonesia).

[6] Tadjoeddin, M.Z. dkk. 2001. *Aspirasi Terhadap Ketidakmerataan: Disparitas Regional dan Konflik Vertikal di Indonesia,* Working Paper: 01/01-1, United Nation Support Facility for Indonesian

IOP Conf. Series: Earth and Environmental Science **79** (2017) 012020 doi:10.1088/1755-1315/79/1/012020

Recovery (UNSFIR), Jakarta.
[7] Marine and Fishery Agency East Java province Final Report (Dinas Kelautan dan Perikanan Provinsi Jawa Timur). 2012. *Penyusunan Rencana Zonasi Wilayah Pesisir Provinsi Jawa Timur 2012-2032*.
[8] Regional Planning and Development Board of East Java Province Final Report. 2016. *Kajian Pengurangan Disparitas Wilayah Pendekatan Pengembangan Wilayah.*
[9] Setiawan, Kusrini. 2010. Econometrics (Ekonometrika). Yogyakarta : Andi.

Water Supply Provision in Sarbagita Metropolitan Area

S Maryati, ANS Humaira, SY Rachmat
Research Cluster for Regional and City Infrastructure System, School of Architecture, Planning and Policy Development, Institute of Technology Bandung
Jalan Ganesha 10, Bandung 40132, Indonesia

E-mail: smaryati@pl.itb.ac.id

Abstract. Sarbagita (Denpasar, Badung, Gianyar, and Tabanan) Metropolitan Area is one of seven metropolitan areas in Indonesia, located in the coastal region of Bali Island. Providing clean water in the coastal region is generally constrained by the limited sources of water. Besides, there is also disparity issue between the core and peri-urban area. The purpose of this study is to explore the conditions of water supply provision in Metropolitan Sarbagita in the context of coastal and peri-urban region. The methods of analysis used are descriptive and association analysis. The analysis shows that the location in the coastal area and peri-urban area does not affect the water supply provision for the case of daily safe water yet it does affect significantly in the specific context of drinking water source.

1. Introduction

Sarbagita (Denpasar, Badung, Gianyar, and Tabanan) Metropolitan Area is one of seven metropolitan areas in Indonesia. Some parts of Sarbagita are located in the coastal region of Bali Island. Water supply provision in coastal region is big issue because generally it is constrained by the limited sources of water in one side and high growth of built up area in another side. In the coastal area of Jiangsu, China, a high density of anthropogenic activities and land use changes affect the ecosystem services value and ecosystem functions [1]. In the case of Jiangsu up to 2020 there will be high probability that the coastline area will experience land transfer. The process of land transfer, mainly into built up are will affect the demand for water and also the availability of water source. Beside the issue of coastal region, in the case of Metropolitan Area there is also disparity issue between the core and peri-urban area. The case of Dammam Metropolitan Area the characteristics of new housing development are exclusiveness, low density, single housing type and poor connectivity with the surroundings [2]. Condition in Indonesia also shows the same result as in Dammam Metropolitan Area, where the development of new housing which is mainly in peri-urban area is characterized by exclusiveness, low density, and poor connection with public infrastructure. Generally, water supply provision in core area is better compared to peri-urban area. The purpose of this study is to explore the conditions of water supply provision in Metropolitan Sarbagita in the context of coastal and peri-urban region.

2. Methods

The method use in this research is association analysis by using crosstab. This kind of analysis was used because the research try to find the relationship between source of water and spatial characteristics (urban, peri-urban) as well as location (coastal, non-coastal). The relationship was analysed one by one. At the first step relationship between source of water and spatial characteristics was analysis, followed by analysis of source of water and location. Besides association analysis, in this research descriptive analysis was also used. Descriptive analysis was used to describe the condition of

IOP Conf. Series: Earth and Environmental Science **79** (2017) 012021 doi:10.1088/1755-1315/79/1/012021

water supply provision in Sarbagita Metropolitan Area. Data reduction techniques as a part of descriptive analysis ware used

3. Result and Discussion

3.1 Case Study Area

Sarbagita Metropolitan Area is one of seven metropolitan areas in Indonesia. It consists of Denpasar City, five districts in Badung Regency, two districts in Tabanan Regency, and four districts in Gianyar Regency as shown in Fig.1. One hundred and seventy sub districts spread over the area with 41 sub districts are located in coastal area and 25 of them are classified into peri-urban area (see Table 1.).

According to the Presidential Decree of Indonesia No.45/2011 regarding Spatial Plan of Sarbagita Metropolitan Area, there are seven core areas with Denpasar –Kuta Area is the main area and Mangupura Area, Jimbaran Area, Sukawati Area, Gianyar Area, Ubud Area, and Tabanan Area as its counter magnet. Other areas remained in Sarbagita Metropolitan Area (outside core area) are categorized as peri-urban areas.

Table 1. Number of villages in Sarbagita Metropolitan Area based on spatial variation and location

	Number of Sub Districts	
	Coastal Area	Non-Coastal Area
Core Area	16	59
Peri-urban Area	25	70

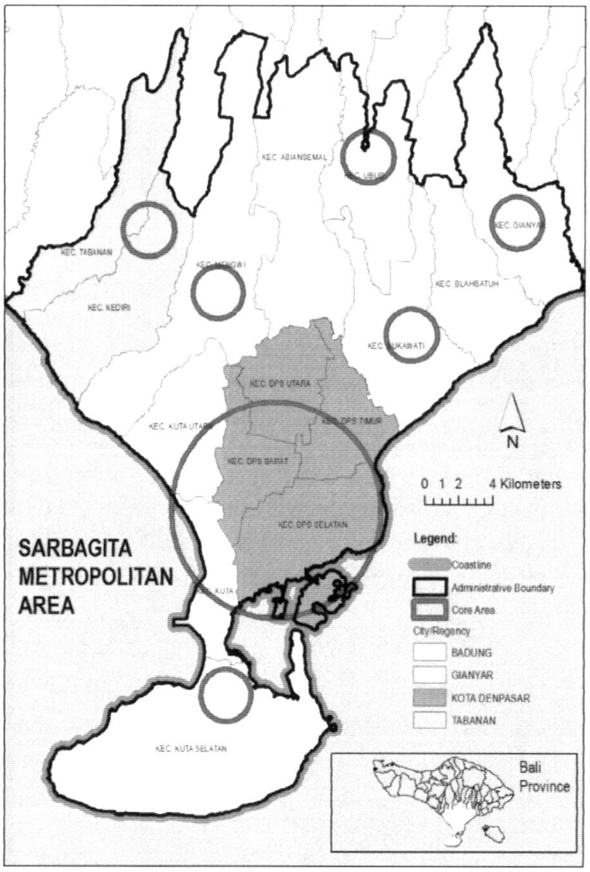

Fig. 1. Sarbagita Metropolitan Area

3.2 Water Provision in Sarbagita Metropolitan Area

Water used by households in Sarbagita Metropolitan Area is distinguished based on the source and its use. Drinking water could be acquired from bottled water, piped water with water meter, well pump, well and wellspring. 42,9% villages acquire drinking water mostly from the piped water with water meter from local water enterprise as known as PDAM (*Perusahaan Daerah Air Minum*) followed respectively by bottled water (23.5%), well pump (19.4%), and wellspring (11.8%). Based on the data gathered from Central Bureau of Statistics, Indonesia (2014), traditional well is still used by households with only 2,4% of villages mostly fulfill their needs of drinking water from this type of water source (see Fig.2.a.).

Drinking water is acquired whether from piped water with water meter, well pump, well, wellspring, and groundwater. The same main water source as for drinking water, piped water with water meter provided by PDAM also takes the biggest percentage (66.5%) among the water sources of daily safe water followed by well pump (27.1%). As illustrated in Fig.2.b., wellspring is scarcely used for daily safe water source shown by only 0.6% village uses this type of water source for fulfilling its daily safe water.

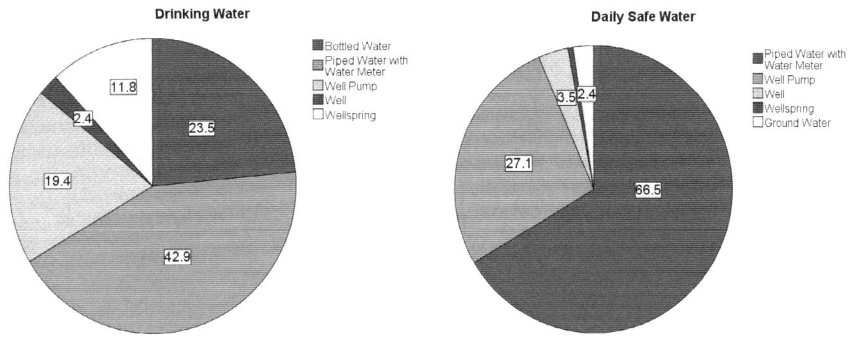

Fig. 2. (a) Percentage of drinking water sources in Sarbagita Metropolitan Area; (b) Percentage of daily safe water sources in Sarbagita Metropolitan Area

PDAM has been widely delivering the needs of drinking water and daily safe water provision and has reached the service coverage in more than 65% villages in this metropolitan area. Each city/regency in Sarbagita Metropolitan Area has its own PDAM that covers its own area. In accordance with the data from BPPSPAM (2014), service coverage of PDAM Tabanan Regency shows the best performance among other PDAM(s) in this metropolitan area, that is 89.8%. Coverage of other PDAMs were also good (Badung Regency: 81.6% and Gianyar Regency: 69.9%), except PDAM Denpasar City (43.3%).

The Chi-Square test is used to determine whether both of location and spatial variation affect or does not affect the water supply provision in Sarbagita Metropolitan Area. If it does affect then we would be able to know in which part it differs according to the information from those following Crosstabs presented. On the basis of the results of Chi-Square test and Crosstab presented in Table 2, 3, 4, 5, 6, 7, 8 and 9, the following interpretations are made.

3.3. Water Provision based on Spatial Variation

The results of Chi Square test reveal that at a confidence level of 95%, there is correlation between water provision and spatial variation in the context of drinking water source (see Table 2.). Nevertheless, it does not show the same result for the test performed on the context of daily safe water source; correlation cannot be found between these two variables (see Table 3.). Some of the reasons of these results can be explained by Crosstabs presented in Table 4 and Table 5 as follows

Table 2. Chi-Square test of drinking water sources and spatial variation in Sarbagita Metropolitan Area

	Value	Df	Asymp. Sig. (2-sided)
Pearson Chi-Square	16.690a	4	.002

a. 2 cells (20.0%) have expected count less than 5. The minimum expected count is 1.76.

Table 3. Chi-Square test of daily safe water sources and spatial variation in Sarbagita Metropolitan Area

	Value	df	Asymp. Sig. (2-sided)
Pearson Chi-Square	4.440a	4	.350

a. 6 cells (60.0%) have expected count less than 5. The minimum expected count is .44.

Correlation between these two variables in the context of drinking water source can be based on two reasons. First of all, as we can see in Table 4., the use of bottled water for drinking is more common in the core area rather than in the peri-urban area. On the contrary, well pump and wellspring are widely used in the peri-urban area due to its typology of geographical condition of the peri-urban of Sarbagita Metropolitan Area that is still characterized by natural condition, so that many villages are still located near to wellspring and water depth in the well is still adequate to be utilized.

In contrast to previous findings, significant differences of water sources used between core and peri-urban area in the context of daily safe water cannot be found distinctly as shown in Table 5. However, piped water with water meter in these two cases (drinking water and daily safe water) is still the main source

Table 4. Crosstab of drinking water sources and spatial variation in Sarbagita Metropolitan Area

		Bottled Water	Piped Water with Water Meter	Well Pump	Well	Wellspring	Total
Spatial Variation	Core Area	27 (36.0%)	31 (41.3%)	12 (16.0%)	2 (2.7%)	3 (4.0%)	75 (100.0%)
	Peri-urban Area	13 (13.7%)	42 (44.2%)	21 (22.1%)	2 (2.1%)	17 (17.9%)	95 (100.0%)

Table 5. Crosstab of daily safe water sources and spatial variation in Sarbagita Metropolitan Area

		Piped Water with Water Meter	Well Pump	Well	Wellspring	Ground Water	Total
Spatial Variation	Core Area	52 (69.3%)	21 (28.0%)	2 (2.7%)	0 (.0%)	0 (.0%)	75 (100%)
	Peri-urban Area	61 (64.2%)	25 (26.3%)	4 (4.2%)	1 (1.1%)	4 (4.2%)	95 100.0%)

3.4 Water Provision in Sarbagita Metropolitan Area

The same method analysis is also applied to identify correlation between variables of water provision and location. In this term, location is differed based on coastal and non-coastal area. In the context of drinking water source, location considerably affects the types of source used for drinking water shown by the statistical result of Chi-Square test presented in Table 6. Conversely, as described in Table 7., we can assume that no correlation can be found between these two variables in the context of daily safe water in this metropolitan area.

Table 6. Chi-Square test of drinking water sources and location in Sarbagita Metropolitan Area

	Value	df	Asymp. Sig. (2-sided)
Pearson Chi-Square	33.748a	4	.000

a. 3 cells (30.0%) have expected count less than 5. The minimum expected count is .94.

Table 7. Chi-Square test of daily safe water sources and location in Sarbagita Metropolitan Area

	Value	df	Asymp. Sig. (2-sided)
Pearson Chi-Square	9.917a	4	.042

a. 6 cells (60.0%) have expected count less than 5. The minimum expected count is .24.

Correlation that occurs between variables of water provision and location in the case of drinking water source can be seen by the pattern of water source mostly used by villages in coastal area (see Table 8.). Villages in the coastal area mostly acquire drinking water from bottled water where piped water with water meter from PDAM is rarely found for drinking water source in these villages. In comparison to the coastal area, villages in non-coastal area mostly used piped water with water meter from PDAM (51.5%) followed by the source from well pump (20.8%) for their drinking water needs.

Even though it is clearly known that for drinking water, there is noticeable difference use of water source between coastal and non-coastal area, yet that matter does not apply in the case of daily safe water. As the matter of fact, there is no major distinction of the use of water source for daily safe water which is described in Table 9. Piped water with water meter followed by well pump is the most widely used water source for daily basis both in coastal and non-coastal area.

Table 8. Crosstab of drinking water sources and location in Sarbagita Metropolitan Area

		Bottled Water	Piped Water with Water Meter	Well Pump	Well	Wellspring	Total
Location	Coastal Area	22 (55.0%)	6 (15.0%)	6 (15.0%)	0 (.0%)	6 (15.0%)	40 (100.0%)
	Non-Coastal Area	18 (13.8%)	67 (51.5%)	27 (20.8%)	4 (3.1%)	14 (10.8%)	130 (100.0%)

Table 9. Crosstab of daily safe water sources and location in Sarbagita Metropolitan Area

		Piped Water with Water Meter	Well Pump	Well	Wellspring	Ground Water	Total
Location	Coastal Area	21 (52.5%)	18 (45.0%)	0 (.0%)	0 (.0%)	1 (2.5%)	40 (100.0%)
	Non-Coastal Area	92 (70.8%)	28 (21.5%)	6 (4.6%)	1 (.8%)	3 (2.3%)	130 (100.0%)

3.5 *Water Provision in* the Coastal and Peri-urban of Sarbagita Metropolitan Are

Turning into more specific case of water provision in the coastal and peri-urban of Sarbagita Metropolitan area, as illustrated in Fig. 3, it can be known that from 25 villages located exactly in the coastal area and classified into peri-urban area, bottled water (37.5%) is the main source for these villages to fulfill their needs of drinking water. From this is also concluded that piped water with water meter from PDAM is the least option for drinking water (16.7%).

As comparison, this piped water with water meter (54.2%) is reversely being the main water source for daily safe water in these villages followed by well pump (41.7%). These matters indicate that drinking water is more vulnerable and cannot be acquire carelessly in the coastal area due to the fact that its geographical condition affects scarcity of drinking water as freshwater aquifers are not available at suitable depths and surface water is highly saline. Households' choice of water source is complex and seasonally dependent (Islam, et al., 2013)

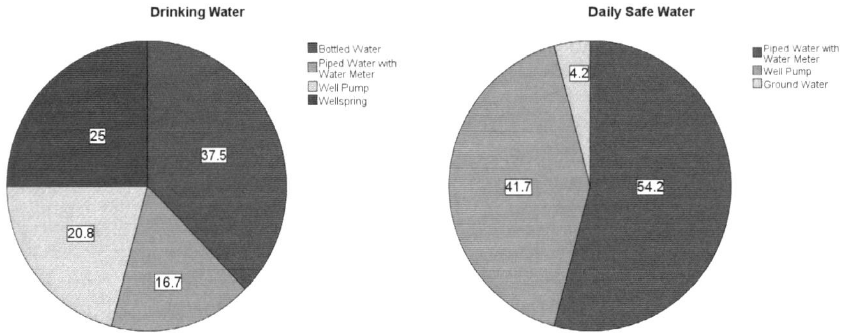

Fig. 3. (a) Percentage of drinking water sources in the Coastal and Peri-urban of Sarbagita Metropolitan Area; (b) Percentage of daily safe water sources in the Coastal and Peri-urban of Sarbagita Metropolitan Area

4. Conclusions

In general, piped water with water meter provided from PDAM is the main water source in Sarbagita Metropolitan Area for both of drinking water and daily safe water with the best performance of service coverage is presented by PDAM Tabanan Regency (89.8%) and the least performs by PDAM Denpasar City (43.3%). To sum up briefly, the findings show that there is noticeable correlation between variables of water supply provision and location (coastal and non-coastal area) as well as spatial variation (core and peri-urban area) in the context water source specifically for drinking water. Bottled water is being the main source for households located in coastal and peri-urban villages whereas piped water with water meter is rarely used for drinking water (least option).

In sharp contrast, this kind of water source (piped water with water meter) is widely used both in core and peri-urban area and even in coastal and non-coastal area if only it is used for daily safe water. From these results, it can be concluded that drinking water especially in coastal area is needed to be prioritized in the context of drinking water supply provision due to its geographical condition that affects to its high salinity. Therefore, some policies, strategies, and programs in providing adequate service of drinking water in coastal area should be formulated and implemented.

Acknowledgements

The research, on which this article is based, is funded by the Ministry of Research, Technology and Higher Education of Indonesia under the scheme of 'Penelitian Strategis Nasional 2016'. The title of the research is 'Kebijakan Penyediaan Infrastruktur Dasar bagi Masyarakat Miskin di Wilayah Peri-Urban (Tahun Kedua)'. The authors would like to thank all the stakeholders involved in this program. However, the authors alone are responsible for any mistakes and shortcomings.

Reference

[1] Chuai, X., Huang, X., Wu, C., Li, J., Lu,Q., Qi, X., Zhang, M., Zuo,T., Lu, J 2016 Land use and ecosystem services value changes and ecological land management in coastal Jiangsu, China. *Habitat International Vol 57*, p.164-174
[2] Al-Shihri, F.S. 2016 Impacts of large-scale residential projects on urban sustainability in Dammam Metropolitan Area, Saudi Arabia, *Habitat International, Vol 56*, p.201-2011
[3] BPPSPAM: Badan Pendukung Pengembangan Sistem Penyediaan Air Minum (2014). *Kinerja PDAM Tahun 2014 Wilayah IV Pulau Bali, NTB dan NTT*. Jakarta: Kementerian Pekerjaan Umum dan Perumahan Rakyat.
[4] Central Bureau of Statistics, Indonesia 2014 *Village Unit Potensial (PODES-Potensi Desa) 2014.*

IOP Conf. Series: Earth and Environmental Science **79** (2017) 012022 doi:10.1088/1755-1315/79/1/012022

Emerging Issues for Bokori Island Development in Konawe Southeast Sulawesi

Ardy Maulidy Navastara
Urban and Regional Planning Department,
Institut Teknologi Sepuluh Nopember,
Surabaya, Indonesia

Email: ardy.navastara@urplan.its.ac.id

Abstract. This article discusses about emerging issues for small island development especially Bokori Island in Konawe Southeast Sulawesi. Indonesia have been known as a big archipelago has more than ten thousand small islands scattered about 5.8 million km2 of oceanic area Sabang in the West end to Merauke at the end East. It is also known as the largest "megabiodiversity" marine nation. However, the weak national policy that handles the development and management of small islands is suspected because it is not accurate in identifying strategic issues that occur in the development and management of small islands. Therefore, this article aims to identify and classify strategic issues related to the development of small islands in Konawe Southeast Sulawesi. A normative and empirical approach is made to discuss this article. It resulted that typology of issues such as internal and external issues, urgent issues and its supporting issues. Then local governments should pay attention to the urgent issues and its supporting issues – it is policy, investment, institutional and technological issues and the endogenous issues such as entrepreneurship and leadership.

1. Introduction

The Government of Indonesia in the Era of President Joko Widodo has placed a vision that Indonesia's future lies in the Sea. So, this government policy is directed at maritime development. As in the following quotation when he was sworn in as president of the republic of Indonesia:

We also want to be present among nations with honour, with dignity, with dignity. We want to be a nation that can make its own civilization. A great creative nation that can contribute to the nobility of global civilization, we must work hard to return Indonesia as a maritime country. Ocean, sea, strait and bay are the future of our civilization. We have too long to turn our backs on the ocean, back to the ocean, back to the strait and bay. Now is the time for us to return everything so that Jalesveva Jayamahe, in the Sea we are victorious, as the motto of our ancestors in the past, can be re-reverberation ... (Speech of President Joko Widodo at Jakarta, 20 October 2014).

In fact, Indonesia's wealth of small islands is potentially very unusual to be developed. Indonesia is an archipelagic country with more than ten thousand small islands scattered about 5.8 million km2 of oceanic area (about 75% of Indonesian territory) from Sabang in the West end to Merauke at the end East, known as the largest "megabiodiversity" marine nation in the world [1]. This fact is easy to understand because with thousands of small islands that make up the archipelago of Nusantara, Indonesia has all the productive tropical marine ecosystems ranging from mangrove ecosystems, sea grass, seaweed to coral reef ecosystems.

With all the productive tropical marine ecosystems dangling around these small islands, aside from being the largest "megabiodiversity" country, Indonesia also has a huge wealth of marine resources as a national asset. But it cannot be denied that the enormous wealth of the ocean is contained in the vastness of the archipelago of the archipelago, apparently on the one hand has not been fully optimized and on the other side is suffered considerable damage. However, the weak national policy that handles the development and management of small islands is suspected because it is not accurate in identifying strategic issues that occur in the development and management of small islands.

The development of small islands in Indonesia is basically one of obligation that should be given special attention by the local government. This is very important considering the potential and challenges of islands owned by Indonesia can help to improve the country's economy. The government policy oriented to the development of small island states in the form of providing information and supporting facilities related to the attractiveness of the islands. The availability of these two things becomes the initial capital that local governments can offer to domestic and foreign tourists.

Each region in Indonesia has its own characteristics that become the attraction to be developed into a tourist area. Southeast Sulawesi is one of Indonesia's provinces and consists of peninsulas and islands with abundant natural resource potential. Some of the natural attractions that are owned include the Buton Fortress, Wakatobi National Park, Moramo Waterfall, and much more.

Some islands in Southeast Sulawesi province have beautiful landscapes and very potential to be developed as a tourist destination, such as Bokori Island. Bokori Island is located near the village of Bajo Tribe off the coast of Kendari. The island has beaches and marine parks that are habitat for Maleo birds. Some of the activities that can be done on the island include: view the habitat of rare animals closely, enjoy the view of the beach and white sand, diving, touring the islands and so forth. Archipelago tour activities can be done because Bokori Island has a close proximity of two surrounding islands, namely Pulau Hari and Pulau Saponda Laut. Access road to this island can be reached from the village of Mekar which is rarely very close to using outboard engine boat. Other tourism activities that have the potential to be developed in this island are activities based entertainment, culinary and art. Provision of accommodation is also very likely to be developed as a supporter of tourism activities, which can be a floating tourist resort. This tourism development approach in Bokori Island is based on sustainable tourism. The complexity of the issues that occur with regard to the development and management of the island of Bokori, demands an accuracy in reading and determining issues and issues map.

Therefore, this article aims to identify and classify strategic issues related to the development of small islands in Konawe Southeast Sulawesi. A normative and empirical approach is made to discuss this article. So, the results that can be explained in this article is the mapping of problems and issues that the better. Please be consideration for local government to put it within the framework of policy formulation.

2. General Information and Characteristics of Bokori Island

2.1. Locus of Bokori Island

The island has an area of ± 8 Ha is administratively located in Soropia, Konawe at coordinates 3o56'29.405 "LS and 122o39'51.686" BT (database of the Ministry of Maritime Affairs and Fisheries). Around the island there are Bokori, Toronipa Beach and Mekar Village (Bajo Indah) located on the shoreline of Regency. Physically, this island is close to the city of Chennai. Distance islands with the nearest port in Kendari (Port Festival) is about ± 10 km. Limit physical Bokori Island region was as

a. North: the Banda Sea and Konawe
b. East: Banda Sea
c. South: Banda Sea and Kendari
d. West: Banda Sea and Kendari

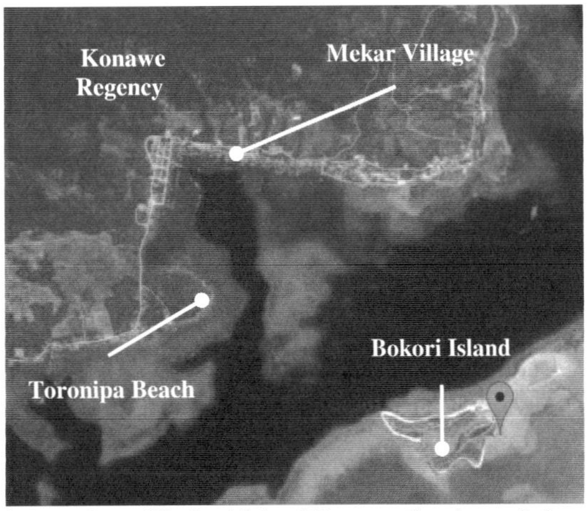

Figure 1. Location of Bokori Island Konawe Southeast Sulawesi

In the Medium-Term Development Plan (RPJMD) Southeast Sulawesi province in 2013-2018 explained that characteristics of Southeast Sulawesi province dominated by the slope 0-40% of approximately 72.23 percent of the land area and water area of 1,114.479 km², has the potential to be developed as cultivation areas, such as fisheries, agriculture, tourism, industry, mining and others by observing the spatial plan. One of the potential that can be developed in the province of Southeast Sulawesi is the tourism sector. With the potential of natural and cultural tourism, the tourism sector is considered as the most potential sector to encourage and accelerate the development of the region in the Southeast.

Based on observation satellite image map series for 3 years (2012-2014), some areas of island shrank significantly. There are three critical points on the northeast side, namely on land connecting the island of A to C, C Island, and the northern tip of the island B. Depreciation coastal erosion caused by the scraping edge of the island to the mainland.

In addition, the suitability of coastal areas is also influenced by oceanographic data. In this case, the adapted from the Guidelines by the Ministry of Maritime Affairs and Fisheries in 2010. Based on these criteria and the results of the physical data compilation in the previous chapter, the island of Bokori meets the following criteria.

Figure 2. The Changing of Bokori Island Period 2012 -2104
Source: Satellite Imagery, 2015

Table 1. Coastal Suitability of Bokori Island

No.	Variables	Suitability Criteria			Existing Conditions
		Good	Moderate	Poor	
1	Slopes (meter)	Ramp (0-8)	Hilly (8-15)	Steep (>15)	
2	Coastal Land Condition	White sand	Sand	Mud	
3	Height of wave (meter)	Low (<1)	Moderate (1-2)	High (>3)	Height of wave maximum= 0 - 1,25 m Significant Height of wave= 0 - 1,25 m Wind Speed= 0 - 5 knot
4	Flow Rate (m/s)	<0,1	0,1-1	>1	Flow Speed= 0 – 0,25 *m/s*
5	Brightness Seawater (meters)	90-100	80-89	<80	Clear water surface
6	The existence of coral reef diversity	Dense and diverse (75%)	Rarely and not diverse (40-75%)	Damaged (40%)	No Data
7	The existence of a special object	Yes and very Special	Yes and Fairly	Nothing	360° view of the sea around the island
8	Disclosure of coastal land	>400	100-400	<100	Island Area A: 26.817,26 m^2 Island Area B: 55.891,09 m^2 Island Area C: 1.046,11 m^2
9	Flood hazards	Nothing	1-2 times	>2 times	Within three months occur continuous tides. Within a week, the tides during the three days and five times high tides for a month
10	Weather changes	Rarely	Moderate	Often	

IOP Conf. Series: Earth and Environmental Science **79** (2017) 012022 doi:10.1088/1755-1315/79/1/012022

3. Exploring Issues of Bokori Island Development in Konawe Southeast Sulawesi

3.1. Mind Mapping the Issues of Bokori Island Development

In Mind mapping issues of Bokori Island development (**see Figure 3.**) showed the relationship of interrelated issues. These issues include those that develop into issues that will affect the development of Bokori Island. The complexity of the problem should be viewed as part of the regional development framework of Southeast Sulawesi. Therefore, a comprehensive approach is needed to understand the issues that are considered for developing Bokori Island. The results of this analysis show that sectoral issues such as the lack of infrastructure development, lack of technology development, local economic potential, cultural social threats, natural environment threats directly and indirectly related to institutional issues, local policies, national policies, Investment even on endogenous issues such as leadership and entrepreneurship.

When viewed from the issue of Bokori Island development efforts, the issues that directly potentially affect development are local government policies (provincial government of Southeast Sulawesi), national policy, local economic potential support, lack of infrastructure development, cultural social threats, natural environment threats as well as the lack of investment in tourism development. These issues can be said to be internal issues of Bokori Island development. While the external issues of development of Bokori Island include issues, whose ramifications are related to internal issues. Issues that include external issues; it is an ineffectiveness of institutions, lack of technological development, lack of entrepreneurship, leadership and local policy of the Konawe district government.

However, the issues of both internal and external issues, when viewed from the urgency of what should be understood by the government of Southeast Sulawesi to be considered as a priority are issues related to local policies of Southeast Sulawesi province and Konawe regency, the lack of investment in tourism development and ineffectiveness of institutions. These three issues require the attention of the government itself, especially in regard to the lack of technology development, entrepreneurship and leadership for local government.

Since Bokori Island began to be revived by the provincial government of Southeast Sulawesi, the surrounding communities especially in Mekar Village get many benefits from the results of motorboat rental used by tourists. Passenger fare to cross to Bokori Island is Rp 1 million (round-trip). Mekar Village residents are a place of relocation Bajo tribe who previously inhabited the island of Bokori. The relocation is done due to the condition of Bokori Island is increasingly unkempt. The condition of Bokori Island is also shrinking due to abrasion and processing of rocks and sand since the island is uninhabited (MySultra.com access 2017).

Since 2014, the provincial government of Southeast Sulawesi has budgeted Rp 1 billion to build tourist facilities in Bokori Island. Then in 2015 the government budgeted funds again about Rp 10 billion Bokori Island will be used as an integrated tourism object of Southeast Sulawesi province that will be connected with another adjacent island, namely Hari Island and Saponda Island. The three coastal tourist areas will be a unified tourist attraction named Bokori Bahteramas Island (BBM Island). In 2015, the government will build 30 international standard cottages on Bokori Island along with other supporting facilities (such as: transportation system and telecommunication network). The area will also begin to accumulate by one-meter high silica sand and planted mangrove trees around the coast, to prevent abrasion that has made the island's area shrink in the past 10 years. (MySultra.com access 2017).

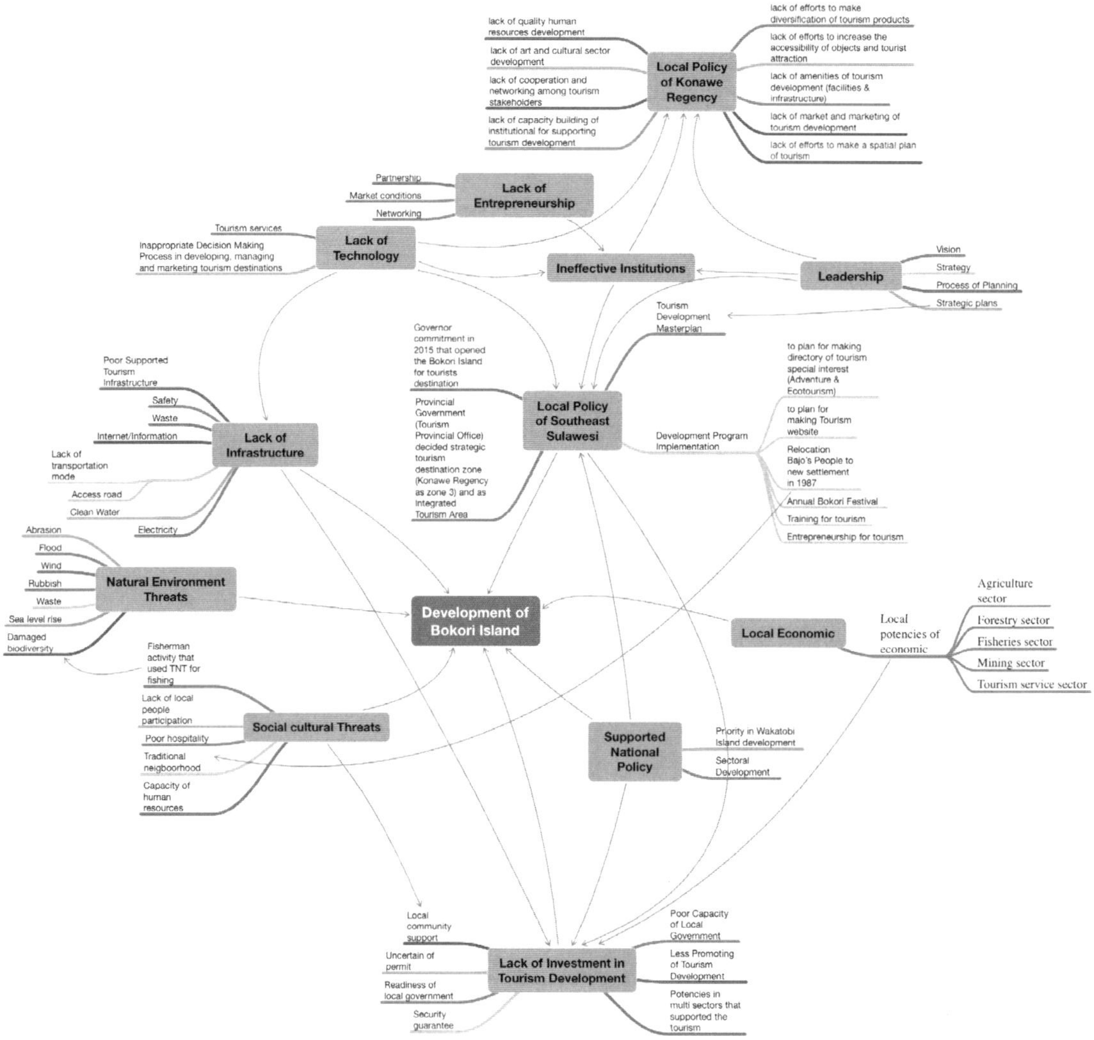

Figure 3. Mind Mapping due to The Issues of Bokori Island Development in Konawe Southeast Sulawesi

3.2. Strategic Issues in Developing of Bokori Island in Konawe Southeast Sulawesi

Furthermore, through mapping the issues are described in **Table 2.** explaining the typology of the issues and grouping strategic issues:

Table 2. Typology of Issues in Developing Bokori Island in Konawe Southeast Sulawesi

Typology of Issues	Strategic Issue	Some Relevant Issues
Internal	Local government policies (provincial government of Southeast Sulawesi),	1. Governor commitment in 2015 that opened the Booker Island for tourists' destination 2. Provincial Government (Tourism Provincial Office) decided strategic tourism destination zone (Konawe Regency as zone 3) and as Integrated Tourism Area 3. Development Program Implementation 4. Tourism Development Masterplan
	National policy	1. Priority in Wakatobi Island development 2. Sectoral Development
	Local economic potential support	Local potencies of economic (Agriculture, Forestry, Fisheries, Mining, Tourism service)
	Lack of infrastructure development	1. Poor Supported Tourism Infrastructure 2. Safety 3. Internet/Information 4. Lack of transportation mode 5. Access road 6. Clean water 7. Electricity
	Social cultural threats	1. Fisherman activity that used TNT for fishing 2. Lack of local people participation 3. Poor hospitality 4. Traditional neighbourhood 5. Capacity of human resources
	Natural environment threats	1. Abrasion 2. Flood 3. Wind 4. Rubbish 5. Waste 6. Sea level rise 7. Damaged biodiversity
	Lack of investment in tourism development	1. Poor Capacity of Local Government 2. Less Promoting of Tourism Development 3. Potencies in multi sectors that supported the tourism 4. Local community support 5. Uncertainty of permit 6. Readiness of local government Security guarantee
External	Ineffectiveness of institutions,	
	Lack of technology development,	1. Tourism services 2. Inappropriate Decision-Making Process in developing, managing and marketing tourism destinations
	Lack of entrepreneurship	1. Partnership 2. Market conditions

		3. Networking
	Leadership	1. Vision 2. Strategy 3. Process of Planning 4. Strategic plans
	Local government policy of Konawe Regency	1. Lack of efforts to make diversification of tourism products 2. Lack of efforts to increase the accessibility of objects and tourist attraction 3. lack of amenities of tourism development (facilities & infrastructure) 4. Lack of market and marketing of tourism development 5. Lack of efforts to make a spatial plan of tourism 6. Lack of quality human resources development 7. Lack of art and cultural sector development 8. Lack of cooperation and networking among tourism stakeholders 9. Lack of capacity building of institutional for supporting tourism development
Urgently Issues	Local Government Policy (Southeast Sulawesi Province and Konawe Regency)	**A. Southeast Sulawesi Province** 1. Governor commitment in 2015 that opened the Booker Island for tourists' destination 2. Provincial Government (Tourism Provincial Office) decided strategic tourism destination zone (Konawe Regency as zone 3) and as Integrated Tourism Area 3. Development Program Implementation 4. Tourism Development Masterplan **B. Konawe Regency** 1. Lack of efforts to make diversification of tourism products 2. Lack of efforts to increase the accessibility of objects and tourist attraction 3. lack of amenities of tourism development (facilities & infrastructure) 4. Lack of market and marketing of tourism development 5. Lack of efforts to make a spatial plan of tourism 6. Lack of quality human resources development 7. Lack of art and cultural sector development 8. Lack of cooperation and networking among tourism stakeholders 9. Lack of capacity building of institutional for supporting tourism development
	Lack of Investment in Tourism Development	1. Poor Capacity of Local Government 2. Less Promoting of Tourism Development 3. Potencies in multi sectors that supported the

IOP Conf. Series: Earth and Environmental Science **79** (2017) 012022 doi:10.1088/1755-1315/79/1/012022

		tourism 4. Local community support 5. Uncertainty of permit 6. Readiness of local government 7. Security guarantee
	Ineffectiveness of institutions	
Supporting Issues due to urgently issues	Lack of technology development	1. Tourism services 2. Inappropriate Decision-Making Process in developing, managing and marketing tourism destinations
	Entrepreneurship for local government	4. Partnership 5. Market conditions 6. Networking
	Leadership for local government	5. Vision 6. Strategy 7. Process of Planning 8. Strategic plans

4. Conclusion

This article concludes that the issues occurring in the development of tourism on Bokori Island can be mapped in the typology of strategic issues covering internal and external issues, urgent issues to be noticed by local governments, both province and regency and the issues being considered to be concerned about issues that are urgent.

In addition to internal and external issues, local governments should pay attention to the urgent issues and its supporting issues – it is policy, investment, institutional and technological issues and the endogenous issues such as entrepreneurship and leadership.

References

[1] Bengen D.Gi., Retrabun A.S.W, Saad S. 2012. *Menguak Realitas dan Urgensi Pengelolaan Berbasis Eko-Sosio Sistem Pulau-pulau Kecil.* Pusat Pusat Pembelajaran dan Pengembangan Pesisir dan Laut (P4L), Bogor, Indonesia.

[2] LPPM ITS Final Report. 2015. *Kajian Pengembangan Investasi Kawasan Wisata Terpadu "Bokori Bahteramas Island" Kabupaten Konawe Provinsi Sulawesi Tenggara*, Indonesia.

[3] RPJMD Southeast Sulawesi province in 2013-2018

[4] UNEP 2014. *Emerging issues for Small Island Developing States.* Results of the UNEP Foresight Process. United Nations Environment Programme (UNEP), Nairobi, Kenya.

IOP Conf. Series: Earth and Environmental Science **79** (2017) 012023 doi:10.1088/1755-1315/79/1/012023

The Sea is Not an Empty Space

P R Satiawan
Urban and Regional Planning Department, Institut Teknologi Sepuluh Nopember, Keputih, Sukolilo, Surabaya 60111, Indonesia

E-mail: puturudy@yahoo.com

Abstract. This research departs from a unique phenomenon, where Indonesian local people, fisherman, in Sapeken sub-district, Sumenep regency, together are able to divide the sea space based on the existence of resources in it with local intelligence inherited from generation to generation. Fisherman in Sapeken sub-district have the spatial intelligence to identify the existence of a sea space called premises "Takat" and "Timpusu". Those spaces however cannot be seen with the eye and merely as an empty space, until the community giving meaning and signing to the space and create the strong perceptual map that prove that the sea is not an empty space. In order to have such a rigorous finding, the researcher using phenomenology methodology that allowed researcher to explore more about the phenomenon.

1. Introduction

Sea, a space on earth that commonly referred as a maritime space is a space in the context of patterns, as the locus of maritime activities, such as resource-rich locations, sailing, cultivation, and marine resource extraction. Indonesia as an archipelago nation which has lots of maritime space, avouch the government to formulate a decision in developing a maritime management in order to utilize existing marine resources. For the government, the sea is a space that should be utilized as much as possible for the sake of Indonesian nation prosperity. But unlike the mainland space, the sea cannot be permanently occupied, fenced or absolutely controlled. Yet, the sea can only be controlled for a limited time. The government's perspective to maritime space leads us to a conclusion that the sea is an integral part of an indivisible territory of the country, but can be distinguished according to the legal regime that governs it.

This research departs from a unique phenomenon, where Indonesian local people, fisherman, in Sapeken sub-district, Sumenep regency, together are able to divide the sea space based on the existence of resources in it with local intelligence inherited from generation to generation. Sapeken sub-district has the most island, which is 42% of the island from a total island that Sumenep Regency has. Sapeken Sub-district located in the east of Sumenep regency and administratively consists of 9 villages: Sabuntan, Paliat, Sapeken, Saseel, Sepanjang, Tanjungkiaok, Pangerungan Kecil, Pangerungan Besar, and Sakala. One of the interesting things in this district is the unique configuration area where small islands stretching around the sea area by forming the letter C. In the middle, there is a group of islands of Saredeng Besar and Saredeng Kecil which is administratively part of Saseel village.

Fisherman in Sapeken sub-district have the spatial intelligence to identify the existence of a sea space called premises "Takat" and "Timpusu". "Takat" is a sea space formed from corals with shallow depths (1-5 meters), and "Timpusu" is formed from corals with depths between 10-15 meters. In relation to the existence of "Takat" and "Timpusu" spaces, revealed that collective agreements enable fishing communities to regulate their management include the types of extractable resources, time or season of extraction, fishing gear used, and extraction methods.

In addition, Sapeken Sub-District is not only used for navigation activities as is commonly occurring in marine areas, but also utilized by the fishermen community for the actualization media in terms of transaction activity both economically and socially, among others commodity trade, diffusion and nursery of all aspects in maritime (Maritime values, maritime ethics, maritime culture, and local skills, etc.), as well as communication. It is interesting to recognize how the fishing communities in this region collectively agree to utilize, manage, interpret, and then formulate conceptions to then produce a spatial order in the region. Research on maritime space has not yet been revealed to reveal a spatial order not only expressed textually but also spatially. Therefore, the researcher makes the research question as "How maritime space construction in Sapeken District of Sumenep regency according to the fisherman perspective?". This study will provide a sea space from the local community's view of fisherman which shows that the sea space is not an empty space, but space that can also be spatially mapped as well as the land.

2. Methods

This research uses phenomenology approach as it is congruence with ontology, epistemology and methodology of research's assumptions. This research conceives ontology realities that shaped by frame of mind construction of community's local and spesific experiences. In epistemology way, researcher has a reciprocal influence with the research object. Hermeneutik and dialecthic method are chosen as the methodology character of this research where the nature of variable and social construction will only can be created by direct communication between researcher and respondents. Therefore, the results of this research are spesific and locally accepted.

Qualitative research method was chosen as the nature of this research is to explore local phenomenon such as the classification of maritime space, takat and timpusu. A qualitative research necessitates the researcher to think inductively by observing phenomenons henceforward analysing and theorizing the observations into concepts and knowledge (Bungin, 2007). This research used inductive paradigm approach to generate answer from research questions. Inductive paradigm of this research is showed in fig. 1.

This research is located in three villages, which are Sassel Village, Sepanjang Village, and Tanjungkiaok Village. The objects of this research are maritime space, fisherman community, and fishing activities where emphasizing on space configuration constructed by cognitive perception of fisherman community.

Information were collected by conducting interview, observation, visualization record, and measurement. The interview was conducted purposively with key respondents, while observation was conducted in order to understand the characteristic of maritime space. Visualization record and measurement were conducted in order to capture the phenomenon thoroughly. The phases of this research also showed clearly in fig. 1

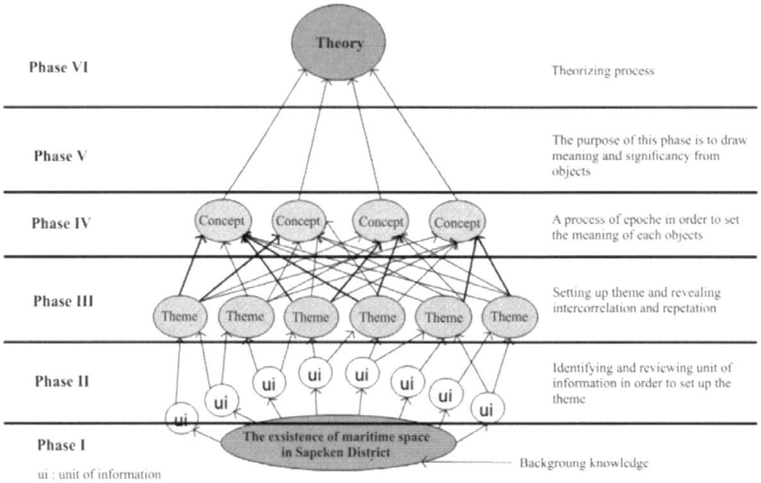

Figure 1. Research Phase
Source: Satiawan, 2017

3. Result and Discussion

3.1 The maritime profile of Spaken Island

The research object is located in Sapeken District where is an archipelago (Figure 2). The total population of Sapeken district in 2012 is 43.455 residents which the population is dominate by female residents with 22.323 residents. Most of the residents of Sapeken district work as fisherman and lived within several communities. This phenomenon can be seen from seperated settlement in different islands. The vernacular architecture such as *rumah panggung* is still exist in Sapeken district. Although for some more prosperous residents are not living in *rumah panggung* anymore.

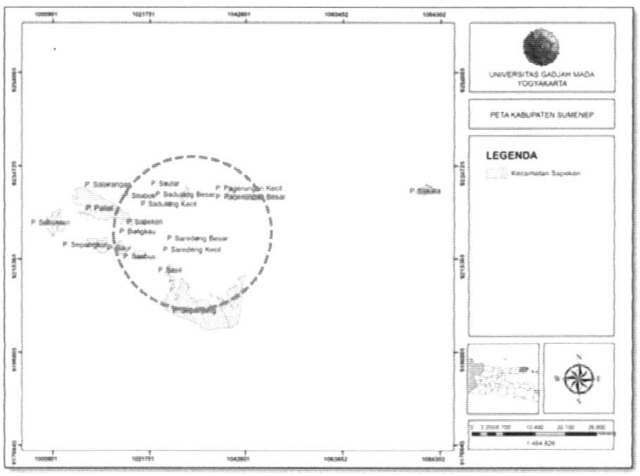

Figure 2. The archipelago of Sapeken district
Source: Satiawan, 2017

The electricity and clean water in Sapeken district is not distributed fairly, only the centre of district that has provided with electricity and clean water. The electricity service is provided individually by using diesel fuel power electricity that starts from 06.00 pm until 05.00 am. Clean water is a big problem in Sapeken district as most of the area still have limited resources of clean water. The only clean water resources located in Saseel island that has become resource for other islands.

3.2 The phenomenon on those maritime space
The initial object within territory observation is the existence of *takat* which is coral reef in terms of physical-geographical, both of which have very shallow depths and extreme depth of approximately 10-20 meters below sea level. In line with its depth, fishing communities in the observed territory tend to categorize in two species of TAKAT. First, which are formed from corals with very shallow to shallow depth (1-5 meters), and second is TIMPUSU which are formed from corals with depths between 10-15 meters. Characteristics of coral depths have implications for the existence of differences in abundance of marine resources possessed, both in type, quantity, and season/time availability (temporal existence). The existence of the takat (and timpusu) in the observed territory is believed by the fishermen communities who inhabit small islands as a place that has strategic value because of its abundance of marine resources. It provides hope and opportunities for fishermen for their life support both in present and in the future by exploit it. Visually identification of takat and timpusu almost have no difference, as shown in Figure 3.

Figure 3. Visual Identification

In the case of marine resources utilization and management, there is a collective agreement where the scheme of utilization is arranged as follows:

1) Utilization of marine resources can only be done in certain seasons by considering the pattern of expansion and migration of marine resources. This provision is set differently for all *takat / timpusu* and types of resources therein. The existence of marine resources and their rotation presence is different between each *takat* and *timpusu* or different among types of resources, because of geographic distinction.
2) Utilization of marine resources can only be done by using certain instruments through consensus process. The types of instruments that can be used for resources utilization in the observed territory is only fishing lines. Fishermen have agreed not to use other instruments such as nets or trawlers.

Life as a fisherman in certain generation brings a loaded of knowledge and understanding to the nature and maritime environment. One of fishermen community capabilities that can be said as one of the local intelligence is the knowledge and understanding of the character of the resource itself, for instance about the availability of natural resources dynamically. In the observation area, another intelligence that needs to be revealed is the ability of fisherman communities to recognize and understand a location, or often termed as spatial intelligence. In aquatic spaces, spatial objects are always dynamic spaces and influenced by natural and environmental process, for example weather and climate. Various findings of spatial intelligence phenomenon by fishermen community in the observed area such as their ability in determining and detecting a location above the water space. From field observations, it was found that the instruments and methods that is used to identify, determine and detect a location in water space include:

1) Wind speed and direction
2) The colour of sea water in various weather and climate
3) The speed of sea currents
4) Wave height
5) Wave character. In general, the wave character is different from wave height. Wave character is the regularity pattern of the wave existence.
6) Detection of sound that arise from seabed (coral), it use to determine the depth of takat / timpusu by putting the ear to the bottom of the boat
7) The use of *Dwi-Referensial* through the use of two hands to create an angle between two hands or to identify the distance between two hands (in *depa* unit). Some objects that is used as reference are islands, hills, highlands, mangrove plants in groups, high buildings (eg BTS towers, light beacons, lighthouse, flames, etc.).
8) Astrological sign, especially at night when the weather is good.

The local spatial intelligence that describe above affects a systemic and intact intelligence. One of manifestation is the ability of fishermen t produce the "map" which is a representation of spatial intelligence. The generated maps are very different from the formal maps geographically that are reconstructed from the formal system. As a result, they have lack of ability to recognize and operate maps that is generated from formal processes. They are only able to recognize and operate the "local" maps compiled by themselves.

3.3 Sapeken Maritime Space Construction (The sea is not an empty space)

There are several interesting activities happen besides the cultivation activities within the maritime space. These activities are economic transaction and social interaction.

Figure 4. Perceptual map of maritime space (left)
and activities in maritime space of Sapeken district (right)
Source: Satiawan, 2017

A. Economic transaction

The catchment of marine resources and seaweed were sold by fishermen directly within the maritime space where basically it was sold in the mainland market. The transaction occurs between fishermen and sellers, either between fishermen and buyers. The reason of this activity is that they can save more time compared if they have done the transaction on the mainland. This activity usually occurs in the cruise lines.

B. Social interaction

Social interaction happens between fishermen and their families. This research found that maritime space is used for maritime educational purpose. The parents will teach their son about maritime activities knowledge such as marine resource cultivation, how to fix the boat, loading and unloading the dock, etc. The second finding is the phenomenon of local knowledge transfer that inherited to the young men in the family about maritime space navigation. This knowledge includes the wind speed and direction, tidal and current measurement, characteristic, etc. The third finding of the research the use of maritime space as a space to interact with other ethnics from different region that usually occur right after the economic transaction finished

4. Conclusion

From the study, we can conclude that the sea space is not an empty space. However, sea is a space that has lots of meanings and spaces for local communities / fisherman who can be considered to have a spatial intelligence to map those space. It also provides new knowledge that from fisherman perceptions of the various spaces of the sea can be expressed spatially, forming a new pattern of seaward. This study is also expected to be input in the marine space planning documents, that local people's perceptions should also be seen to develop a fairer public policy for all concerned

References

[1] Bungin B 2007 Qualitative Research: Communications, Economics, Public Policy and Other Social Studies. Kencana. Jakarta

[2] Satiawan P R 2017 Disertation draft. Unpublished. Gadjah Mada University.

IOP Conf. Series: Earth and Environmental Science **79** (2017) 012024 doi:10.1088/1755-1315/79/1/012024

Green Infrastructure Concept for JABODETABEKJUR Metropolitan Area

Gunawan Tanuwidjaja[1] **and Bill Gates Chang**[2]

[1] Architecture Program Study, Petra Christian University, Jl, Siwalankerto 121-131, Surabaya, Indonesia
[2] Student of Architecture Program Study, Petra Christian University, Jl, Siwalankerto 121-131, Surabaya, Indonesia

Abstract. Sixty "Mega Cities" would emerge by 2015 catering of 600 million populations, and were threatened by the climate change, because of cyclones, flooding, etc. Jakarta became a metro region covering Jakarta, Bogor, Tangerang, Bekasi, Depok and Cianjur. Jakarta metropolitan faced the very high population growth, urban sprawling, traffic jams, flooding, green open space reduction, environmental degradation, urban slums and illegal street hawkers. Flooding and traffic congestions were the two most important issues to solve.

SWOT analysis and urban design solutions were produced to create a sustainable solution. Related to transportation issues, Singapore Mass Rapid Transport (MRT) concept was evaluated. Meanwhile the Netherlands' polder concept as well as Singapore's Integrated Water Management were also analyzed.

The development of above ground MRT as well as Busway could be developed to connect Jakarta Metropolitan Region. The networks were developed on the main toll road networks. The MRT and Busway would eventually replace the need of automobile use in the future. The Transit – Oriented - Development (TOD) with high density can be suggested to be concentrated nearby the MRT and Busway interchange stations. The Netherlands' polder and were adopted for urban' low-lying lands in Jakarta Metropolitan Region, A polder system was defined as the Integrated Man-made Drainage System consisting Dikes, Drains, Retention Ponds, Outfall Structures or Pumping Stations. The polder system was proposed to be extended to Tangerang and Bekasi area.

1. Introduction

The world's population increased exponentially from 2.521 billion to 6.782 billion from 1950 to 2009. Furthermore, it was predicted that the population could reach 9 billion by 2040. The urbanisations rate also had increased the urban population from 30% in 1950 to 50% in 2007 from total population. The World's urban population was predicted to reach 60% in 2030. The unsustainable development phenomena, such as exponential urban population growth, rapid urbanisation, limited urban planning, less stringent development control, and land speculation, further caused urban sprawling in the "Mega Waterfront Cities." The Mega Cities with more than 5 million populations each city. Sixty "Mega Cities" would emerge by 2015 catering of 600 million populations, i.e. Singapore, Hong Kong, Jakarta, Mumbai, Bangkok and Manila (Schultz, 2006).

The Climate Change made the "Mega Cities" more vulnerable to cyclones, high winds, flooding, coastal erosion and deposition, and sea-level rise (Nicholls 1995, Rosenzweig & Solecki 2001). And combination of climate change and other aspects would cause the urban lowlands more susceptible to floods and tidal waves.

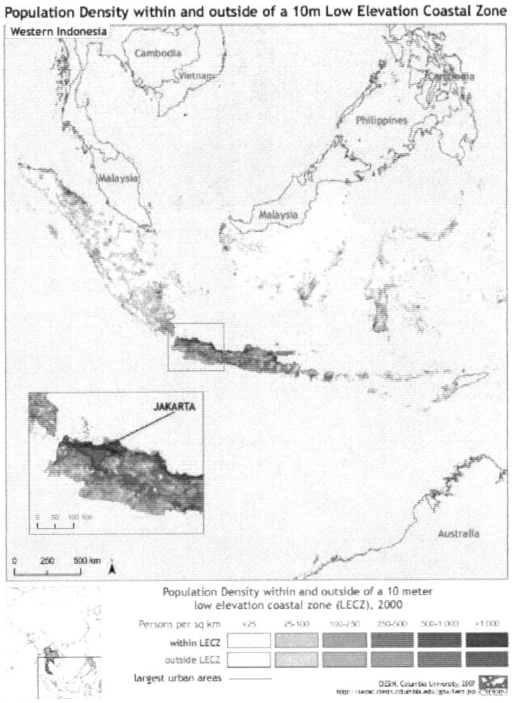

Figure 1. Locations of Mega Cities in the Coastal Areas of Indonesia.
Sources: CIESIN, Columbia University, 2007 http://sedac.ciesin.columbia.edu/gpw/lecz.jsp.

Jakarta as one of the Mega Waterfront Cities was developed unsustainably (Goldblum & Wong 2000). The growth and development of spatial Jakarta City was developed driven by the Economic Globalization. Jakarta suburb area development followed the pattern of development in Western countries. Zone - Industrial area sprung up in the Jakarta suburb of developments had shifted the center of downtown. While the region - the area that has a high price, or "prime lands" has been converted into Office, Retail and Shopping Complex. It was recognized that poor management and coordination of the implementation of ineffective planning led to poor quality of urban planning. And the Jakarta Development Model City was illustrated by Ford (1993).

Jakarta has evolved becoming a metro region covering Jakarta, Bogor, Tangerang, Bekasi, Depok, Puncak and Cianjur. The Jakarta Metro Region was referred as Jabodetabekpunjur region and formed based on Government Regulation (*PP*) No.26 Year 2008. This area also was identified as National Strategic Area (*Kawasan Strategis Nasional*) in the Indonesian National Spatial Plan (*Rencana Tata Ruang Wilayah Nasional*) due to the economic potential. And due to that Central Government should integrate the mega city development. In the Jakarta metro region, a gap between elite – planned - housings and unplanned - urban - village (kampung kota) existed. Jakarta metropolitan city faced similar issues of very high population growth, urban sprawling, traffic jams, flooding, green open space reduction, environmental degradation, urban slums and illegal street hawkers (Dardak & Poerwo, 2007).

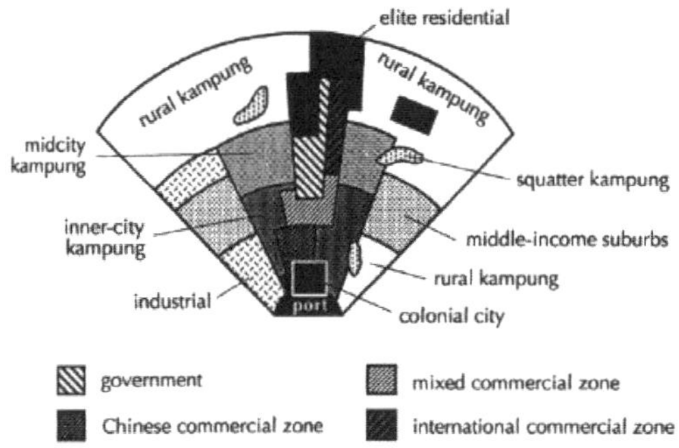

Figure 2. A Model City in Indonesia, like Jakarta Metropolitan City described by Ford (1993).

Figure 3. Development of Urbanization in the region Jabodetabekjur of the year 1972-2005, The red areas showed the urbanisation areas. Source: (Dardak & Poerwo, 2007)

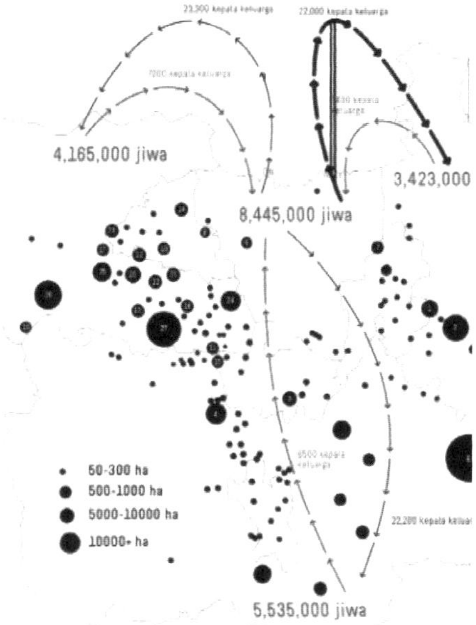

Figure 4. Urban Migration in Jabodetabekjur that showing Urban Sprawling. Source: Sutanudjaja & Arifin, (2011)

The issue of flooding and traffic were two important issues to discuss. Because of urban sprawling, chronic traffic congestions were found and wasting of Rp. 10 Billion per year fuel cost and Rp. 17.2 Billion vehicle operation cost per year (the city government (Sutanudjaja & Arifin, 2011).

Floods were the regular disasters for Jakarta City. In 1619, Jakarta or previously named Batavia was developed by Jan Pieters Z. Coen with waterfront city concept or urban polder concept following the city of Amsterdam. But unfortunately, Jakarta was inundated by major floods in 1621, 1654, and 1918. Further, major inundations recorded happened in 1976, 1997, 2002, and 2007.

The 1997's Jakarta floods occurred city-wide and created the national tragedy which attracting public even worldwide attentions. The flood covered the area of 4 sub-districts (Kelurahan), 745 houses, displacing 2640 persons with 80 cm water level. Further, the Jakarta flood got worse in 2002 affecting 60% of Jakarta. It also impacted Tangerang and Bekasi area. It was reported killing 142 persons and displacing 114,441 persons. Further, Rp 4 trillions economic losses were estimated caused by the flood.

In 2007, Jakarta flood affected more than 60% of Jakarta. It also affected Bogor, Depok, Tangerang and Bekasi. 80 persons were reported killed, while 340,000 persons had been displaced. Furthermore, 74,000 houses were submerged; 670,000 people lived without electricity; and 82.150 sq km of roads were damaged. Further The Director of Urban, Spatial Planning and Land Issues of National Planning

Board of Indonesia (BAPPENAS), Mr. Luky Eko Wuryanto, estimated that Rp 8 trillions economic losses were caused by the floods. The breakdown were: Rp 1,7 trillions of housings damage; Rp 600 billions of infrastructures damage; Rp 2 trillions of Industrial, Financial and Small Medium Enterprises (SMEs) sector losses; Rp 4,8 billions of institutions buildings damage; and Rp 3,6 billions other losses. This data actually illustrates the worsening floods in Jakarta, and increasing economic losses due to the flooding.

In general, Jakarta floods are caused of two groups of factors, which are: natural factors and human factors. They are:

1. Natural Factors
   - Approximately, 40% of Jakarta areas are low-lying land below the highest astronomical tide.
   - Flat topography of Jakarta (even concave in some area sop potential for flooding).
   - 40% of Jakarta are located within the flood prone areas of the main rivers.
   - High rain intensity in the upstream of Jakarta Metro Region.
   - Extensive catchment affecting Jakarta (850 sq km).
   - Limited water bodies capacity (drains, rivers, retention ponds and lakes).
   - Geological and soil types of the Jakarta Metro Region are susceptible to land subsidence and erosion.
2. Human Factors
   - Disintegration of Spatial Plan and Drainage Master Plan (macro and micro level) of Jakarta Metro Region.
   - Overpopulation in several areas (strategic areas and slums).
   - Extreme groundwater extraction.
   - Conversion of forests, wet agricultural lands
   - Conversion of water bodies (wetlands, agricultural lands, drains, rivers, retention ponds, lakes).
   - Unsustainable land development practice causing increasing soil sedimentation.
   - Solid waste and waste-water pollution in the water bodies.
   - The unclear system role-sharing among in environmental management especially in the drainage management.

Jakarta Spatial Plan (RTRW) of 2030 recommended several solutions such as (Planning Agency of Jakarta Special Province/ BAPPEDA DKI Jakarta, 2009) :

- Systems and transport infrastructure
- Regulation of water and flood control
- Green open space
- Global warming and energy limitations
- Economic and financial world crisis
- Socio-demographic pressure and urbanization (urban sprawl)
- Need of facilities and infrastructure provision such as: clean water, liquid and solid waste, telecommunications, energy
- The need of disaster mitigation
- Limited public funding

Because of similar finding by Jakarta Planning Agency, we proposed a conceptual infrastructure framework for sustainable development in Jakarta Metropolitan Region.

2. Methods

The research was conducted with SWOT analysis and urban design solutions. To create a sustainable solution, a literature review on Singapore as well as other cities were conducted. And the solution was discussed with the architects and proposed to the Jakarta's Metropolitan Urban Design Competition. The paper also proposed to disseminate the concept again.

3. Mass Rapid Transportation

For transportation issues, Singapore Mass Rapid Transport (MRT) concept was evaluated. Meanwhile the Netherlands' polder concept as well as Singapore's Integrated Water Management were also analysed for further implementation.

The MRT system in Singapore could answer the need of urban sprawling of the Jakarta Metropolitan Region. The Urban Redevelopment Authority (URA) proposed Master Plan Singapore-based mass transportation system MRT (Mass Rail Transit) in 1970-1980 and the integrated planning of the Housing Development Town board (HDB Town) which is a high density in various Locations in Singapore. (http://www.ura.gov.sg).

The development of above ground MRT as well as Busway could be developed to connect Jakarta Metropolitan Region. The networks were developed on the main toll road networks. The existing connectivity should be expanded to the suburban of Jakarta Metropolitan. The MRT and Busway would eventually replace the need of automobile use in the future. The Transit – Oriented - Development (TOD) with high density can be suggested to be concentrated nearby the MRT and Busway interchange stations. Additional drive - and - park structures could be proposed in several areas of the urban fringe to reduce the traffic to the city.

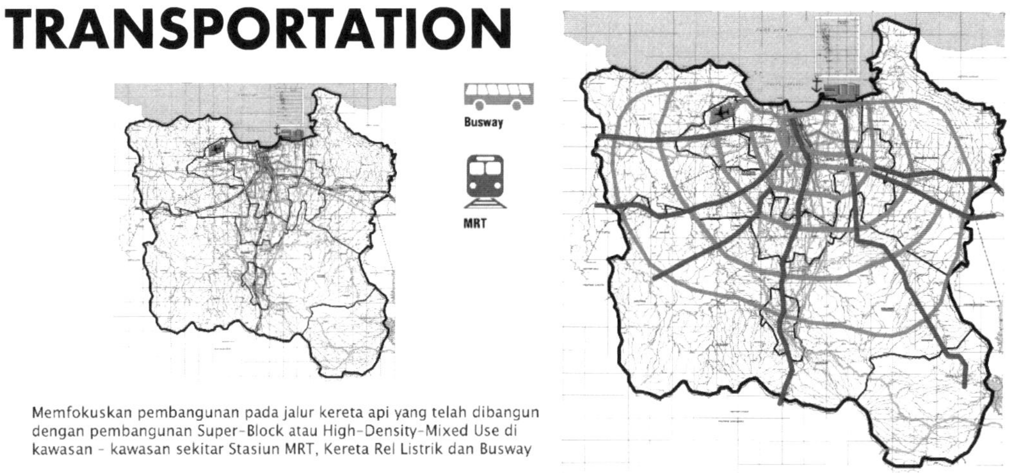

Figure 5. Proposed Networks (Above - the - Ground - MRT and Busway) were proposed to connect the Jakarta Metropolitan Regions. The Dark Purple Lines represented the MRT line on top of the toll road, while the light blue lines represented the Busway routes utilising the toll road.

Figure 6. Above - the - Ground – MRT line developed on top of toll road

Figure 7. Busway line developed on top of the toll road

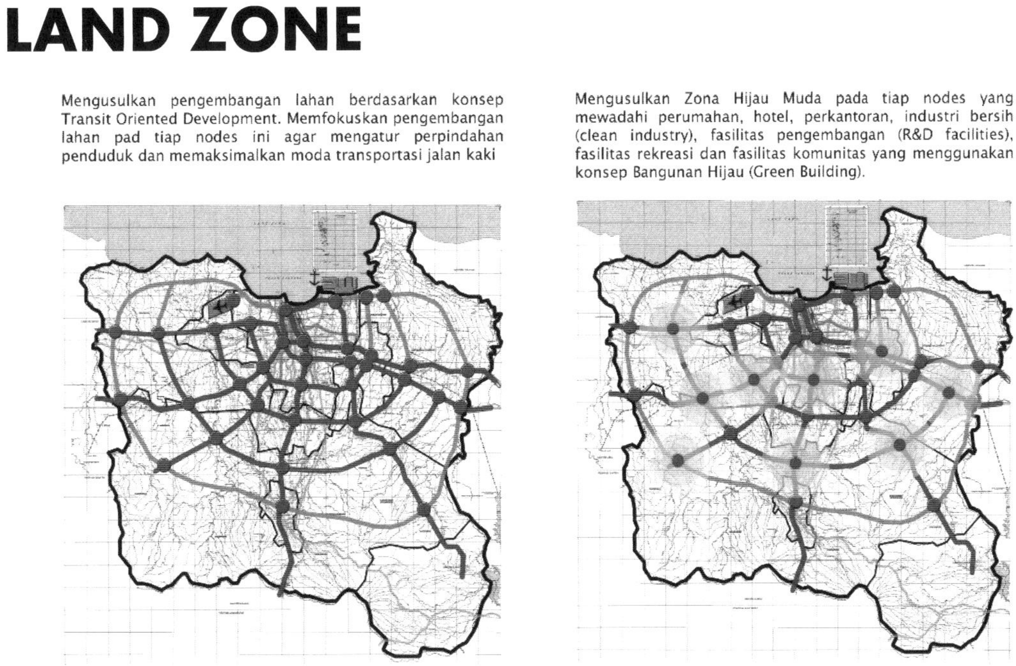

Figure 8. The Transit Oriented Developments adjacent to Proposed Networks Nodes (Above - the - Ground - MRT and Busway). The Nodes were the interchange stations of MRT and Busway.

Figure 9. The MRT and Busway interchange station

The Netherlands' Polder was also adopted for urban' low-lying lands in Jakarta Metropolitan Region, A polder system was defined as "An Integrated Man-made Drainage System consisting Dikes, Drains, Retention Ponds, Outfall Structures and Pumping Stations. Designed Landscape in the Polder is required to ensure its effectiveness. Polder Dikes should be planned and designed considering potential of soil strength, land subsidence, social conflict as well as accessibility issue. The maintenance of infrastructures becomes a critical point for successful polder operation. Lastly, the Polder system must be planned in integration from macro spatial plan, urban design and water management of the macro (river basin) system."

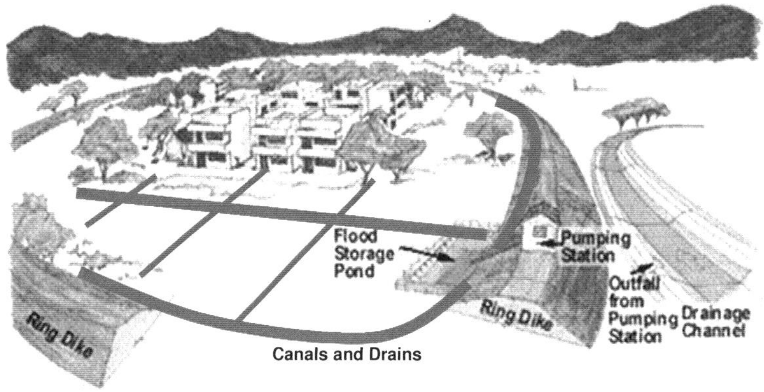

Figure 10. Illustration of Polder System. Sources: Ministry of Public Works et.all. Urban Polder Guideline

Because of the current polder system was not integrated, the polder system should be evaluated and expanded to Tangerang and Bekasi area. The polder system should be integrated with the current connectivity. And some roads could be utilised for drainages and dikes.

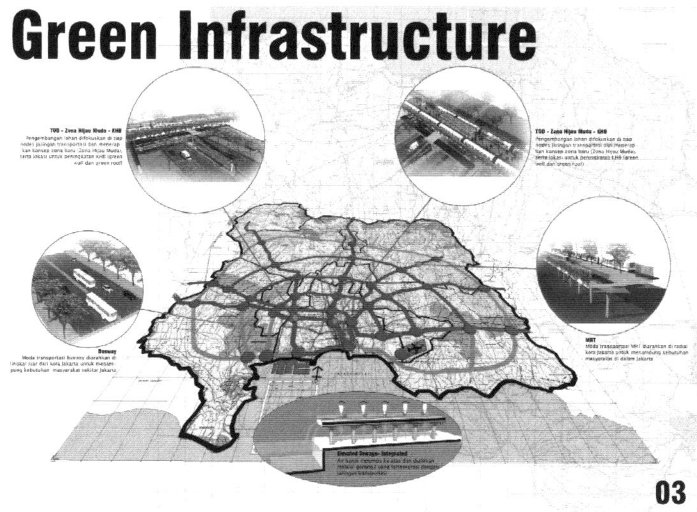

Figure 11. The Polder System on the Northern Area of Jakarta Metropolitan Region (sympbolised in the yellow area)

The research was conducted with SWOT analysis and urban design solutions. To create a sustainable solution, a literature review on Singapore as well as other cities were conducted. And the solution was discussed with the architects and proposed to the Jakarta's Metropolitan Urban Design Competition. The paper also proposed to disseminate the concept.

4. Conclusion

Jakarta metropolitan city faced three main issues of urban sprawling, traffic jams, flooding. The issue of flooding and traffic were two important issues to discuss. The development of above - ground - MRT and Busway networks could be developed to serve the Region. The networks were developed on the main toll road networks. The existing connectivity should be expanded to the suburban of Jakarta Metropolitan.

Meanwhile Polder system should be adopted for urban' low-lying lands in Jakarta Metropolitan Region. The new polder system should be expanded to Tangerang and Bekasi area with integration to the Metropolitan Drainage System. These systems could become the green infrastructure of the Greener Jakarta Metropolitan area.

References

[1] Suhaida M S, Chua K H and Leong Y P 2011 *Sustainable Development in the Building Sector Building Green Framework* in Malaysia 15th International Conference on ISO & TQM (Malaysia) 1–8. Retrieved from http://www.uniten.edu.my/newhome/uploaded/admin/research/centres/iepre/2012/papers/Sustainable Development in the Building Sector Green Building Framework in Malaysia.pdf

[2] Tan C S, Suhaida M S and Leong Y P 2011 *Sustainability Development through Energy Efficiency Initiatives* Malaysia 15th International Conference on ISO & TQM (Malaysia) Retrieved from http://www.uniten.edu.my/newhome/uploaded/admin/research/centres/iepre/2012/papers/Sustainability Development through Energy Efficiency Initiatives in Malaysia.pdf

[3] U.S. *Environmental Protection Agency, Basic Information Green Building.* Available from http://www.epa.gov/greenbuilding/pubs/about.htm_Accessed on 1 March 2013.

[4] Voss K, Musall E and Lichtmeß M 2011 *From Low-Energy to Net Zero-Energy Buildings: Status and Perspectives Green Building.* **6(1)** 46–57.

[5] Luis P L, José O, Rocío G and Ismael R M March 2009 *A review of benchmarking, rating and labelling concepts within the framework of building energy certification schemes Energy and Buildings.* 41(3) 272-278 ISSN 0378-7788.

[6] Andrew P, Arno K and Raymond L April 2012 A roadmap for navigating voluntary and mandated programs for building energy efficiency *Energy Policy.* 43 415-426 ISSN 0301-4215.

IOP Conf. Series: Earth and Environmental Science **79** (2017) 012025 doi:10.1088/1755-1315/79/1/012025

Sustainable Agro-Industrial Ecology Concept of the Madura Island

Joyce Martha Widjaya[1] **and Gunawan Tanuwidjaja**[2]

[1] *Research Institute of Policy and Technology Implementation (Pusat Penelitian dan Pengembangan Kebijakan dan Penerapan Teknologi). Jl. Raya Waru no 20, Surabaya Indonesia*
[2] *Architecture Program Study, Petra Christian University, Jl, Siwalankerto 121-131, Surabaya, Indonesia*

Abstract. Madura as one small island in East Java Province, Indonesia faced many challenges due to limited transportation connectivity, limited water resources and karst geology. Due to this reasons, the Government of Indonesia proposed a strategic plan to improve the development of the island to Surabaya, the largest port for the Eastern of Indonesia. It was started with building the Surabaya – Madura (SuraMadu) Bridge with 5.7 km length in 2003. The bridge was finally completed in 2009, improving the traffic flow into the island and development of Madura Island.
Unfortunately, the strategy would not be comprehensive without strategic development of the Madura Island, especially in Bangkalan District (Kabupaten Bangkalan). The Central Government has proposed a Green Industry with zero waste and clean energy concept. This industry and port would process the agriculture products from Madura for the export and Eastern part of Indonesia market. Therefore, an industrial ecology concept was needed to achieve the sustainable green industry for Eastern of Indonesia.

.

1. Introduction

Sustainable Coastal Development was crucial for developing countries like Indonesia. The Madura Island was found as one exotic island in the East Java, Indonesia. Comprising of 5'304 sq km area, the Island was located adjacent to Surabaya, the largest Port City of Eastern Indonesia. The Island became very important to Indonesia due to Industrial and Port expansion need from Surabaya which serving the Eastern part of Indonesia and the export market. And before 2009, the island faced development hindrance due to limited transportation connectivity, limited water resources and soil fertility. These phenomena were related to the karst ecology that dominated the Island.

Because of that, the Central Government initiated building the Surabaya – Madura (SuraMadu) Bridge witn 5.7 km in lengths connecting the Largest East Port City and the island. The Bridge was completed in 2009, serving the industrial traffic as well as domestic traffic between the areas. The bridge was designed with middle span width of as 424 m and height of 35 m allowing the sea transportation to travel under the bridge.

The impact of the bridge development would be less effective without integrated strategy for regional development in the Island. The strategy would comprise of the integrated development of Commercial in the Bridge Foot, the Bangkalan Integrated Industrial and Port area, as well as integration of Spatial Planning of 4 Districts (Kabupaten) in the Madura Island. And the paper was prepared for recommending "the Bangkalan Integrated Industrial Port Area"

The "green industry, zero waste and clean energy" concept were proposed for integrated industrial and port of Madura Island. The area was actually already proposed in the Bangkalan District (Kabupaten Bangkalan) directly connecting with the SuraMadu Bridge. Unfortunately, the

IOP Conf. Series: Earth and Environmental Science **79** (2017) 012025 doi:10.1088/1755-1315/79/1/012025

implementation of "Green Industry" had not been evaluated based on the carrying capacity of the karst ecology of Madura.

2. Methods

The research was conducted with SWOT analysis of Natural Resources and Socio-Economical Aspect of the Madura Island. The analysis was done based on the secondary data as well as primary survey with assistance of Research Institute of Policy and Technological Implementation of Public Works and People Housing Department, Central Government of Indonesia. This was conducted based on Adaptive Landscape Evaluation Tool that was tested in Bintan Island, Indonesia (Tanuwidjaja dan Malone - Lee, 2009). [1]

3. Biofuel Production

The increasing biofuel production, had created the devastations of tropical forests in Indonesia, especially Sumatera, Kalimantan, Sulawesi and Papua. For instance, the 64 million ha Indonesian tropical forests were deforested since 1950 for forests products. In addition, 1,871,500 ha of forests were slashed every year for agriculture development including the palm oil and other biofuel plantation. This actually had caused the extinctions of 147 species of mammals, 114 birds, 28 reptiles, 91 fishes and 28 invertebrates. (WRI et al. (2002) - State of the Forest Indonesia; WRI et al. (2000) - Trial by Fire; FAO Global Forest Resources Assessment 2005). [2]

Therefore, more sustainable biofuel production was still in discussion to answer the greener energy challenge worldwide. Therefore, for reducing environmental damage, some crops produced in dry-plantation in Madura could substitute the palm oil. Off course, further study regarding the supply – demand ration was needed to ensure the food supply adequacy in the local context. After fulfilling the food demand, the crops could be used for sustainable biofuel for Madura, Eastern part of Indonesia or the export market.

The karst landscape of Madura the island was suitable only allowing the non-irrigated agriculture and plantation to survive in the past. Observing the existing crops of non-irrigated plantation in Madura and the business potential, the biofuel industry was found possible to be developed in the "Green Industrial Area of Bangkalan. One solution for the concept was The Industrial Ecology. It was defined as a "systems-based, multidisciplinary approach seeking the understanding of behaviour of integration of human and natural systems". Therefore, the industrial ecology could describe the sustainable industry using the analogy of natural systems, and with consideration of social, environmental, economical and technological aspects. [3]

The industrial ecology was proposed by Frosch & Gallopoulos in 1989. Their idea was to propose industrial systems that perform following the nature of ecosystem, where wastes could become resources to another industries reducing use of raw materials, pollution, and saving on waste treatment?" One successful project of implementing the industrial ecology was the Kalundborg industrial Park. The industrial park was located in Denmark. The industries followed Frosch & Gallopoulos idea to reuse each other's waste in the form of by-products and heat. A power plant, an oil refinery, a pharmaceutical plant, a plasterboard factory, an enzyme manufacturer, a waste company as well as the city itself were included in the system. For instances, the gypsum produced by the Power Station as the by-products of the electricity generation process would become a resource for the BPB Gyproc A/S for producing plasterboards. [4]

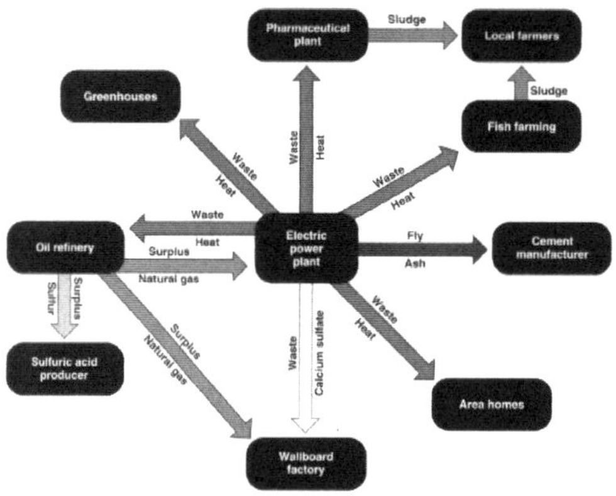

Figure 1. The Kalundborg Industrial Ecology System, Source: (Miller, 2003). [5]

3.1. The Industrial Ecology for Biofuel or Energy Production

Christensen and Kjær (2009) conducted further research on the Industrial ecology in the energy sector in Denmark. [6] The paper explained the implementation of industrial symbiosis concept for fuelling energy systems with the biomass and waste materials and its potential development. It also highlighted the ethical issue of bio-fuel. The food crises related to the biofuel productions in recent year led to criticism from NGOs and institutions world widely. The European Union responded with including the so-called sustainability criteria's in the renewable energy directives (EU 2009). [7]

Therefore, The Danish Government (2007) decided adopting fibrous biomass waste, and not food crops for bio-fuel. Consequently, the food crops lands in Denmark should not be converted. It also implied that imported crops from developing countries should be grown in sustainable manners. Therefore, industrial ecology models for sustainable biofuel productions were needed. [8]

Christensen and Kjær (2009) also presented industrial symbiosis concepts for biomass waste resource utilisations. One concept proposed was the integration of biogas facility and the agricultural land. The system utilised sugar beet as the raw material because of the high biomass yield per acre and long tradition of sugar beet farming in the region. The sugar beet was grown in a crop rotation system (4-6 crops), which included the clover. The clover would ensure a high biogas production and later used as fertilizers. [9]

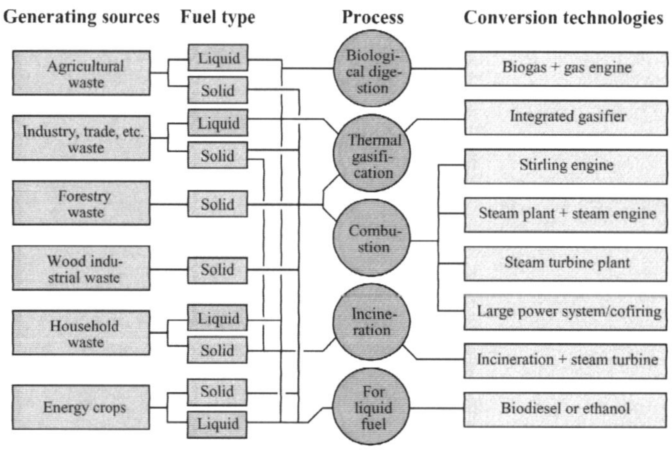

Figure 2. Existing Technologies of Biofuel Utilisation. [10]

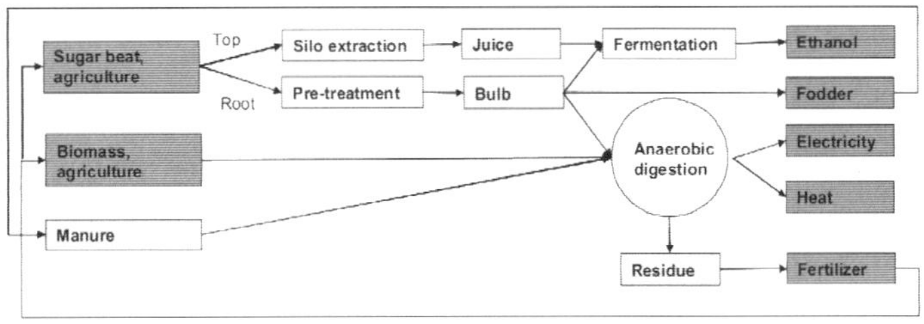

Figure 3. The 2nd Christensen and Kjær (2009) Industrial Ecology Concept (with Medium-sized Biogas Facility linked to Agriculture Facility) for Energy Sector in Denmark.[11]

The Biofuel study by Nexant (2008) for Brazil has shown the competitiveness of biofuel productions among countries. Nexant collaborated with Expetro, Brazilian consultancy with extensive experience in the biofuels industry in Brazil to provide the feedstock options and issues; direction of technology developments; and competitive position of Brazilian biofuels in the global markets. [12] The research found that the success key for successful biofuel production, careful integrated study was needed in 5 aspects, such as: policy, markets, feedstock, production economics, and technological developments. Further, the relative global production economics for Ethanol was showed in Figure 4. [13]

The graph described that the biofuel productions costs were still competitive with maximum wholesale gasoline price. And it showed that corn, switch grass and sugarcane were economically viable to be developed. And it could be concluded that the potential development of biofuel was promising.

Another study titled "Classification of Industrial Symbiosis Synergies: Application in the Biofuels Industry" by Martin, et.al. (2009) was conducted to identify synergies between biofuel industries in the Östergötland, Sweden. The study was carried out with workshop involving the regional biofuel industries and academic participants. The research identified synergies within the industry and with external industries. Finally, 70 industrial synergies were recorded. [14]

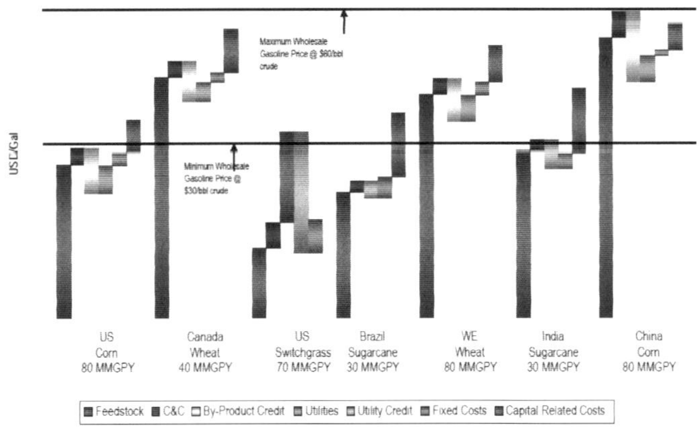

Figure 4. The Relative Global Production Economics for Ethanol (Cost Axis removed). [15]

The study explained the 3 types of synergies with biofuel industries such as:

- Biofuel ↔ Biofuel Synergies
- Biofuel ↔ External Synergies

- External ↔ Biofuel Synergies. [16]

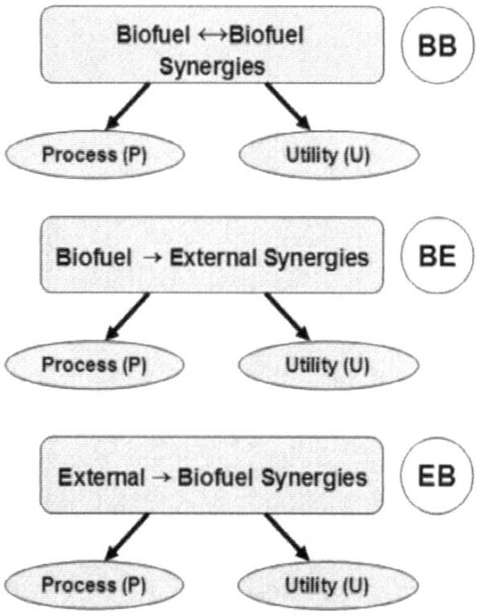

Figure 5. Industrial Symbiosis (IS) Models in the Biofuels Industry. [17]

The First Industrial Symbiosis (IS) model was the biofuel to biofuel production synergies (BB). Exchanging goods, materials and services between biofuel industries were conducted with sharing the services, utilities, by-products and products in both biofuel production systems. Two examples of BB were:

- Ethanol used in biodiesel trans-esterification reaction
- Biomass/stillage from ethanol production used for biogas production.

The second IS concept in biofuel industry was utilizing goods, materials, services from biofuel productions for external industries, or could be described as BE model. Some of BE examples were:

- Stillage used for animal feed
- Biogas digestive used as bio-fertilizer.

The third IS of was utilizing goods, materials, services from external industries for biofuel productions, or named as EB model. This would reduce the waste disposal costs of the external industries as well as reducing the biofuel cost productions. Few EB examples were:

- Waste vegetable oil (WVO) from snack food industries used for biodiesel production
- Animal wastes used for anaerobic digestion (biogas production).

Beside the 3 IS models, the research also presented the utilization of the Products (or Process) and Utilities between Industries were also identified. Van Berkel (2006) identified that synergies could be developed from utilisations of by-products (or named "by-product synergies"). [18] For the biofuel productions, some materials/ products/ "energy carriers" were found suitable for other conversions such as: raw oil for biodiesel production, cereal grains for ethanol production, and CO2 produced during fermentation. This was represented with "P" notation in IS Models. [19]

Further, Martin, et.al. (2009) presented several examples of Industrial Symbiosis in biofuel productions in the Östergötland, Sweden in the following table. [20]

Table 1. Industrial Symbioses in Biofuel Productions in Östergötland, Sweden

Synergies	Classification	Information
Glycerol from Biodiesel production used in Biogas Production	BB-PP	Glycerol, a major by-product of biodiesel production was used in many anaerobic digestion for methane production.
Exhaust from Ethanol Producer used to Dry Biogas Digestate	BB-UU	Exhaust from heating (utility) was used for drying of biogas digestate to optimize transportation.
Ethanol Stillage used for Biogas Production	BB-PP	The biogas was produced from stillage as the main input material
Wastewater from Biodiesel or Ethanol production used for Algae Production	BE-PU	Waste water (utility) was used for external industry (Algae production)
Glycerol from biodiesel production used for healthcare and cosmetics industry	BE-PP	Glycerol was used as an important raw material in cosmetics
Ethanol stillage used to produce Energy Pellets	BE-PU	Ethanol stillage (by-products of Ethanol production) was used as energy pellets for subsequent industries.
Animal Fats from Slaughtering used for Biodiesel Production	EB-PP	Animal fats (by-products of the slaughtering industry) were used in biodiesel production.
Algae Harvested from Baltic Sea used for Biogas Production	EB-UP	Natural algae as the pollution of Baltic Sea (utility) was used for biogas production.
Synthetic diesel production produces alcohol as a by-product, this can be used for biodiesel production	EB-PP	Alcohol (by-products from synthetic diesel production) was used for trans-esterification process in the biodiesel production.

These symbioses actually showed that the industrial ecology in biofuel productions would bring long term benefits to the local economy. Further, this would become the environmental endowment, in the form of pollution reduction that would improve the environmental quality (Kneese, Ayres & D'Arge, 1970; and Thampapillai D.J. 2002). [21]

4. Natural Resources and Socio-Economic Aspect

Analysis of Natural Resources and Socio-Economic Aspect of the Madura Island was conducted based on secondary and primary data. It was concluded there was a relationship of limited natural resources (water resources constraints, the limited fertility of the soil because of karst geology) and the development in Madura. This was supported by Adib (2009) that the non-irrigated plantations of Madura were subsistent because the less fertile soil types (the formation of limestone and marl deposits) and the limited rainfall volume. [22]

The average monthly rainfall varies between 0-276 mm/ month. This later caused constraint to agricultural activities between June to September. And it also caused the intermittent pattern of the Maduranese Rivers (the rivers were only flowing only during the rainy season) (The Office of Statistics of Bangkalan, 2009). [23]

Maduranese Geological condition consisted of several formations such as Alluvium, Pamekasan Formation, Pucangan Formation, Madura Formation, Pasean Formation, Bulu Formation, Tawun

Formation Ngrayog Formation and Menuran Formation (Geological Research and Development Centre, 1992). [24] Madura Formation and Alluvial were less suitable for industrial development because of potential land subsidence and sinkhole collapse (Suyanto and Sugalang 2003). [25] Further, the Maduranese Sedimentary Geology consisted of sediment clastic medium sandstone, sediment clastic fine claystone, sediment clastic limestone, sediment clastic alluvium, sediment clastic fine marl, sediment reef limestone. The sediment clastic limestone, sediment clastic alluvium, and sediment reef limestone were found less suitable for development for the industry. Meanwhile, the agriculture could be developed in areas with the clastic sediment alluvium. [26]

The Maduranese Aquifer Productivity, were mainly low (Poespowardoyo & Setiadi, 1985). It caused by high permeability and limited storage capacity of limestone. The permeability varied between of 0.2 to 380 litres/sec m (Madura groundwater Study, 1977). Contrary, the soil's storage capacity was very low, around 1-5%, resulting in unfertile soil. Thus, agricultural engineering was required for the agricultural development. [27] Some limestone aquifer springs were present, such as Pucung Spring in Bangkalan, Omben Spring in Sampang, Pamekasan Spring, and Kermata Spring in Sumenep. Unfortunately, the aquifer potential would not meet the needs of existing water according to the study (Research Institute of Water Resources, 1995). [28] The research found that water usage deficit had happened in Madura in the dry season of 1995. This was shown in the average ratio between annual water demand and supply (Q80%) achieving 58%. The data recorded of specific deficits in the three sub-watersheds such as Bangkalan (164 %), Semajid (102%), and Sabuntar (101%). Research Institute of Water Resources (1995) predicted that the average ratio between annual water demand and supply (Q80%) in dry season in 2020, based on the 1993 District Spatial Plan would reach deficit. The deficit were even worse in Bangkalan (428%), Sabuntar (101%), and Semajid (102%). It showed that the need of infrastructure provision of for Madura development. [29]

Soil Type of Madura comprised of Alluvium, Andisols, Gleysols, Cambisols, Lithosols, Non-Cultivated Bare Soil and Regosols in Madura. The Alluvium, Andisols and Gleysols were found less suitable for industry. While Cambisols, Lithosols, Non-Cultivated Bare Soil and Regosols were less suitable for agricultural development (Research Institute of Agriculture Soil Resources, 2009). [30]

Bangkalan Land Use showed that non-irrigated plantation comprised of 2.463 km2 (or 50%) from the island of Madura. This showed that non-irrigated plantation contributed enormously to the Regional Gross Domestic Production (Office of Statistics Bangkalan, 2009). [31] The non-irrigated plantations normally produced corn, peanut, rice, tapioca, coconut, greenpeal, kapok, soya bean, sweet potato and fan palm. Office of Statistics Bangkalan, (2009) [32] reported that 50% the Regional Gross Domestic Production of Bangkalan were contributed by the agriculture sector.

The main producing area of the Madura Island were reported as followed (Office of Statistics Bangkalan, 2008)

- corn – 80'012 ha,
- peanut – 36'587 ha
- rice – 40'027 ha,
- tapioca – 4'616 ha,
- coconut – 4'465 ha,
- green peal – 3'539 ha,
- kapok – 3'268 ha,
- soya bean – 1'449 ha,
- sweet potato – 1'295 ha,
- and fan palm – 669 ha. [33]

The development of livestock production was recorded possible in Bangkalan, Madura. There were 900 horses, 137'017 cows, 1.985 buffaloes, 60'360 goats and 4.024 sheeps in 2008 (Office of Statistics Bangkalan, (2008). [34] It showed great potential of implementation of industrial ecology similar to Christensen & Kjær (2009) concept in Denmark [35] or Martin ,et.al. (2009) in Sweden. [36]

Limited transportation network of Madura Island hindered the industrial and agricultural development in the past. Based on site reconnaissance, many of the Regional Road were too narrow for catering regional transportation. Thus, it caused inefficiency of the transportation. Furthermore, the subsistence agriculture was found in the island causing farmers poverty. BPS predicted the population working in agriculture reached 53% of the total labour force, meanwhile 60% of Bangkalan District population were found poor (BPS 2009). This showed the clear relationship between the limited Natural Resources, subsistence agriculture, and poverty of Maduranese. [37]

Therefore, Concept of Industrial Ecology of Integrated Industrial and Port of Bangkalan, Madura was suggested. The area was designated for this purpose in Bangkalan and Socah. This would be incorporated with the surrounding non-irrigated plantations and livestock farms. The plantations would supply some amount of corn, peanut and tapioca for biofuel productions. This amount would not reduce the initial volume required for regional food supplies. Meanwhile, farmers could enjoy higher sales prices because of biofuel production. Eventually, the biofuel could support the regional electricity production or exported to Eastern Indonesia or other countries. On the other hand, fertilisers as the by-products of anaerobic digestions, could be used by the farmers to improve the soil fertility. On one condition, that the agricultural land development should not be developed in the Madura Formation or other karst geological condition.

The Sustainable Industrial Ecology Strategy required integrated industrial area planning. The planning process must comprise suitable land selection; the selection of prospective industries and logistics; and integration with transportation master plan. Then the master planning process should be conducted with environmental impact assessment. Furthermore, Logistics Park was suggested to built in adjacent to the industrial and port areas. They could be developed in areas with excellent access to the Port and the SuraMadu Bridge.

Infrastructures were also needed to be provided such as water, electricity, sewage treatment, solid waste treatment, etc. Three water supply options were available for the Integrated Industrial and Port of Bangkalan, Madura. They were the piping water supply from Surabaya, the ground water usage, or the development desalination plant. Of the three options, the first and second options were not possible because of bridge structure limitations as well as karst phenomenon that could easily collapse in Madura. And the seawater desalination plant is was the only solution for the raw water supply for Madura development. Off course, this option should be accompanied with careful planning, Environmental Impact Assessment (EIA) and Detailed Engineering Design (DED).

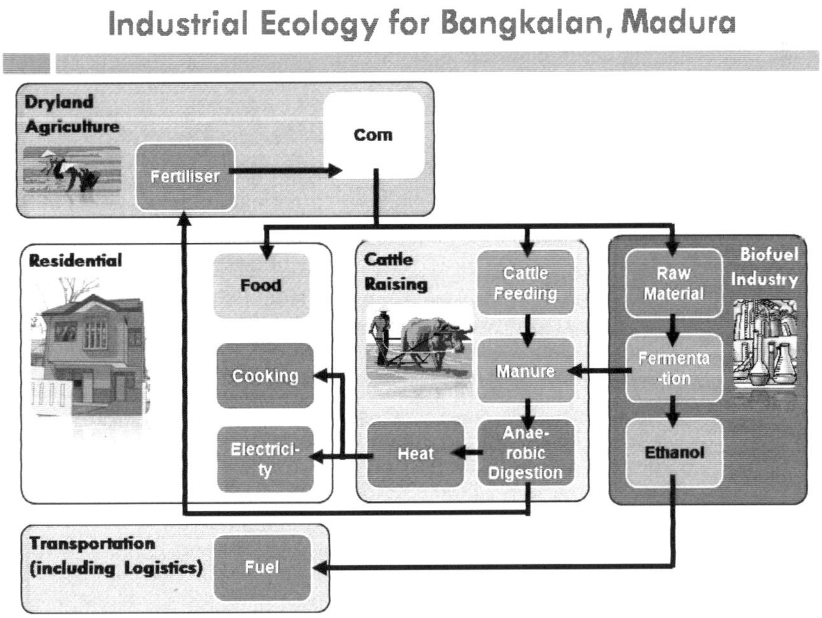

Figure 6. The Industrial Ecology Concept for Integrated Industrial and Port Park, Non-irrigated Agricultural Lands, Livestock Farms, and Communities in Bangkalan Madura.

Other improvement such as the agricultural revitalisation, human resource development (HRD) were needed collaboration with Maduranese and Islamic culture involving Moslem Religious Leaders (*Kyai*), Local Leaders (*Kepala Desa*) and Islamic Boarding Schools (*Pesantren*). The Vocational Training Centers (*Balai Latihan Kerja/ BLK*) could be set up independently or integrated with existing boarding schools.

5. Conclusion

The island of Madura faced challenges in coastal development because of limited transportation services, limited water resources and soil fertility due to karst phenomenon. This led to the unique social – economic condition. Consequently, the industrial ecology concept was needed for Bangkalan, Madura Island. The industrial ecology concept was also important for other coastal cities for the more sustainable development with careful analysis and planning.

Acknowledgements

We thanked

- Prof Marian Chertow, Profesor of Industrial Ecology, Yale School of Forestry and Environmental Studies;
- Petra Christian University;
  - Eunike Kristi Julistiono, S.T., M.Des.Sc.(Hons.) Head of Architecture Department;
  - Agus Dwi Hariyanto, M.Sc., ST., Former Head of Architecture Department;
- Surbana - Jurong Pte Ltd;
- Green Impact Indonesia

References

[1] Adib M 2009 *Ethnografi Madura (Maduranese Ethnography,*(Surabaya:Pustaka Intelektual Surabaya)

[2] Allenby Brad 2006 *The Ontologies of Industrial Ecology.* Progress in Industrial Ecology (Inderscience Enterprises Ltd.) **3** (1/2): 28–40. http://enpub.fulton.asu.edu/cesem/paper/PIE.pdf.

[3] Danish Government 2007 *National report on the implementation of directive 2003/30/EC of 8 May 2003 on the promotion of the use of biofuels or other renewable fuels for transport for 2006*, report submitted by the Danish government to the European Union

[4] De Joonge H *1988 Madura dalam Empat Zaman: Pedagang, Perkembangan Ekonomi dan Islam (Madura in the Four Ages: Trader, Economy Development and Islam)* (Jakarta: PT Gramedia)

[5] Eddy Juwono Slamet 1999 *Madura, Masa Lalu, Masa Kini dan Masa Yang Akan Datang: Sebuah Tinjauan Perilaku Ekonomi (Madura, The Past, The Present and The Future: The Economical Review),* presented in National Seminar Electrical Engineering Department 1999 (Surabaya: ITS)

[6] EU 2009 *Directive of the European Parliament and the Council on the promotion of the use of energy from renewable sources amending and subsequently repealing Directives 2001/77/EC and 2003/30/EC* (Brussels)

[7] FAO Global Forest Resources Assessment 2005. Retrieved from: http://www.fao.org/forestry/fra/en/;

[8] Frosch R A and Gallopoulos N E 1989 Strategies for Manufacturing. *Scientific American* **261** (**3**): 144–152. "International Society for Industrial Ecology". http://www.is4ie.org/Default.aspx?pageId=215327. Retrieved 1/08/2009.

[9] Geological Research and Development Centre, Bandung 1992 Peta Geologi Bersistem Indonesia, Lembar Surabaya & Sapulu 1608-4 & 1609-1 dan Lembar Tanjungbumi & Pamekasan 1609-2 & 1608-5 skala 1:100.000, (Indonesian Systematic Geological Map,

Surabaya & Sapulu Page - 1608-4 & 1609-1 and Tanjungbumi & Pamekasan Page - 1609-2 & 1608-5 scale: 1:100.000)
[10] Kneese A V, Ayres R U and D'Arge R C 1970 *Economics and Environment: A Material Balance Approach, Resources for the Future* (Washington D.C)
[11] Martin M, Ivner J, Svensson N and Eklund M 2009 *Biofuel Synergy Development: Classification and Identification of Synergies using Industrial Symbiosis*, (Sweden: Linköping University)IEI Report Number- LIU-IEI-R--09/0063--SE (2009).
[12] Martin M, Ivner J, Svensson N and Eklund M 2009 *Classification of Industrial Symbiosis Synergies: Application in the Biofuels Industry* (Sweden: Environmental Technology and Management, Linköping University)
[13] Miller G T 2003 *Environmental Science, Working With Earth, 10th edition*, Brooks/Cole Thomson Learning USA.
[14] Nexant Chem Systems 2008 P*rospectus Brazil's Biofuels Industry*, Outlook for a Global Leader, November 2008 (New York) accessed in http://www.chemsystems.com/reports/search/docs/prospectus/MC07_Brazil_Biofuels_Pros.pdf
[15] Office of Statistics Bangkalan, Madura 2009 *Kabupaten Bangkalan dalam Angka 2008 (2008 Bangkalan District Statistics).*
[16] Poespowardoyo R S and Setiadi H 1985 Geological Research and Development Centre, Bandung, Peta dan Laporan Hidrogeologi Lembar VIII – Surabaya, Jawa, Skala 1 : 250.000 (Hydrogeological Maps and Reports Page VIII – Surabaya, Java, Scale 1 : 250.000).
[17] Research Institute of Agricultural Soil Resources 2009 GIS Map.
[18] Research Institute of Water Resources, Public Works Department, Indonesia 1995 *Studi Neraca Air SWS Madura (Water Balance Study of Madura Island).*
[19] Sugiyanto and Sugalang 2003 *Geological Research and Development Centre, Bandung, Inventarisasi Geologi Teknik Kabupaten Bangkalan Jawa Timur* (Inventory of Geological Engineering Condition in Bangkalan District, East Java) ; www.dgtl.esdm.go.id
[20] Tanuwidjaja G and Malone-Lee L C 2009 *Applying Integrated Ecological Planning and Adaptive Landscape Evaluation Tool for Developing Countries in the Framework of Sustainable Spatial Planning and Development, Study Case Bintan Island, Indonesia*, In International Seminar Positioning Planning in Global Crises (Bandung: Institut Teknologi Bandung)
[21] Thampapillai D J 2002 *Environmental Economics: Concepts, Methods, and Policies* (Melbourne: Oxford Universities)
[22] Van Berkel R 2006 *Regional Resource Synergies for Sustainable Development in Heavy Industrial Areas: an overview of opportunities and experiences.*
[23] World Resources Institute, Forest Frontiers Initiative, WWF-Indonesia & Telapak Indonesia Foundation 2000 Trial by Fire, Forest Fires and Forestry Policy in Indonesia's Era of Crisis and Reform. Retrieved from: http://www.globalforestwatch.org/english/indonesia/index.htm;
[24] World Resources Institute, Forest Watch Indonesia, Global Forest Watch 2002 *State of the Forest Indonesia.*

Typologi of Island City in Indonesia

Alvaryan Maulana[1] **and Tania Benita**[1]

[1] *Rural and Regional Planning Research Group, Institut Teknologi Bandung, Bandung 40135, Indonesia*

Abstract. As an archipelagic country, Indonesia consist of thousands of island. Some of them are big enough to contain growth and become the center of settlement and activity in Indonesia. However, growth and habitation are not only mushrooming in main island. Several small island also experience growth and become densely population places and simply become a city within island. This study aims to identify island city in Indonesia and creating the typology of the island city. This study is using exploratory approach and heavily rely on statistical figure of every single autonomous region as data sources. Eventually, this study found twelve (12) island cities in Indonesia, and three distinctive typology of island cities.

1. Introduction

Indonesia is one of five archipelagic countries which have been approved by United Nation on the Law of the Sea or UNCLOS along with Fiji, Papua New Guinea, the Bahamas and the Philippines. This approval issued within the UN Convention on the Law of the Sea which held on 1982. Based on the convention, article 46, Archipelagic State means "a State Constituted wholly by one or more archipelagos and may include other islands" (UNCLOS, 1982).

Around 300 million of Indonesian lived in 17.500 islands in Indonesia. Among them, there are five main island; Sumatera, Java, Kalimantan, Sulawesi and Papua. Besides, there are also group of island such as Maluku archipelago and Lesser Sunda. Apart from main island land and group of island, there are also smaller group of island, spread across Indonesia, from Natuna Archipelago, *Kepulauan Riau, Kepulauan Seribu, Kepulauan Aru,* and *Kepulauan Talaud* and dozens of other small islands.

Java Island, home of 141 million of Indonesian (BPS 2012), is the densest and most developed island in Indonesia, possessing six metropolitans; Jakarta Raya, Bandung Raya, Cirebon Raya, Kedungsepur, Kartamantul and Gerbangkertasusila. Java is growth center of Indonesia.

However, center of growth are not limited in main-island. Several smaller island, apart from main island, also experience advancement because of various push factors, whilst it populations are still not comparable with main island. However, some of them are granted administrative status such as City of Sabang in Weh Island (granted municipality status in 1965), City of Batam in Batam island (1999), City of Ternate in Ternate Island (2010), City of Tarakan in Tarakan Island (1997), City of Tual in Tual archipelago (2007), City of Mataram in Lombok Island (1993) City of Ambon in Ambon Island (1958).

This phenomenon indicates that smaller island in Indonesia have tendency to developed as urbanized places. However, Island tend to have limited natural resources and land area, complex geography and more vulnerable to disasters. (Bass & Dalal-Clayton, 1995). Those limitation means that smaller island have distinguish urban process from other places, along its extreme density since it located in smaller are, an island. (Grydehøj et al., 2015)

Purpose of the study is to acquire a comprehension of urbanized small-island in Indonesia by assessing every local autonomous regions based on their spatial, economic and demographic characteristic. This study will be a platform to develop new approach and perspective in managing cities and island. Therefore, several objectives are designed in order to comply the main purposes. Objectives in this study, are:

- To explore and identify island city in Indonesia
- To identify the typology of island city in Indonesia

2. Methods

The To understand small but urbanized island which have considerable amount in Indonesia, this study reflect on similar but not identical concept called island city as the root theory, but not applicable entirely. This study is using explorative approach, while the information and data for the study are collected from statistical document issued by Badan Pusat Statistic (National Statistical Bureau), published in 2015. Every region

But what is Island Cities? Is it a city located in island, or an island which already urbanized? (Tutt, 2014) argue that the concept of island city could be an oxymoron concept since it possess contradictive and paradoxical components ; A city, which are huge, expanding progressively and chasing into the future; and an island which are obsolete, constrained and secluded. Yet, this paradox needs to be challenged since most of the densest place in the world are located on small island. (Grydehøj et al., 2015).

Table 1. Island City Classification

No	Source	Island City Classification	Example
1	(Tutt, 2014)	Cities which located in smaller island contagious to larger land mass, expand and diffuse with it surrounding, leaving insular characteristic	Singapore, Hong Kong, Montreal
		Historical cities which bounded by the time and dying without tourism attraction which most of them are walled cities in the past	Venice, Visby
		Modest cities with heavy limitation to expand, literally and metaphorically. Have subsistence activity	(examples are not given)
2	(Grydehøj et al., 2015)	Cities which are contagious with one or more small island	Abu Dhabi, Lagos, Mumbai
		Cities which are located one or more dense and urbanized Small Island	Lagos, Montreal
		Cities located within mainland	Shanghai, Tokyo, Toronto
		Cities within small island or archipelagos that cannot be regarded as an urban zone, but already meet the requirement of urban function	Bahrain, Malta, Penang

Sources: Tutt (2014), Grydejoj et al (2015)

These classifications mainly talks in spatial context. The way city located, and city relation with main land are common variable in both classification. First analysis is identifying the spatial attributes of each region which classified based on their "islandess" characteristic or simply measuring how island an island, and its distance from main land. In broader context, every places in Indonesia have insular characteristic since Indonesia is a large group of island. However, in this study, five main

IOP Conf. Series: Earth and Environmental Science **79** (2017) 012026 doi:10.1088/1755-1315/79/1/012026

islands are not considered as an island. Rather, these five main islands are considered as a landmass area which don't possess insularity. Remote, rural and distantly are true traits of island. Hence, in order to be eligible island city, a region must located outside Indonesian main island; Sumatera, Java, Kalimantan, Sulawesi and Papua island.

The second assessment is measuring urban characteristic of each region. Several definition of city that being used are shown in table 2. In summary, city is defined as a region which possess a certain amount of population and certain types of economic activity.

Table 2. Definition of City

No	Source	City Definition
1	Law No 26/2007 on Spatial Planning	Regions which main activities are non-agriculture and have functions as a residential urban neighborhood, and possess concentration and distribution of government services, social services, and economic activities.
2	National Statistical Bureau (Rule No 37th 2010)	Total Population is more than 10.000 persons
		Population Density is more than 5000 person/Km^2
		Less than 25% of total population work on primary sectors

Sources : Law No 26/2007

Therefore, the definition of island city in Indonesia is "a region which located outside of Indonesian Main Island and fulfill urban characteristic; population density and economic structure".
To identify island cities in Indonesia, 506 regions (Kabupaten/Kota) are assessed. Although Indonesia recognized two type of autonomous region; Kota (municipalities), and Kabupaten (regencies) based on their legal status and capacities, these two classifications are considered in the same position, as a city. Facts that there are actual cities classification in Indonesia is ignored in order to accommodate regions which not have cities status but already fulfill characteristic of a city.

3. Typologi Island

For Based on analysis, there are only 12 autonomous region, from total 506 autonomous region which meet the features of island city. List of the regions is shown in Table.3

Table 3. Typologi Island of The City in Indonesia

No	Name	Population	Density (Person/Km^{2})	% of Primary Sector	Typology
1	Kota Batam	1,037,187	1,450	1	Single zone on group of small island
2	Kota Denpasar	863,600	6,758	1	Part of small island, sharing with other region
3	Kota Mataram	450,226	7,345	3.12	Part of small island, sharing with other region
4	Kota Kupang	390,877	2,168	3.64	Part of small island, sharing with other region

5	Kota Pangkal Pinang	327.167	1,955	5.77	Part of small island, sharing with other region
6	Kota Ambon	411,617	1,145	5	Single urban zone, located on one small island
7	Kota Ternate	212,997	1,314	8	Single urban zone, located on one small island
8	Kota Tanjung Pinang	203.153	848	3.15	Part of small island, sharing with other region
9	Kota Sabang	30,653	250	8	Single urban zone, located on one small island
10	Kota Bima	159,736	719	13	Part of small island, sharing with other region
11	Kota Tarakan	239,787	956	17.67	Single zone, located on one small island
12	Kabupaten Bintan	151,123	77.65	25.32	Single urban zone on group of small island

Source : Analysis 2015

Most of the island cities in Indonesia already have municipality status, leaving Kabupaten Bintan as the only region with regency status. Kota Batam is the most populous island city with number of population is more than 1 million person, while Kota Sabang is the least populous island city. However, the densest island city is claimed by Kota Mataram, followed by Kota Denpasar as the second densest island city in Indonesia.

There are three typologies of island city based on the way they located. First typology is island city which located on group of small island, scattered. Kota Batam and Kabupaten Bintan are classified as this group of island city.

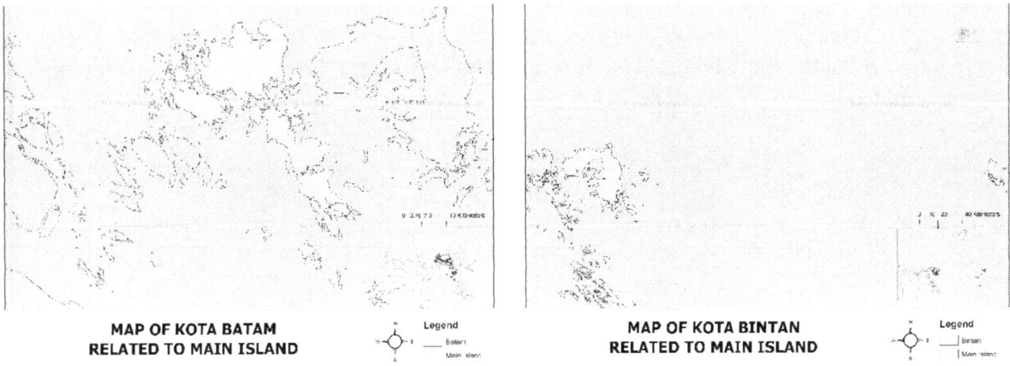

Figure 1. (Left) Map of Kota Batam, (Right) Map of Kabupaten Bintan

Second typology is island city which located on a single island. This island city occupy entire/ most of the island as its territory. Kota Ambon, Kota Ternate, Kota Sabang and Kota Tarakan are classified into this typology.

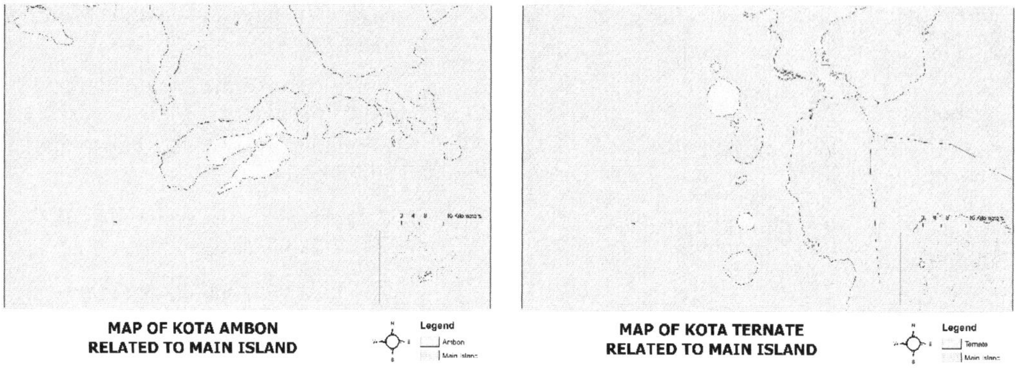

Figure 2. (left) Map of Kota Ambon, (right) Map of Kota Ternate

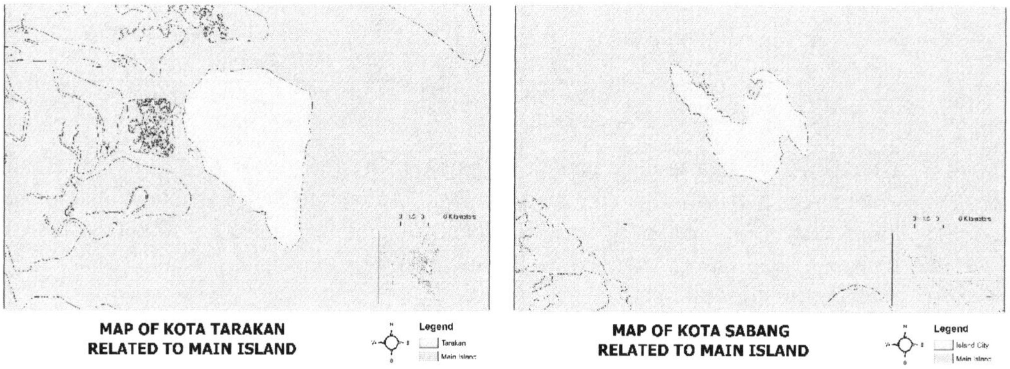

Figure 3. (left) Map of Kota Tarakan and (right) Kota Sabang

Third group is island city which located in relatively-large small island and share the ownership of the island with other regions. This typology is the most common island city in Indonesia. This group consist of Kota Denpasar, Kota Bima, Kota Kupang, Kota Mataram, Kota Tanjung Pinang and Kota Pangkal Pinang.

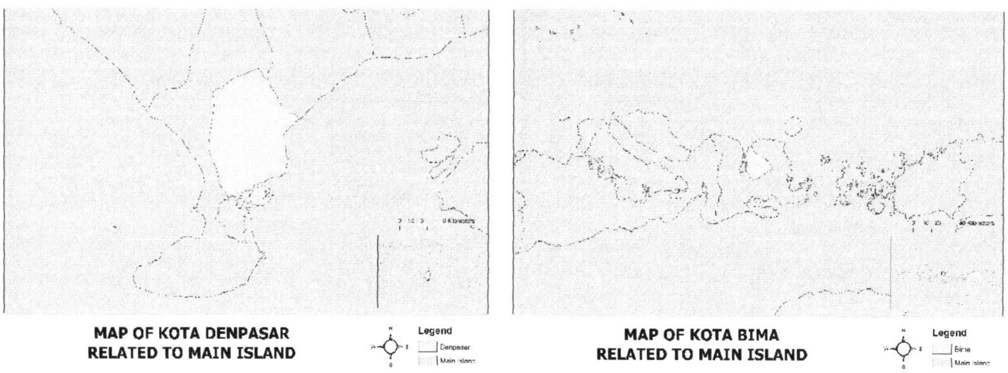

Figure 4. (left) Kota Denpasar, (right) Kota Bima

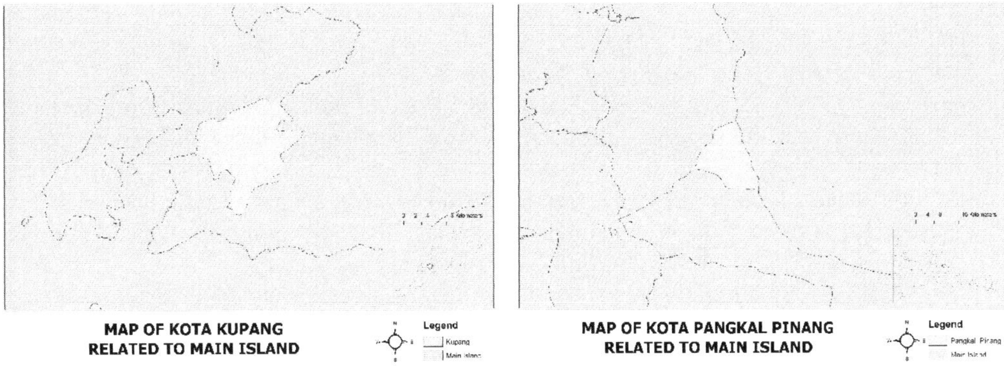

Figure 5. (left) Kota Kupang, (right) Kota Pangkal Pinang

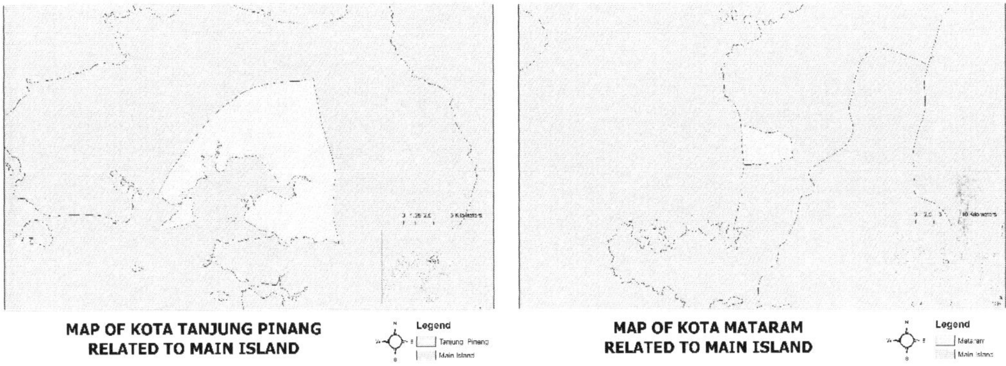

Figure 6. (left) Tanjung Pinang, (right) Kota Mataram

4. Conclusion

Concept of island city arise because of densely populated city which have economic shifting from primary activity to secondary and tertiary activities located on land scarce area. As insularity, or islandess are strongly related to heavy limitation on land and natural resources. Consequently, the combination of growth rate and the limitation create a new characteristic of places and obviously have significant differences with other in Indonesia

Furthermore, this study conclude that there are three distinctive type of island city in Indonesia and leads to different characteristic and traits in managing island city. This distinction means that each type of island city need differences managing approach. Assuming every island cities are similar might leads problems in foreseen future. Hence, the understanding of island city and the typology of island city are necessary.

Acknowledgements

This research was supported by Delik Hudalah from Regional School of Architecture, Planning and Policy Development, Institut Teknologi Bandung who also help the substantial aspect. We would also like to show our gratitude to fellow researcher, Yustina Octifanny, and Nabilla Adharina who provide insight and assistance for comments that greatly improved the manuscript.

References

[1] Bass S C and Dalal-Clayton D B 1995 *Small Island States and Sustainable Development: Strategic Issues and Experience-7755iied.* IIED.

[2] Grydehøj A, Pinya X B, Cooke G, Doratlı N, Elewa A, Kelman I and Swaminathan R 2015 Returning from the horizon: introducing Urban Island Studies. *Urban Island Studies.* ***1*****(1),** 1–19.

[3] National Statistical Bureau 2015 *Regional Statistical Figure* 2015, Indonesia.

[4] Tutt P A 2014 Defining the island city: ancient right versus modern metropolis, as considered at Peel, Isle of Man. *Island Studies Journal*, ***9***, 2, 196–204**.**

[5] United Nation 1982 *United Nation Convention on the Law of the Sea.* (Jamaica)

Conjoint Analysis for New Service Development on Electricity Distribution in Indonesia

D L Widaningrum, Chynthia, L D Astuti, M A B Seran
Industrial Engineering Department
Faculty of Engineering
Bina Nusantara University
Jalan KH Syahdan 9, Jakarta 11480, Indonesia

E-mail: diwdaningrum@binus.edu

Abstract. Many cases of illegal use of electricity in Indonesia is still rampant, especially for activities where the power source is not available, such as in the location of street vendors. It is not only detrimental to the state, but also harm the perpetrators of theft of electricity and the surrounding communities. The purpose of this study is to create New Service Development (NSD) to provide a new electricity source for street vendors' activity based on their preferences. The methods applied in NSD is Conjoint Analysis, Cluster Analysis, Quality Function Deployment (QFD), Service Blueprint, Process Flow Diagrams and Quality Control Plan. The results of this study are the attributes and their importance in the new electricity's service based on street vendors' preferences as customers, customer segmentation, service design for new service, designing technical response, designing operational procedures, the quality control plan of any existing operational procedures.

1. Introduction

Illegal use of electricity in Indonesia is still rampant, especially for activities where the power source is not available, such as in the location of street vendors. It is not only detrimental to the state, but also harm the perpetrators of theft of electricity and the surrounding communities. These conditions trigger state electricity company of Indonesia to deliver new services to support many activities which location are uncertain, called SPLU, an electric charging stations. This product is expected to be the solution to reduce electricity theft and illegal electricity usage. This product will be launched first in DKI Jakarta, as one of coastal area in Indonesia, with the main target are the street vendors. This service is expected to help the street vendors to gain legal electricity and avoid electricity theft, since it may harm the state and endanger the street vendors as a user.

The first generation of SPLU is designed and made without customers' involvement. The product has been designed without knowledge of preferences and expectations from customers. Lacks of knowledge about customer preferences were causing a perception gap between institution and customers. For example, 5500 VA power which was previously planned were over supply for street vendors, some of them when the preliminary project ask electricity power in range 450-1300 VA.

Learning from experience of production the first generation of SPLU, the company requires the development of a new service or new service development that systematically and accordingly to customers preferences service SPLU second generation which will be launched. It was common that new services were solely based on the experience of a team of developers. The subjective process will be difficult to implement the new service development competitive and sustainable [1]. Meanwhile, ideas for new service development preferably gained from customer suggestion, customer concern, and customer demographics [3]. With the design of the development of a new service or new service development to suit the preferences of these customers, expected service SPLU second generation can meet the wishes of our customers and maximize the benefits of the SPLU.

New service development has been widely recognize, especially in academic community. Not always for profit business, NSD also effective to design a non-profit service, such as the emergency management services, to prevent another disaster for obtaining unhealthy foods [4[highlight the importance of all organization members' involvement for a successfuly NSD, such as front line employees [5] and also tied with the nature of dynamics team, including conflict intensity and frequency, and conflict management styles [6].

[7] Conducted a study to revealed important parameters in NSD process, that was grouped as locality, timing, and human resources internal factor. In NSD process, one important technique in determining the service attributes is conjoint analysis. Conjoint analysis is a powerful method to design both product attributes such as food product, packaging, or automotive [8],[9],[10],[11],12], and service attributes, such as hotel services, tourism package, ecosystem services, and online business [13],[14],[15],[16].

The problems discussed in this research are: The customer's considerate attributes to use SPLU; The design of the SPLU service according to customer's preferences if analyzed using conjoint analysis; Customer segmentation based on similarity of services customers preferences SPLU if analyzed using cluster analysis; The technical response that must be carried out by PT PLN Disjaya to meet the expectations and needs of the customer; How to design the right operations to do PT PLN Disjaya to realize the technical response that has been designed before; Quality control plan that was formed to be able to control the operational activities that have been designed.

According [17], New Service Development that is systematically done with 5 stages, consists of opportunity identification, customer understanding, concept development, process design, and refinement and implementation.

2. Methods

This study, the process of New Service Development (NSD) is done in five steps as shown in Figure 1.

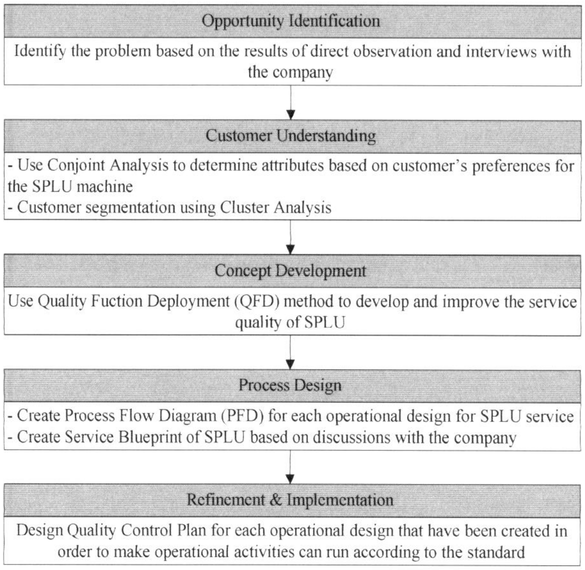

Figure 1. New Service Development Methodology

Distributing questionnaires at this step to determine the most favoured stimuli by respondents based on the respondent's preferences. The object of this research is the street vendors who trade on the official site based on the local government of DKI Jakarta (West Jakarta , Central Jakarta, South Jakarta, East Jakarta and North Jakarta).

3. Result and Discussion

3.1. Opportunity Identification

To create the new service development for second generation of SPLU service. The following diagram in Figure 2 is the interrelationship diagram that used as the basis to create the new service development for next generation of SPLU service. Figure 2 shows the existence of a discrepancy between the SPLU machine (1st generation) that will be placed in 16 locations in East Jakarta to the needs of the target customers. Before the next generation of SPLU machine place in the whole area in Jakarta, special research needs to find the customer's preferences regarding the SPLU service. By creating the new service development for SPLU service next generation, the company can determine the customer's preferences, customer segmentation, technical response to fulfill the customer's needs, and control any operational activities.

3.2. Customer Understanding

To create the new service development for second generation of SPLU service. The following diagram in Figure 2 is the interrelationship diagram that used as the basis to create the new service development for next generation of SPLU service. Figure 2 shows the existence of a discrepancy between the SPLU machine (1st generation) that will be placed in 16 locations in East Jakarta to the needs of the target customers. Before the next generation of SPLU machine place in the whole area in Jakarta, special research needs to find the customer's preferences regarding the SPLU service. By creating the new service development for SPLU service next generation, the company can determine the customer's preferences, customer segmentation, technical response to fulfill the customer's needs, and control any operational activities.

Conjoint Analysis

In conjoint study, the determination of attributes is done by reviewing the service design of SPLU next generation that has been designed by the development teams from PT PLN (Persero) Disjaya and examine the results of the survey that has been done by the company against target of customers, i.e. street vendors. Table 1 presented the attributes and levels of service on the SPLU machine.

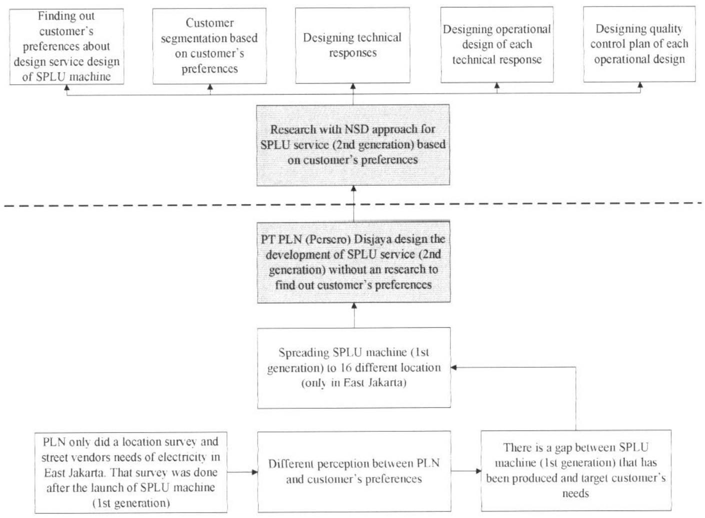

Figure 2. Interrelationship Diagram

Table 1. Comparison Between Supply and Demand of Clean Water in Ambon Bay

Attributes	Levels
Payment method	Co-Branding with Bank
	Electronic Money
	Prepaid Token
Instruction to use	Written Instruction
	Audio Instruction
And an entry	450 VA
	900 VA
	1300 VA
	2200 VA
	3500 VA
	4400 VA
	5500 VA
Paid Wifi as additional feature	Need
	Do Not Need

Before analysis was conducted on the importance value and utility of each attributes and levels, an evaluation of the goodness of fit model of the correlation value that is has the Pearson's coefficient < 0,05, then the model has accurately so that the calculation results worthy to be analyzed more.
From 180 data that has been collected, 180 respondents have seen that the Pearson's coefficient < 0,05. This means all the answers given to respondent when filling the questionnaire has been accurate and consistent so it can analyze more. The following Table 2 is presented the conjoint analysis result that consist importance value and utility of each attribute and level from the preference analysis from 180 respondents on the year 2037, more sub-districts are having the land availability status of deficit. Thus, it is concluded that the land carrying capacity in Ambon Bay, is decreasing time to time, which is shown by the increasing land deficit in the sub-district units.

Table 2. Attributes and Levels of SPLU Service

Attributes	Levels	Utility Estimate	Std. Error	Importance Value
Payment Method	Electronic Money	-,296	,079	33,951
	Co-branding with Bank	-,283	,079	
	Prepaid Token	,580	,079	
Instruction of Use	Written Instruction	-,067	,059	10,678
	Audio Instruction	,067	,059	
Power Capacity	450 VA	,076	,116	47,386
	900 VA	,717	,116	
	1300 VA	,621	,153	
	2200 VA	-,162	,153	
	3500 VA	-,343	,153	
	4400 VA	-,408	,153	
	5500 VA	-,501	,153	
Paid Wifi as Addional	Need	,088	,059	7,986

Attributes	Levels	Utility Estimate	Std. Error	Importance Value
Facilities	Do Not Need	-,088	,059	
(Constant)	2,053	,065		

From the four existing attributes, the respondents considered the most important attributes of power capacity of SPLU service with a percentage of 47,386%. After that, followed by the attribute that has the priority to 2, 3, and 4 in the SPLU service are the payment methods (33.951%), the instructions for use (10.678%), and the attributes of paid wifi as additional facilities (7.986%).

The payment method attribute, respondents prefer prepaid token with the utility value of 0.58 as a medium through which consumers in the payment transaction before using electricity from the SPLU machine. The majority of respondents has last education in junior and senior high schools was minimal to determine the payment with electronic method. In addition, the determination of the prepaid token due to the comfort level of street vendors who have been using prepaid tokens in daily life.

The instructions for use attribute, respondents prefer the audio instructions from machine with a utility value of 0.67 as a way of delivering to consumers about steps when using SPLU machine.The power capacity attribute, respondents were more likely to use the power of 450 VA, 900 VA, and 1300 VA with utility value respectively 0.076; 0.717; 0.621. This is because the street vendors have different type of merchandise and uses several electronic tools such as lights, fans, rice cooker, TV, music player, and blender so that the power of 450 VA, 900 VA, and 1300 VA can be used to power the electronic tools.

The attribute of additional facilities paid wifi, respondents consider additional facilities should be installed such as paid wifi. This is because the majority of street vendors aged between 21-30 years and 31-45 years. The age range was relatively young and understand the use of the internet.

Cluster Analysis

To classify the 180 prospective customers based on the similatiry of customer's perception then do some steps. The steps is called two stage clustering that using Ward's method and K-means [3]. The first step, Ward's method is used to determine the optimal number of clustering [3]. The output of the Ward's method was shown to the coefficient agglomeration schedule. Coefficient algorithm of 180 respondents causes elbow on the graph. Thus the number of the clustering can be found by the total number of respondents was reduced by the number of respondents who experienced elbow [3], namely (180-176) = 4 clusters

The second steps is used the K-means method, this method will show the number of members in each cluster and have been classify each respondent based on group respectively and show average values of SPLU service combination to all customers and each segment. To validate the clustering of potential customers, it can use the Analysis of Variance (ANOVA). Results of the F-test one way ANOVA shows p-value for all combinations which was obtained p-value $< 0,05$. This means that there are significant differences between each cluster on the service combinations on a machine SPLU.

The majority of respondents in cluster 1, aged between 25-30 years with the last education is high school. This cluster has the highest average value on a combination of services SPLU with details as follows: payment method - prepaid token, instructions for use – audio instruction from machine, the power capacity - 1300 VA, and the need to paid wifi as additional facilities. Respondents in cluster 2 consists of respondents aged between 31-45 years and graduated from high school. This cluster has the highest average value on a combination of services SPLU with details as follows: payment method - prepaid token, instructions for use – audio instruction from machine, the power capacity - 1300 VA, and the need to paid wifi as additional facilities.

Respondents in cluster 3 between the ages of 31-45 years old and have a high school education and the last is already trading more than 5 years. This cluster has the highest average value on a combination of services SPLU with details as follows: Payment method - electronic money, instructions for use – written instruction, the power capacity - 1300 VA, and need to paid wifi as additional facilities. In cluster 4, majority of respondents were in the age range 31-45 years with the last education middle and high school and has been trading for 3-5 years. This cluster has the highest

IOP Conf. Series: Earth and Environmental Science **79** (2017) 012027 doi:10.1088/1755-1315/79/1/012027

average value in combination with the following services: payment method - prepaid token, instructions for use – written instruction, the power capacity - 450 VA, need to paid wifi as additional facilities.

3.3 Concept Development

At this stage, the method using Quality Function Deployment for translating customer needs regarding services SPLU (2nd generation) to be able to design the technical response to be made by the company to meet customer needs.

Identification Of Customer Needs

After observing directly to the target customers and reviewing the results of a survey that has been done by PT PLN Disjaya, there are some attributes that need to be considered in the design of SPLU services are payment method, usage instructions, power capacity and additional facility paid wifi. The priority attribute of the desire desire level based on customers obtained through the process of conjoint analysis in aggregate and individual.

Determine The Technical Response

Technical response is the things that should be done by PT PLN (Persero) distribution of Jakarta Raya to be able to develop and improve the service quality of the SPLU will be given to the customer. Some technical responses that must be done by the company include:

1. The sale of prepaid tokens with a smaller nominal
2. Designing the energy return system
3. Forming a special division to take care the entire operations of SPLU machine
4. Doing quality checking of SPLU machine before released
5. Designing instruction of using SPLU machine
6. In 1 SPLU machine will be provided some meters with different power capacity
7. Cooperating with the PT Icon+ to provide wifi service
8. Maintenance on a regular basis against SPLU
9. Cooperation with vendors that manufacture SPLU machine
10. Surveying on the target location
11. Introducing SPLU directly to target consumers

Determine The Technical Response

House of Quality (HoQ) is a matrix that represents the Quality Function Deployment (QFD). HoQ represents relationship between technical responses that has been designed by the Corporation with the needs of customers within the service SPLU (2nd generation) as can be seen in Figure 3.

Process Design

At this stage will be explained about the service of process that will be given to customers who presented in the draft blueprint and the operational service of the technical response that has been designed in the form of process flow diagrams. The process visualize by service blue print to show the design of the SPLU service system that will be given to customers ranging from entering the number token to use electricity from SPLU machine, and the design of the SPLU service starts from the customer's finished using electricity from SPLU machine to design the remaining energy return system.

Operational design for doing quality checking of SPLU machine before released are consist of operational design for maintenance on a regular basis against SPLU, operational design for partnership with vendors that produce SPLU machines, and operational design for location survey.

Refinement and Implementation

At this stage will be designed Quality Control Plan for each operational design which have been prepared so that all operations can run according to standard. Quality Control Plan consist:

a. *Quality Control Plan for Doing Quality Checking of SPLU Machine Before Released.* Quality checker teams must have a high degree of accuracy in order to avoid errors as there are components of the SPLU machine are not checked and errors in filling checklist.
b. *Quality Control Plan for Maintenance on a Regular Basis Against SPLU.* In regular maintenance of SPLU machine there are five processes, namely the checking condition of SPLU machine, report the results of checking, assessing the extent of damage and the cost involved, approving repair damage to components, repair damaged components, create reports on the repair components, review the components has been improved, making SPLU machine maintenance reports to the Distribution Office.
c. *Quality Control Plan for Cooperation with Vendor.* In cooperation with vendors, there are four processes that establish the criteria for selection of vendors, determine the appropriate vendor criteria, producing SPLU machine by vendor, evaluate the vendor performance.
d. *Quality Control Plan for Location Survey.* There are four processes of location surveys, namely in cooperation with the local government to find the official location for laid SPLU, conducted a survey directly to the proposed location, making the survey report of appropriate location to be paired SPLU to the Distribution Office, and issued the approval letter.

Level	Customer needs	Technical Response 1	2	3	4	5	6	7	8	9	10	11	The Relative Importance: aggregate	CL 1	CL 2	CL 3	CL 4
L11, L12, L13	Payment Method	9	9	1							9	9	33.95	26.50	36.68	25.22	40.33
L21, L22	Audio Instruction		9	1	9	9			9	9	9	9	10.68	12.68	10.44	10.35	9.58
L31, L32, L33, L34, L35, L36, L37	450 VA, 900 VA, 1300 VA	9		1	9		9		9	9	9	9	47.39	50.48	45.94	53.30	44.03
L41, L42	Paid Wifi			1	9			9	9		9	9	7.99	10.35	6.94	11.13	6.06
Importance Absolute	1	692.75	352.60	100	661.53	114.13	454.28	93.12	661.53	568.40	900	900					
	2	743.54	424.06	100	569.90	93.96	413.44	62.50	569.90	507.40	900	900					
	3	706.72	320.15	100	672.98	93.13	479.70	100.1	672.98	572.83	900	900					
	4	759.24	449.17	100	537.03	86.21	396.28	54.54	537.03	482.49	900	900					
Importance Relative	1	23.87	22.81	25	27.10	29.46	26.05	30.01	27.10	26.67	25	25					
	2	25.62	27.43	25	23.34	24.25	23.71	20.14	23.34	23.81	25	25					
	3	24.35	20.71	25	27.57	24.04	27.51	32.27	27.57	26.88	25	25					
	4	26.16	29.05	25	22.00	22.25	22.73	17.58	22.00	22.64	25	25					
Cluster 1 ranks		10	11	9	4	2	6	1	3	5	7	8					
Cluster 2 ranks		2	1	3	10	6	8	11	9	7	4	5					
Cluster 3 ranks		9	11	8	3	10	4	1	2	5	6	7					
Cluster 4 ranks		2	1	3	10	8	6	11	9	7	4	5					
Importance Absolute Aggregate		732.03	401.66	100	594.45	96.10	426.47	71.87	594.45	522.58	900	900					
Importance Absolute Aggregate (%)		13.71	7.52	1.873	11.13	1.80	7.99	1.35	11.13	9.79	16.86	16.86					

Figure 3. House of Quality

4. Conclusion

Based on the direct observation and survey by PT PLN Disjaya, here are several customer's considerate attributes to use SPLU machine are payment method (with levels: co-branding with bank, electronic money, and prepaid token), instruction of use (with levels: written instruction and audio instruction), power capacity (with levels: 450 VA, 900 VA, 1300 VA, 3500 VA, 2200 VA, 4400 VA, and 5500 VA), and paid Wifi as additional facility. Based on data processing and analysis that has been done using conjoint analysis, design of the SPLU to meet the preferences of the customer is using the prepaid token as a method of payment, audio instructions from the engine, the capacity of power of 450 VA, 900 VA, and 1300 VA, and customers require paid wifi as additional facility.

The segmentation is done based on similarity on customer's preference of SPLU service. Customer segmentation is done by using cluster analysis. Results from cluster analysis this is customers are divided into four clusters. After knowing the customer's needs to the services of the company, then the SPLU designing technical response to meet the needs of the customers. The technical response among other prepaid tokens with nominal sales are smaller, making locations and the introduction of the SPLU products directly to the target consumer.

From the technical response then would be designed to run the operations of the technical responses is structured. Some of the operational design that has been designed, among others, the design of the operational quality of the SPLU machine checked before the draft was circulated, the routine maintenance operations against SPLU, draft operational cooperation with vendors and operational design of the survey locations. From some operational design, the quality control plan will be designed for each process in the design of such operations in order to ensure that all operational activities already underway in accordance with standards and provide the results in the appropriate standard.

References

[1] Boukis, Achilleas, & Kaminakis, Kostas. (2014). Exploring the Fuzzy Front-end of the New Service Development Process – A Conceptual Framework. *Procedia - Social and Behavioral Sciences, 148*, 348-353. doi: 10.1016/j.sbspro.2014.07.052
[2] Claret, Anna, Guerrero, Luis, Aguirre, Enaitz, Rincón, Laura, Hernández, Mª Dolores, Martínez, Inmaculada, . . . Rodríguez-Rodríguez, Carmen. (2012). Consumer preferences for sea fish using conjoint analysis: Exploratory study of the importance of country of origin, obtaining method, storage conditions and purchasing price. *Food Quality and Preference, 26*(2), 259-266. doi: 10.1016/j.foodqual.2012.05.006
[3] Crumbly, Jack, & Carter, Lemuria. (2015). Social Media and Humanitarian Logistics: The Impact of Task-technology Fit on New Service Development. *Procedia Engineering, 107*, 412-416. doi: 10.1016/j.proeng.2015.06.099
[4] Fitzsimmons, J. A., & Fitzsimmons, M. J. . (2006). *Service Management Operation, Strategy, Information Technology*. New York: McGraw-Hill.
[5] Gounaris, Spiros, Chatzipanagiotou, Kalliopi, Boukis, Achilleas, & Perks, Helen. (2016). Unfolding the recipes for conflict resolution during the new service development effort. *Journal of Business Research, 69*(10), 4042-4055. doi: 10.1016/j.jbusres.2016.03.046
[6] Haeberle, Dieter, Imran, Saed, van Husen, Christian, & Droll, Carsten. (2016). A New Approach for the Development of Services for Industrial Product-Service Systems. *Procedia CIRP, 47*, 353-357. doi: 10.1016/j.procir.2016.04.079
[7] Kim, K. J., & Meiren, T. . (2010). New Service Development Process. In W. K. G. Salvendy (Ed.), *Introduction to Service Engineering* (pp. 253-267). New Jersey: John Wiley & Sons, Inc.
[8] Lebeau, Philippe, Macharis, Cathy, & Van Mierlo, Joeri. (2016). Exploring the choice of battery electric vehicles in city logistics: A conjoint-based choice analysis. *Transportation Research Part E: Logistics and Transportation Review, 91*, 245-258. doi: 10.1016/j.tre.2016.04.004
[9] Lee, Seung Hyun. (2016). How hotel managers decide to discount room rates: A conjoint analysis. *International Journal of Hospitality Management, 52*, 68-77. doi: 10.1016/j.ijhm.2015.09.014
[10] Nuraeni, Shimaditya, Arru, Arlavianyssa Pradiva, & Novani, Santi. (2015). Understanding Consumer Decision-making in Tourism Sector: Conjoint Analysis. *Procedia - Social and Behavioral Sciences, 169*, 312-317. doi: 10.1016/j.sbspro.2015.01.315
[11] Pentus, Kristian, Mehine, Tanel, & Kuusik, Andres. (2014). Considering Emotions in Product Package Design through Combining Conjoint Analysis with Psycho Physiological Measurements. *Procedia - Social and Behavioral Sciences, 148*, 280-290. doi: 10.1016/j.sbspro.2014.07.044
[12] Shen, Zhonghua, Wakita, Kazumi, Oishi, Taro, Yagi, Nobuyuki, Kurokura, Hisashi, Blasiak, Robert, & Furuya, Ken. (2015). Willingness to pay for ecosystem services of open oceans by choice-based conjoint analysis: A case study of Japanese residents. *Ocean & Coastal Management, 103*, 1-8. doi: 10.1016/j.ocecoaman.2014.10.016
[13] Surjandari, Isti. (2010). *Conjoint Analysis: Konsep dan Aplikasi*. Jakarta: Penerbit Universitas Trisakti.
[14] Widaningrum, Dyah Lestari. (2014). The Importance of Take-Out Food Packaging Attributes: Conjoint Analysis and Quality Function Deployment Approach. *EPJ Web of Conferences, 68*, 00036. doi: 10.1051/epjconf/20146800036
[15] Wu, Wann Yih, Liao, Ying Kai, & Chatwuthikrai, Anon. (2014). Applying conjoint analysis to evaluate consumer preferences toward subcompact cars. *Expert Systems with Applications, 41*(6), 2782-2792. doi: 10.1016/j.eswa.2013.10.011

[16] Yang, Yefei, Lee, Peter K. C., & Cheng, T. C. E. (2016). Continuous improvement competence, employee creativity, and new service development performance: A frontline employee perspective. *International Journal of Production Economics, 171*, 275-288. doi: 10.1016/j.ijpe.2015.08.006

[17] Yusuf Dauda, Samson, & Lee, Jongsu. (2015). Technology adoption: A conjoint analysis of consumers' preference on future online banking services. *Information Systems, 53*, 1-15. doi: 10.1016/j.is.2015.04.006

IOP Conf. Series: Earth and Environmental Science **79** (2017) 012028 doi:10.1088/1755-1315/79/1/012028

Principles of Waterfront Renovation to Decisive Spaces for Local Identity: A Study Case of Port Adelaide, South Australia

S Taher[1], F I Haqi[2]
[1] Urban and Regional Planning, University of South Australia
Adelaide, Australia
[2]Department of Architecture, State Islamic University of Sunan Ampel
Surabaya, Indonesia

E-mail address: sufia.taher05@gmail.com

Abstract. If governments have an objective to participate in a globalized economy, they should address the challenges of blighted inner city and particularly older water front areas. In doing the latter there is often a degree of consistency in place making. While the processes and principles behind urban design and planning need to have an internal logic across different scales to create quality public realms, the outcome of these interventions are frequently unresponsive to regional diversities in culture and identity. This indifference can lead to changes in the built form through urban regeneration activities that do not respect, celebrate or represent the history and traditions of the local community associated with an inner harbour or other forms of urban waterfronts. This research was undertaken to develop a set of design principles for the Inner Harbour renewal project in Port Adelaide, South Australia that would be sensitive to the historical role of the water front and in particular to incorporate the sentiments and needs of long term residents for whom the character of the area and their attachment to it shapes their sense of identity. The research established a theoretical framework for the development of design principles such as functionality, accessibility, natural environment and culture, history and memory. The findings showed that the redevelopment did not significantly disrupt residents' sense of place or sense of local identity however some more additional design attributes would nevertheless strengthen the capacity of the project to preserve local identity. This paper concludes with recommendations for preserving local identity for residents adjacent to waterfront renewal projects.

1. Introduction

In the recent past, the most noticeable feature of city revitalization is waterfront redevelopment [3]. Cities have grasped the prospect to generate new images of harbour by developing highly visible waterfront sites that have become vacant as a result of changes in traditional port and industrial functions [5]. As a result of the rise of globalization, the flow in waterfront redevelopment has occurred within distinctive political and economic environments. Cities are now faced with the challenge of having to compete in the global market or the risk of falling in-to decline. In this circumstance, waterfronts offer a priceless opportunity to market cities on the global stage. Thus, the community aspirations for public access and open space are often contrasted against political ambitions seeking to utilize the marketing value of waterfront with flagship-projects, while commercial interests seek to exploit maximum financial gain [13]. Thus, the improvement of waterfront contributes clearly to the value of urban life [10].

Renewal of urban waterfronts is unavoidable and vital [19]. Conversely, the successful economic waterfront generation may not be successful culturally or socially [8]. Port Adelaide is characterized with such specific qualities, background and maturity. Port Adelaide waterfront indicates a drastic revolution in the discursive identity, the economic function and the social relations through the redevelopment of it as a future -oriented waterfront landscape [13]. The redevelopment of Port Adelaide waterfront is located at 14 km north-west of the city of Adelaide and is expecting to accommodate over 2000 new up-market dwellings to supply over 4, -500 new residents. This paper presents an Australian perspective of this inclusive trend from a particular position and discussed local identity from the perspectives of planning and design. The findings will help planners identify how the development and growth of Port Adelaide Waterfront can influence the identity of the area. Therefore, this paper has a significant potential to reduce the knowledge gap on how planning and design of waterfront redevelopment project can preserve the identity of a city and also its adjacent neighbourhood.

2. Methods

2.1. Study case

A case study approach is used due to its ability to explore a case in detail and depth [4],[11]. [20] claims that a case study is preferable when "how" and "why" questions are being asked in condition where researcher has less control in the actual living situation. The most important strength of a case study method is the use of many different sources of evidences such as participant-observation, questionnaires and interviews to allow researcher to address a broader range of behavioural issues [20]. The key components of research design involved analysis of policy and planning documents, site visits and interviews with a sample of residents.

2.2. Questionnaire survey

A questionnaire survey was conducted of 180 residents (number of properties) of both sides of 'The Minorities street', Ethelton- Port Adelaide and district (Figure 1). The site was chosen based on the nearest redevelopment area, also as the research aims to evaluate the involvements of adjacent community around the waterfront redevelopment in Port Adelaide and the impacts of redevelopments among the adjacent residential neighbourhood through resident's satisfaction in terms of behavioural response towards it. Below is the image of boundary of McLaren Wharf and Cruickshank's Corner Precincts Master Plan frame work site and the Port Adelaide Centre vision and frame work boundary which embraces a significant relation or impacts in this research study area.

The study area covers approximately 2.25 ha with 360 residents. Among those 360 residents, 180 residents were chosen randomly for the survey. In the North side of the study area is St. Vincent St. with church St and Port Mall in East side. In the West side of the area is Old Port Reach (water body) and Port Princess Dolphin cruises in the South.

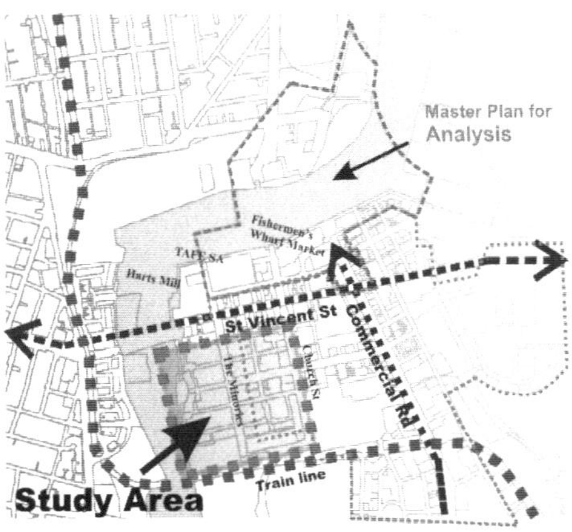

■ ■ : Boundary of McLaren Wharf and Cruickshank's Corner Precincts Master Plan frame work

■ ■ : Boundary of Port Adelaide Centre vision and frame work

■ ■ : Boundary of Study area

Figure 1. Study area and nearest redevelopment boundary.

As a quantitative approach, a survey was conducted among the residents who live near the redevelopment project (Figure. 1) of Port Adelaide and somehow are affected (positively or negatively). The survey was accomplished between medium samples of 36 adults (age 18+) from local community. At first, the survey questionnaire was piloted to 5 respondents. After the pilot study, the questionnaire was reviewed again; then researcher distributed survey questionnaire to participants; these include information sheet, research description, aim and objectives, which were put in envelopes and placed in-to randomly 180 residents' post box with a small note about how the questionnaire will be collected and the time of collection. The survey was carried out from May 2015. Among them, 36 residents responded and returned the completed questionnaire. Questionnaire is not only useful for gaining factual data such as ages, genders and living period, but also gathering data like opinions and values (Neuman 2007). The questionnaire was designed to investigate the residents' perceptions and preferences towards the nearest redevelopment and feelings about their own neighbourhood.

2.3. Data analysis

A statistical analysis process was followed to analyse the quantitative data collected from the survey questionnaire [11]. In order to work with data more carefully and successfully, raw data were coded. The questionnaire survey data were coded using Statistical Package for the Social Science (SPSS) software. The statistical techniques used in the analysis were descriptive statistics including, Chi square (X^2) to test statistical significance between gender and perceptions, frequency distribution and cross tabulations (crosstab). Descriptive statistics might present a strong impending for the understandings of the data where very minor is known about particular subjects [16]. Standard deviation and Mean of data using Likert Scale were required to identify the degree of participants' replies and the allocation of these replies [6].

Descriptive and inferential statistics were used to analyse the data from the findings. Furthermore, a thematic analysis assisted in comprehending and describing the meaning of the data. For thematic analysis, comments from the survey questionnaire were written in Microsoft excel sheet. Then, those comments were categorized under some main topics or ideas and presented by a frequency table. To focus on the overarching themes and categories, the thematic analysis was presented in the following chapter by using diagrams or specific quotations from the data.

Finally, the research method is followed by synthesis and analysis of the findings in terms of the residents' behavioural responses and satisfaction (attitudes, values, and feelings) with the analysis of Port Adelaide Redevelopment Master plan. Some conclusions are then drawn from the analysis and synthesis of both quantitative and qualitative findings in the research. Figure 2 is describing different phases of the data analysis.

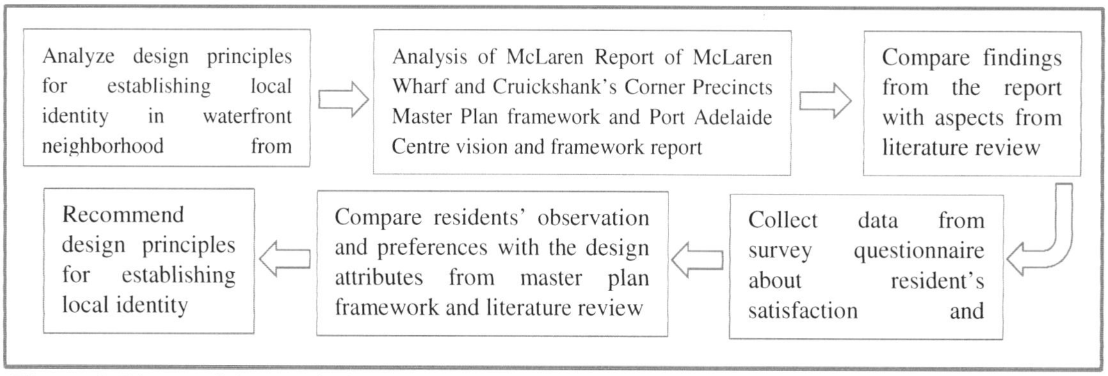

Figure 2. Phases of data analysis.

3. Findings and discussions

IOP Conf. Series: Earth and Environmental Science **79** (2017) 012028 doi:10.1088/1755-1315/79/1/012028

3.1. Synthesis of design guidelines

The strategies are intended to facilitate planners to inform the planning and design of waterfronts in order to create a local identity by providing them with design phases that assist to foster a local identity and to assist professionals to identify key characteristics to consider when analysing and assessing a waterfront development project, particularly, on how to produce a public place for local communities through design and planning. The key strategies, namely functionality, accessibility, natural environment and culture, history and memory, offer a brief logical implementation for studying the place identity for any waterfront improvement. The synthesis of design attributes is presented in Table 1 below where green is defining the attributes are provided, red is defining not provided and orange defining the attributes are partially provided.

Table 1: Synthesis of design guidelines, source: Author 2015.

Design strategies for the establishing local identity through planning and design in Waterfront Development (Port Adelaide)		
Key aspects	Design Attributes	Outcome
Functionality	Mixed land uses and different services to attract more people	
	Affordable housing for low- income groups to promote social equality	
	Diverse retail services within walking distance	
	Recreational functions on the water edge with basic services	
	Design resolution that balance the privacy of the waterfront neighborhoods and the street level activities for tourists	
	More job opportunities and volunteering openings	
	International hotel and long term consolidation and redevelopment of townhouses	
	Increase car parking facilities	
	Student services and student accommodation	
Accessibility	Well design public transit system connecting to city core and other regional area- extended tram services	
	Water based transport along the river	
	Opportunities for people accessing the water	
	Improve way finding signage to key attractions in Port Adelaide	
	Pedestrian and biking road connection to local paths with proper security	
	Connection between St Vincent St and the waterfront	
	Frequency bus and train service and alternate location for bus interchange	
	View corridors from inner city to waterfront	
Natural Environment	Find a home for the 'Falie' and 'Nelcebee' that complements the renewal of the area	
	Restoring and showcasing the vessels in an Interpretive Precinct	
	Protecting and restoring natural attributes in the waterfront	
	Creating habitats for wildlife	
	Bioengineering approaches to soften banks and landscapes	
	Landscape features for natural and educational activities	
	Managing storm water	
	Local plants and plants with cultural meanings	
Culture, history and memory	Public spaces that highlight the culture and history of the site such as, plaza, parks	
	Cultural facilities, such as theatre, museum, library to enrich the spiritual life of local residents	
	Celebrate Kaurna Culture and History through public art	
	Increase the connection between maritime museum and Cruickshank's Corner	
	Restoration of heritage vessels	
	Restoration of heritage building for future uses	
	Incorporating existing histories and themes such as, the maritime history, the indigenous history, the transport history or themes into the hard and soft built form, open spaces, landscape treatments etc.	
	A building for historic boat repairs, boat building and other maritime trades	
	Providing public place for cultural gathering or cultural events	

From the above review, it can be seen that there are some draw backs in the proposal of McLaren Wharf and Cruickshanks Corner Master Plan in terms of functionality, environment and culture, history and memory such as, in the phase of functionality-affordable housing for low-income groups to promote social equality, the design resolution that balances the privacy of the waterfront neighbourhoods and the street level activities for tourists, the phase of nature and environment creating habitats for wildlife, the bioengineering approaches to soften banks and landscapes, the landscape features for natural and educational activities, the local plants and plants with cultural meanings, and lastly in the culture, history and memory section providing public place for cultural gathering or cultural events. However, the maximum level of the design attributes been considered in the proposal. In next section of the case study, the residents' perceptions and preferences will be identified.

3.2. Residents' characteristics and perceptions

3.2.1. Respondents' age group and gender

From the 36 returned questionnaires, all of them had complete data and therefore, nothing was excluded from the statistical analysis. Table 2 illustrates respondents' age groups and gender.

Table 2: Age group and gender; Source: Author 2015

Age group	No. of males	% males	No. of females	% females	Total	% total
36- 45 yrs	3	8.3%	3	8.3%	6	16.7%
46- 55 yrs	2	5.6%	1	2.8%	3	8.3%
56- 65 yrs	5	13.9%	7	19.4%	12	33.3%
66- 75 yrs	6	16.7%	2	5.6%	8	22.2%
76 yrs and over	6	16.7%	1	2.8%	7	19.4%
Total	22	61.1%	14	38.9%		

Table 2 describes the distribution of samples of 36 people. The samples include comparatively more men than women: 61.1% men and 38.9% women. There are more men and women in the middle age groups of 56- 65 years. The number of men and women in the age distribution of youngest age group 36- 45 years were equal: 8.3% males and 8.3% females. However, there were differences in other age groups, such as 12 people in the 56- 65 years age group and of those 5 were male and 7 were female. Also, there were 6 men and 1 women of the respondents' age group of 76 years and over. All of the respondents were from the adjacent neighbourhood of the redevelopment project and had a high probability of being affected (positively or negatively) by the drastic changes.

3.3. Observations and preference toward the existing neighbourhood

According to [2], place attachments are positive bonds to physical and social surroundings that hold identity and offer other emotional settlement. Thus, it is assumed that most of the respondents have a strong place attachment with Port Adelaide and also significant character of sense of belongingness. Responses from questionnaire survey showed almost similar results and pattern in observation and preference. Differences between East side and West side of the Minories street in regards to residential satisfaction, sense of community and sense of belonging were examined to determine if the different characters of two sides of the streets (East side: Low socio economic & West side: High socio-economic area), influenced the degree to which the residents experienced each variable. The results indicated very minor differences between the two sides of the street on these variables.

Additionally, the participants responded negatively on both quantity and quality of existing situation of neighbourhoods such as, parking facilities in East side of Minories, security of parking near the shopping Centre, lack of opportunities, poor accessibility to the waterfront. Conversely, they responded positively on the approach of adjacent redevelopment project which they assume will bring more opportunities, more people, and will increase overall functionality, accessibility, and environment of the locality. There is a strong connection between the participants' observation and preferences where most preferable aspects were those who received negative responses. According to [9], observation is a key aspect in inclination. It is assumed that preference of principle aspects in this study reveals an anticipation of development of the existing character.

3.4. Observation and preference toward the redevelopment projects

The majority of the respondents were very positive and expectant about the nearest redevelopment project. Most of the respondents believed that the redevelopment project will benefit the community in future. However, some responses replicate the issue of losing identity of actual Port. Such as, approximately total 30% respondents from questionnaire survey assumed that there could be an issue of losing identity after the redevelopment. Moreover, 14% respondents considered the redevelopment is an inefficient and uneconomical use of land. Still, the majority of the respondents disagreed with these. Several respondents from both sides of the Minories Street commented that the redevelopment will bring the opportunity of using more facilities in their own area rather than traveling elsewhere. As an extension of the participant's response, development is very necessary for bringing life back in Port Adelaide.

Results from the analysis of Master Plan and the residents' observation and preferences indicate that inadequate design attributes that contribute to establishing a local identity in waterfront development were achieved. Insufficient provision of supporting facilities such as, poor accessibility, slight esteem for the local history, value and memory, and superseding the green environment lead to an inaccessible and under- utilized waterfront. Moreover, a place that creates an attachment to local communities will benefit the economic in the long term.

The results indicate that the residents' satisfaction level towards the existing neighborhood condition, with respect to functionality, accessibility and culture, history and memory aspects is significantly low. The primary reasons identified are: lack of public place for cultural gathering, lack of concern about protecting and restoring natural attributes in the waterfront, poor accessibility towards the waterfront from East side of the Minories, lack of concern about restoration of heritage building for future uses, lack of recreational functions on the water edge with basic services and most importantly lack of job opportunities not only for residents also no initiatives for welcoming more students in Port Adelaide neighborhood. Based on the key aspects, design attributes were collected to outline the design strategies. The strategies are proposed as recommendations to support planners to inform their design of waterfronts to try to create a local identity. The design attributes present valuable references and can act as a checklist for designers for assessment of their proposals.

4. Conclusion

This paper concludes with recommendations for preserving local identity for residents adjacent to waterfront renewal projects and are based on the developed design guidelines, the current status of built environment, resident's observation and preferences and the Master Plan frame work of McLaren Wharf and Cruickshanks Corner precinct (Table 3).

Table 3: Recommendations for future improvement of waterfront; Source: Author 2015

Key Aspects	Recommendations
Functionality	Integrating different retails and facilities into the greenbelt along the water edge, with café or tea house, enough public toilets and small shops
	Incorporating more recreational facilities, incorporating bikeways in to the boulevard along the water edge, adding enough seating options and spots to enjoy the riverside scene, adding playground and community gardens.
	Affordable housing for low- income groups to promote social equality
	Design resolution that balance the privacy of the waterfront neighbourhoods and the street level activities for tourists
	Mixed land uses and different services to attract more people
	More job opportunities and volunteering openings
Accessibility	Increase car parking facilities
	Pedestrian and biking road connection to local paths with proper security
	Water based transport along the river
	Frequency bus and train service and alternate location for bus interchange
	Well design public transit system connecting to city core and other regional area- extended tram services
Natural environment	Creating habitats for wildlife
	Restoring and showcasing the vessels in an Interpretive Precinct
	Bioengineering approaches to soften banks and landscapes
	Landscape features for natural and educational activities
	Local plants and plants with cultural meanings
Culture, history and memory	Providing gathering places for community activities such as, dancing, playing, celebrating religious festivals and others
	For preserving the historical style of the buildings, apply and consider the design in new construction
	Restoration of heritage vessels and heritage building for future uses
	Incorporating existing histories and themes such as, the maritime history, the indigenous history, the transport history or themes into the hard and soft built form, open spaces, landscape treatments etc.
	A building for historic boat repairs, boat building and other maritime trades
	Public spaces that highlight the culture and history of the site such as, plaza, parks
	Cultural facilities, such as theatre, museum, library to enrich the spiritual life of local residents
	Celebrate Kaurna Culture and History through public art

However, related studies for the perceptions of planning and design are not enough. The study contributes to the argument by cantering on establishing local identity at a city waterfront level through planning and design. The developed strategies might not be able to assure the social and cultural success of the waterfront development. Yet, literature studies prove that not addressing design issues listed in the strategies will direct to a loss of a sense of place attachment, local community and place identity.

In conclusion, participants responded negatively aspects on the existing facilities and attributes such as lack of functionality, poor accessibility towards the waterfront from the East side of the Minories, lack of respects on local history, value and memory, and an inaccessible and under- utilized waterfront. On the other hand, the participants responded positively in terms of the expected benefits from the redevelopment project. Recreational facilities, accessibility towards the waterfront, more job opportunities, and maintenance of the all historical buildings become the most preferable attributes that respondents considered should be provided through the waterfront redevelopment project. Several factors were considered regarding framework for the master plan; however, a balance needs to be achieved between the global and the local. Waterfront developments that keep local in mind need to offer functionality that attracts local residents from different backgrounds, support social equality and develop value of life, also be conversely reachable: produce ambience that relates inhabitants to local mores and ethnicity, celebrate local culture and history; and significantly respect the natural environment. These qualities create a sense of local identity, creating an attachment among local community and their waterfronts.

References

[1] Brown G, Brown B, and Perkins D 2004 New housing as neighborhood revitalization: place attachment and confidence among residents' *environment and behavior* 36(6) pp. 749- 771, DOI: 10.1177/0013916503254823.
[2] Brown B, Perkins, Douglas D, and Brown G 2003 Place attachment in a revitalizing neighborhood: Individual and block levels of analysis *Journal of Environment Psychology* 23(3) pp.259- 271
[3] Chang T, and Huang S 2011 Reclaiming the city: Waterfront development in Singapore *Urban Studies* 48(10) pp 2085- 2100.
[4] Creswell and John W 2003 *Research Design: Qualitative, quantitative, and mixed methods approach* 2nd ed (Sage, London)
[5] Giovinazzi O and Moretti M 2010 Port Cities and Urban Waterfront: Transformations and Opportunities *TeMA: Journal of Land Use, Mobility and Environment* 3 pp 57- 64, DOI: 10.6092/1970-9870/123
[6] Gob R, McCollin C, and Ramalhoto MF 2007 Ordinal methodology in the analysis of likert scales *Quality & Quantity* 41(5) pp 601-626.
[7] Hauge AL 2007 Identity and place: a critical comparison of three identity theories, *Architectural Science Review* 50(1) pp 44- 51, DOI: 10.3763/asre.2007.5007.
[8] Hakim A 2014 Post- development evaluation of user's satisfaction on lakefront in Dhanmondi neighborhood across its residents, Degree of Masters of Architecture, Department of Architecture, Bangladesh University of Engineering and Technology (Bangladesh)
[9] Kaplan R and Kaplan 1989 *The experience of nature: a psychological perspective*, 1st edn, Cambridge, New York: Cambridge University Press)
[10] Liu H 2013 Establishing local identity through planning and landscape design in urban waterfront development (Ontario, Canada: Thesis in Master of Landscape Architecture, University of Guelph)
[11] Matthews B and Ross L 2010 *Research methods: A practical guide for the social sciences*, (Longman, England)
[12] Neuman WL 2007*Basics of social research: qualitative and quantitative approaches* (Boston: Pearson Education, Inc).
[13] Oakley S 2007 Discourses of community in urban waterfront regeneration: the case study of the Port Adelaide waterfront redevelopment', *in refereed paper at the State of Australian Cities: National Conference* 7 pp. 360- 369, viewed 8 October 2014, < http://www.fbe.unsw.edu.au/cityfutures/SOAC/discoursesofcommunity.pdf>.
[14] Oakley S 2001 Cities as spaces and places: the politics of urban governance in Australia (South Australia Unpublished PhD thesis Department of Sociology, Flinders University,)
[15] Oakley S 2005 Working Port or Lifestyle Port? A preliminary analysis of the Port Adelaide waterfront redevelopment *Geographical Research* 43(3) pp. 319-326.
[16] Ozguner H 2011 Cultural differences in attitudes toward urban parks and green spaces, *Landscape Research* 36(5) pp 599-620
[17] Paumier Cy 2004 *Creating a vibrant city center: Urban design and regeneration principles*, (Washington, DC: Urban Land Institute)
[18] Sepe M and Pitt M 2014 The characters of place in urban design', *Urban Design International 19(3)* pp. 215- 227.
[19] Smith H and Ferrari MSG 2012 *Waterfront regeneration: Experience in city- building*, (Routledge, New York)
[20] Yin RK 2003 *Case Study Research: Design and Methods* 4th (Sage, London)

Factor Influencing Creative Industries Development in Kenjeran Surabaya

V K Siswanto, E B Santoso
Urban and Regional Planning Department
Institute Technology of Sepuluh Nopember

E-Mail: velykazu@gmail.com

Abstract. Indonesia needs to improve the competitiveness of local products to optimize its contribution to the ASEAN Economic Community (AEC) for the welfare of the population. This can be done by improving the development of creative industries. Kenjeran Sub District is one of the coastal villages in the city of Surabaya, which has a large number of poor people. The potential of the creative industries is still not visible in efforts to increase public welfare. Therefore, it is necessary to investigate the factors that influence the development of creative industries in Kenjeran. This study using in-depth interview were analysed using content analysis and Cartesian diagram to determine the important factors that affect the development of creative industries in Sub district of Kenjeran. By using five respondents from the government shows that the factors that have a high level of importance and advantages are low among other regulatory factors of raw materials, import and export, extortion, and administrative procedures.

1. Introduction

Indonesia needs to improve the competitiveness of local products to optimize its contribution to the ASEAN Economic Community (AEC) by developing creative industries. The economic contribution of the creative industry is very significant for the economy.

The economic contribution of the creative industry is very significant for the economy. In 2013, the creative economy accounted for 7.05 percent of Indonesia's GDP, or approximately 641,815.4 billion rupiah of the total GDP reached 9,109,129.4 billion. Creative industries are also able to absorb the labour force amounted to 11,872,428 votes or 10.72% of the total employment of 110 801 648 votes [3]. The economic potential of the creative was developed with micro-based small and medium enterprises (SMEs). SMEs are able to absorb labour force large enough and provide opportunities in an effort to grow and compete with companies that are more likely to use large capital (capital intensive) [2]. SMEs proved to survive and become a cog economy, especially after the economic crisis [2].

Surabaya City is the largest city in the province of East Java, which has great potential in the development of creative industries. Therefore, it is necessary to study the factors that influence the development of creative industries in Kenjeran Sub District. Research conducted by the creative industries was expect to be able to compete and be able to capture all the opportunities of the ASEAN Economic Community (AEC).

1.1. Creative Industries

The creative industry is an industry that is derive from the utilization of creativity, skill and talent of individuals to create wealth and jobs through the creation and utilization of data creation and the individual copyright. According to the government definition of Creative Industries is an industry that

relies on human creativity, by utilizing the talents and skills possessed so as to improve the standard of living through the creation (or ideas) and exploitation of IPR (intellectual property rights) [4].
Creative industry in Indonesia was divided into several aspects including Advertising, Architecture, Goods Art, Craft, Design, Fashion, Video films and photography, interactive games, Music, Art Shows, Publishing and Printing, Computer Services and software (software), television and radio, Research and Development and Culinary [1].

1.2. Creative Industries in Kenjeran Sub District

The creative industry is an industry that should be develop, especially in the city of Surabaya. There are wide varieties of creative industries in the city of Surabaya. However, in Kenjeran sub district there are only a culinary creative industries and crafts as shown in the following table.

Table 1. Creative Micro Business Industries in Kenjeran Sub District 2016

No	Name of Businesses	Type of Business	Sales Result	Start in Business
1	Hj.Coriyah	Craft Shells	Rp. 50.000.000,-	1985
2	Fakturohman	Craft Shells	Rp. 5.000.000,-	2000
3	Supiyah	Craft Shells	Rp. 5.000.000,-	1993
4	Hasan Su'ud	Craft Shells	Rp. 5.000.000,-	1993
5	Kardi	Craft Shells	Rp. 10.000.000,-	1990
6	Chozin	Crackers Seafood	Rp. 10.000.000,-	2001

Based on the table shows that there are five small micro craft shells and a small micro crackers seafood in Kenjeran with a turnover between 5 million to 50 million. There is a creative micro small business, which started in 1985. It shows that the creative industries are located in this district can survive for a long time.

2. Methods

The first phase to obtain data about the factors that influence the development of creative industries in Sub District of Kenjeran with conducting in-depth interview with five stakeholders from the government. After that, results of in-depth interview will be further analysed using content analysis.
The qualitative content analysis [5] is consists in a bundle of techniques for systematic text analysis. The main idea of the procedure of analysis is thereby, to preserve the advantages of quantitative content analysis as developed within the science communication and to transfer and further develop them to qualitative-interpretive steps of analysis [6]. Characteristics as a method of research that attempts to capture the symbolic meaning of the messages [7]. The symbolic meaning of the messages expressed from data found in books, manuscripts, or documents studied. Furthermore, the results of the content analysis was processed and illustrated in in Cartesian diagram to get cluster influential factor in the development of creative industries in the Kenjeran Sub District.

Table 2. Respondents of Indepth Interview

Respondent Code	Name	Institution
A	Benny Iriawan	Surabaya City Development Planning Board
B	Fatmawati	Department of Cooperatives Surabaya
C	Devy	Department of Community Empowerment and Family Planning
D	Raditya	Department of Community Empowerment and Family Planning
E	Fidelia	Department of Industry and Trade of Surabaya

3. Result and Discussions

The results of in-depth interviews with five respondents on the importance and advantages of the factors that influence the development of creative industries in Sub Kenjeran will described in the following table.

Table 3. Factors that influence the development of creative industries

Code	Factor	Code	Factor
A	Human Resources: 1) Skill labor 2) Managerial Capabilities 3) Number of Workers 4) Innovation of Labor 5) Skills / labor experience 6) Education workforce	E	Technology 1) The use of technology production process 2) The use of technology marketing 3) Use of Technology administration 4) The use of technology product design 5) Control Technology
B	Raw material 1) Raw material Origin 2) the availability of raw materials 3) alternative raw materials 4) Selection of raw materials 5) Infrastructure Support	F	Capital 1) Sources of Capital 2) The level of ease in obtaining capital 3) Use of Funds
		G	Legality 1) Have Permission
C	Product 1) Quality of Production 2) Price of Production 3) Competitiveness of Production	H	Business Climate and Competition 1) Raw Materials Regulations 2) Regulation of Import and Export 3) Administrative Procedures 4) Piracy 5) Illegal Payments 6) A healthy competition 7) Taxes 8) The role of industry associations and professional associations 9) Institutional which encourage the growth of creative industries 10) Relationship triple helix (government, academia and business people)
D	Marketing 1) Promotion 2) The product market (market) 3) Cooperation with other businesses		

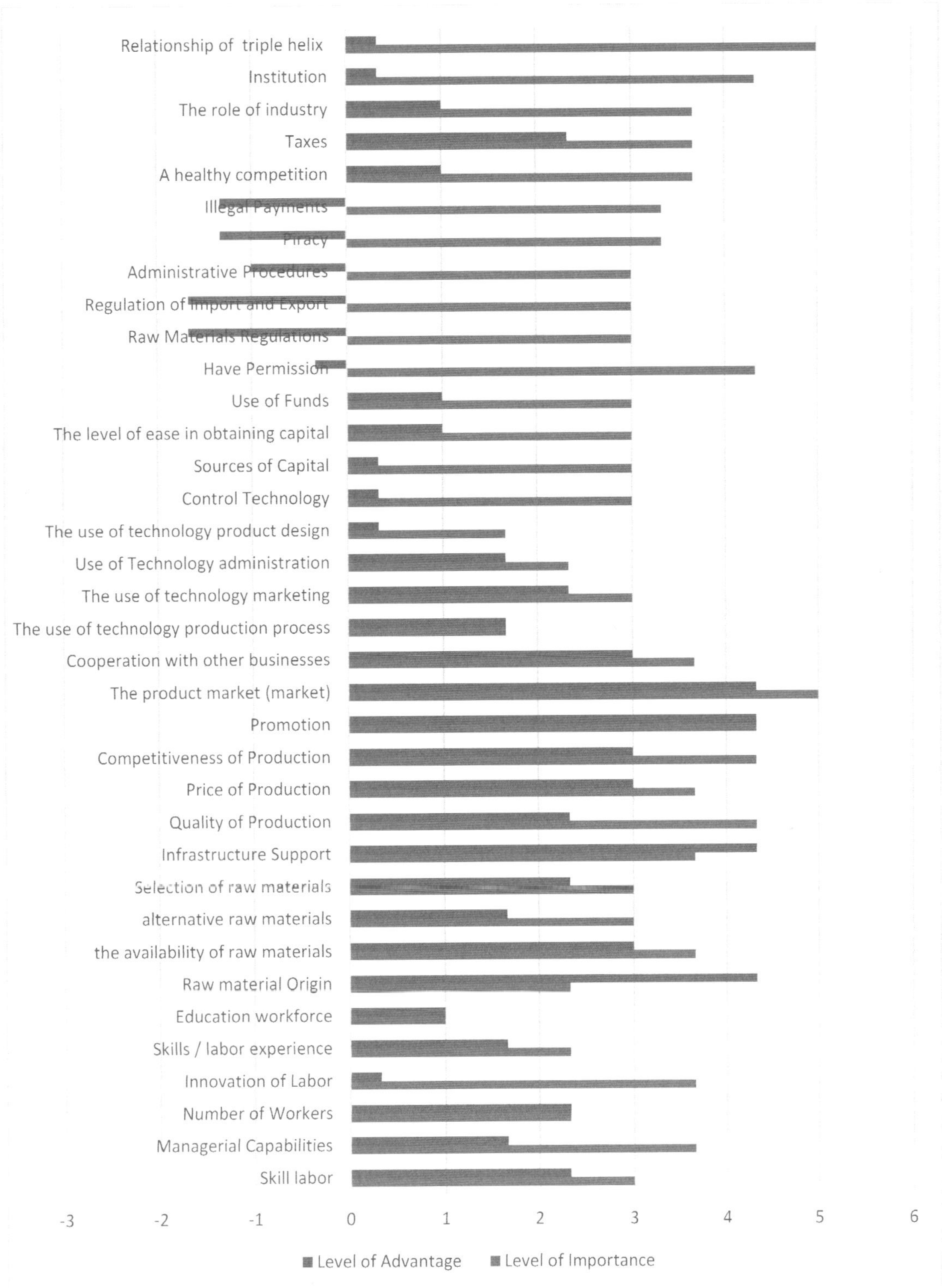

Figure 1. Summary of Interview Regarding the Importance and Advantage Factor develop Creative Industries

Based on the figure 1. shows that the important factors for develop the creative industries in Kenjeran Sub-District are (1) the innovation workforce (2) infrastructure (3) quality of production (4) the price of production (5) Competitiveness of production (6) promotion (7) product market (8) cooperation with other businesses (9) protection against piracy (10) Extortion (11) fair competition (12) Tax (13) Role of industry (14) Institutional (15) Relationships of triple helix.

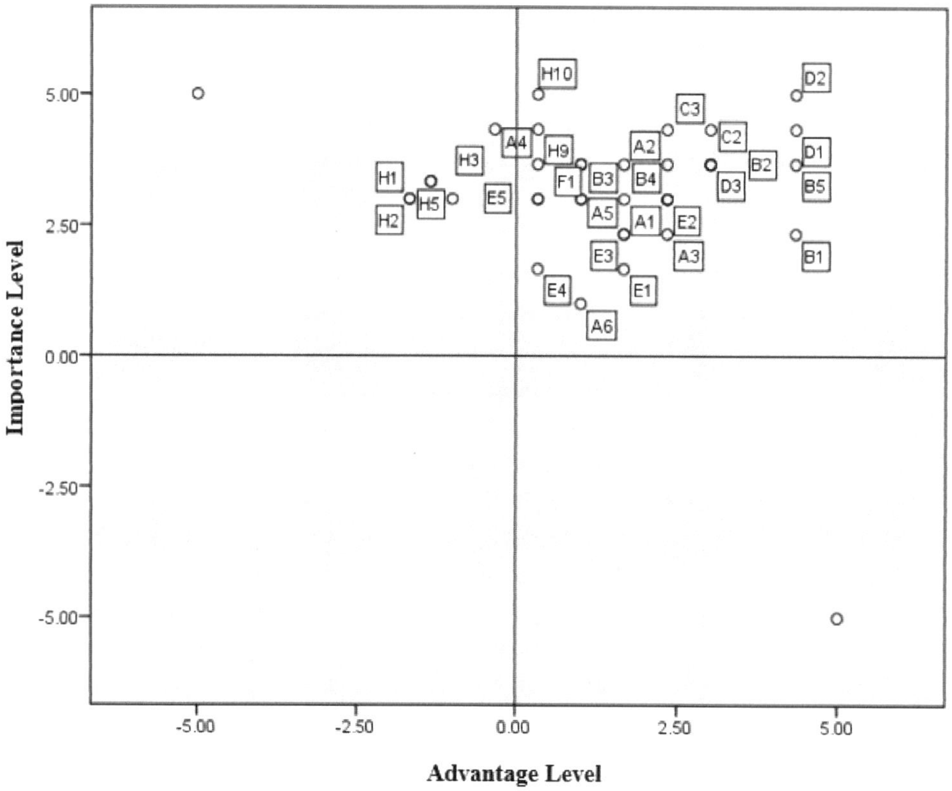

Figure 2. Results Summary of interviews using Cartesian diagram Regarding the Importance and Advantage Factor to develop the Creative Industries

Based on the graph shows that the critical factors required for the development of creative industries in the city of Surabaya are factors that have a high level of importance and advantages are low among other regulatory factors of raw materials, import and export, extortion, and administrative procedures. This occurs because the license management is still consider very difficult for the executors of creative industries. In addition, the regulations regarding the export and import of raw materials still not applied for protection products in the city of Surabaya. Therefore, this needs to be restyled.

4. Conclusion

Factors that have a high level of importance and advantages are low among other regulatory factors of raw materials, import and export, extortion, and administrative procedures. Policies to develop creative industries in the city of Surabaya is a creative human resource development, improvement of production quality advantages, the use of local raw materials, improvement of infrastructure and the development of triple helix cooperation.

Acknowledgements

Institute Technology of Sepuluh Nopember supported this research in 2016. We thank to our colleagues for full support related with preparation and finishing this paper..

References

[1] Departemen Perdagangan Republik Indonesia. 2009. Development of Indonesia's Creative Economy 2010-2014. Jakarta: Departemen Perdagangan.

[2] Sudayanto. Ragiimun and Wijayanti, Rahma Rina. 2014. Strategy of Small and Medium Enterprises Empowerment Facing ASEAN Free Market.

[3] Syarif, Mahmud et al. 2015. Developmental Analysis And The Role Of The Creative Industry To Face The Challenges Of The Asian Economic Community 2015. Seminar Nasional Inovasi dan Tren (SNIT). United Nations Development Programme

[4] Widiasari, Muchlison. 2015. Identification of Creative Industries in Blangkon Handicraft Subsector To Facing Asean Economic Community (AEC) Case Study: Serengan District, Surakarta. Muhammadiyah Surakarta University.

[5] Altheide, D.L. (1996). Qualitative media analysis. Qualitative Research Methods Vol. 38. Thousand Oaks: Sage.

[6] Ballstaedt, S.-P.; Mandl, H.; Schnotz, W. & Tergan, S.-O. (1981). Texte verstehen, Texte gestalten. München: Urban & Schwarzenberg.

[7] Krippendorff, K. 1980. Content Analysis: An Introduction to its Methodology. Beverly Hills: Sage Publication

IOP Conf. Series: Earth and Environmental Science **79** (2017) 012030 doi:10.1088/1755-1315/79/1/012030

The Evaluation of Facilitation Process in Building Community Capacity about OVOP Concept in Kenjeran Coastal Area, Surabaya

K D M E Handayeni, E B Santoso, V K Siswanto
Urban and Regional Planning Department
Institute Technology of Sepuluh Nopember

E-Mail: erli.martha@urplan.its.ac.id

Abstract. The concept of One Village One Product (OVOP) is an approach to the development potential of the area in the region to produce products that can compete in the global market, while still having unique characteristics of the area. Bulak District is one of the Kenjeran coastal area in Surabaya, Indonesia. Bulak District has had a great potential of marine products, but still contribute greatly in improving the people's welfare. Total activities of SMEs in the District of Bulak quite a lot, but the resulting product unknown to the wider community and the global marketplace. Activity of facilitation for SMEs society do to build community capacity in the implementation of the concept of OVOP. Based on the results of the evaluation assistance through Wilcoxon Signed Ranks Test result an increased understanding of the community regarding to the five subjects related OVOP concept. There are six factors to note in mentoring activities that need to be considered for the sustainability of community capacity building programs on OVOP.

1. Introduction

Concept of One Village One Product (OVOP) was first initiate in Oita, Japan. OVOP is an approach to potential development areas in the region to produce products that can compete in the global market, while still having unique characteristics of the area. The resulting product is a product that utilize local resources, both natural resources, and human resources [3]. One Village One Product (OVOP) is a government effort to increase the value-added products in the regions in order to improve the welfare of society in a container cooperatives or SMEs [1]. Program of one village one product (OVOP) which has been successfully develop in some Asian countries such as Japan and Taiwan are now proclaim as a national movement in Indonesia. Application of OVOP in Indonesia implemented through the Ministry of Industry program since 2008 to develop the potential of small and medium industries in various sectors. Implementation of this program is expected to improve the welfare of local residents by optimizing the potential for a global market.

OVOP containing the spirit of empowering the community really rely on micro, small and medium enterprises (MSMEs) and cooperatives as the spearhead. SMEs are able to absorb 99.9 percent of the workforce in Indonesia. Average economic contribution of SMEs to GDP for Indonesia is 59 percent. For the improvement of the SME sector is very important and strategic. However, as happened during the time, SMEs often hampered classic problem, namely the marketing gap. Products are already good often cannot be sold-or if it can be sold cheaply valued-because small businesses have not had sufficient access to market it. Therefore, it is necessary to develop a concept of One Village One Product to solve the problems. Hopefully, it will create economic growth in the region which is the increase of public revenue that occurred in the region, which is the entire value added (added value) that occurred [2].

Bulak District is one of the Kenjeran coastal area in Surabaya. District of Bulak have less welfare. This is evident from the number of poor people amounted to 10% of the overall population in the District of Bulak. The potential of the marine products still does not contribute significantly to the efforts to increase the welfare. Therefore, it is necessary to do the mentoring activities in order to optimize all the advantages with the concept of OVOP based PEL order to improve the welfare of residents in the District of Bulak.

2. Methods

The method used in this study is a non-parametric inference method by Wilcoxon Signed Ranks Test. This method aims to test the effectiveness of assistance activities in the introduction to the public OVOP concept SMEs in District of Bulak. Effectiveness from the absence of understanding of the concept of OVOP change in conditions before and after their mentoring activities. The use of non-parametric inference method based on a conclusion reached on the number of population samples that are not too many and not the assumption of normality on the tests performed. The samples used in this study were 30 respondents from SME representative types of food products processed in the District of Bulak, Surabaya. A number of respondents follow through mentoring activities concerning the material OVOP. Therefore, all respondents follow the pre-test and post-test to measure their understanding of the concept of OVOP. Then the pre-test and post-test will be assessed and measured into a Likert scale with a value scale of 1 means that it is not understood; scale of 2 means do not understand; scale 3 means moderate; 4 scale means understand; 5 scale means that it is understood. Through this method of testing the hypothesis, that consists of the null hypothesis (H0) and the alternative hypothesis (H1) as follows:

- H0 : understanding of the concept of OVOP in conditions before and after the assistance was the same, meaning that no effective mentoring activities affect the level of public understanding of SMEs
- H1 : understanding of the concept of OVOP in conditions before and after the assistance is different, which means that effective mentoring activities affect the level of public understanding of SMEs

Conclusion on the hypothesis testing using a confidence level of 95% or 5% alpha. The null hypothesis (H0) is rejected if the probability value <0.05 and vice versa H0 if the probability value> 0.05.
In this study also used a descriptive method to evaluate how the mentoring activities / facilities can effectively increase the capacity of MSME community. The success of this facilitation activity was also influence by the role of facilitator in presenting the material as well as design assistance activities of interest.

3. Result and Discussions

Implementation of assistance activities carried out by 6 teams of facilitators and 30 SMEs. This activity begins with a pre-test to determine the extent of SMEs to understand and apply the concept of OVOP. Furthermore, at the end of the program, 30 SMEs to follow the post-test to measure whether there is any change in the understanding of the concept of OVOP and implementation. In the five subjects, related OVOP concept tested to 30 participants, the results showed an average increase understanding of the pre-test (average score of 39.19) to post-test (average score of 63.75), as in chart 1 below, this means that the assistance activities undertaken an impact increase public understanding of OVOP. People are realizing that the product has been produced should be based locally but globally competitive.

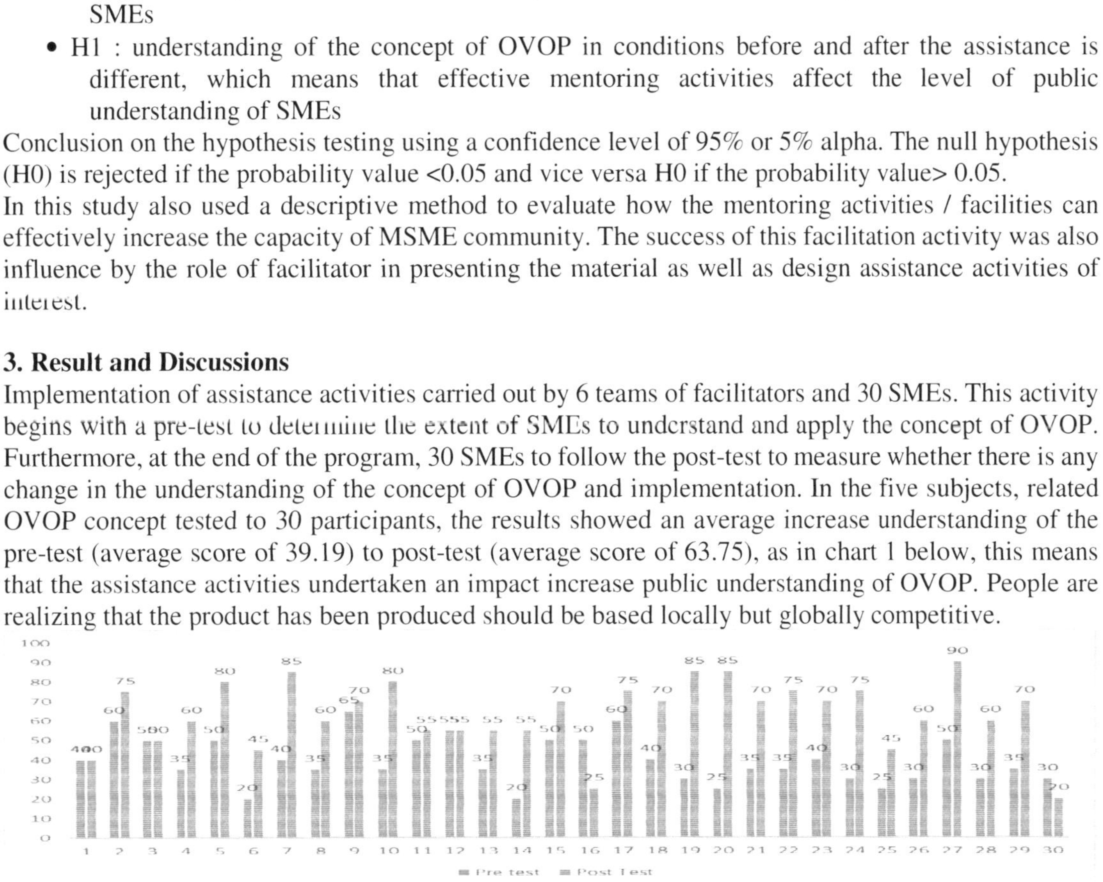

Figure 1. Comparison of Results Assessment between Pre-test and Post-test

Results of the assessment pre-test and post-test is measure into a Likert scale of a scale of 1 to 5. Further, the measurement results were teste by Wilcoxon Signed Rank method to see if mentoring is effective or not. In Figure 2 and 3 below shows the results of testing with Wilcoxon Signed Rank method. In Table 1 is explained that there is one participant with a negative value which means the ranking of post-test value is lower than the pre-test, 23 participants with a positive ranking meaning that post-test value is greater than the pre-test, and 6 participants with the post- test equivalent to the pre-test. Overall, the process of mentoring activities provides an understanding change for SMEs on OVOP.

Ranks

		N	Mean Rank	Sum of Ranks
post - pre	Negative Ranks	1[a]	6,00	6,00
	Positive Ranks	23[b]	12,78	294,00
	Ties	6[c]		
	Total	30		

a. post < pre

b. post > pre

c. post = pre

Figure 2. Rank Results between Pre-test and Post-test

Test Statistics[a]

	post - pre
Z	-4,198[b]
Asymp. Sig. (2-tailed)	,000

a. Wilcoxon Signed Ranks Test

b. Based on negative ranks.

Figure 3. Statistical Tests with Method of Wilcoxon Signed Rank

Statistical tests with methods Wilcoxon Signed Rank also showed a significant difference in the value between the pre-test to post-test, as shown in Table 2. This shows that there is a significant influence on the activities of the community capacity building assistance to SMEs in understanding the concept of OVOP. The effectiveness of assistance activities is influence by the role of the facilitator team and the mentoring process. Five factors influence the success of the mentoring activities in improving the capacity of the community about OVOP understanding, namely:

a) Knowledge SMEs on how the public perception and acceptance of the product produced. In this mentoring activity played back video that contains interviews with people outside Surabaya and related products typical Kenjeran. The result most people do not know the typical products Kenjeran. Through this video, SMEs aware that the products that has been made by the community in Kenjeran still not yet widely known. The market share of products produced only a local scale. Based on this, SMEs feel intrigued by the concept of OVOP that can offer product development strategies of local but global competitiveness. Participants are becoming increasingly interested in mentoring activities to participate in mentoring activities.

b) The facilitator team play a huge role in providing analogies and examples to improve competitiveness of local scale product. It provides a clear and easy for participants to understand OVOP. In addition, the facilitator gives examples of the practice in the form of pictures and videos that are interesting and easy to understand.

Figure 4. Comparison of Results Assessment between Pre-test and Post-test

c) The team of facilitators provide assistance in groups, so that the delivery of content OVOP easily accepted. Participants were divided into 4 groups each accompanied by one facilitator. Each group have a task to identify the potential and problems of product development efforts. As shown in the picture below.

(a) (b)

(c) (d)

Figure 5. (a) Facilitator in Group 1; (b) Facilitator in Group 2; (c) Facilitator in Group 3; (d) Facilitator in Group 4

d) It should be their reward on the activity of the group in the process of mentoring. Each group ultimately competing to give the best results to obtain the reward. It is very effective in encouraging group and individual levels of participation in mentoring process. As a result, all of the group following the mentoring process with great enthusiasm.

IOP Conf. Series: Earth and Environmental Science **79** (2017) 012030 doi:10.1088/1755-1315/79/1/012030

Figure 6. Group Rewarding

e) It is necessary to reward the achievements of participants to increase their participation in the group discussions. The mentoring process done by selecting the most active participants in following the mentoring process.

Figure 7. Participant Rewarding

f) Need any assistance activities to disseminate the results of local government publicity efforts of the community capacity building process OVOP so that these activities can be sustain.

Figure 8. Publication of Facilitation Result to the Local Government

4. Conclusion

The concept of One Village One Product (OVOP) is an approach to the development potential of the area in the region to produce products that can compete in the global market, while still having unique characteristics of the area. Bulak District is one of the coastal area in Surabaya, Indonesia. Bulak District has had a great potential of marine products, but still contribute greatly in improving the people's welfare. Total activities of SMEs in the District of Bulak quite a lot, but the resulting product unknown to the wider community and the global marketplace. Activity of facilitation for SMEs society do to build community capacity in the implementation of the concept of OVOP. Based on the results of the evaluation assistance through Wilcoxon Signed Ranks Test result an increased understanding of the community regarding to the five subjects related OVOP concept. There are six factors to note in mentoring activities that need to be considered for the sustainability of community capacity building programs on OVOP. This activity of facilitation needs to be done sustainably to strengthen the capacity of SMEs community in the implementation of OVOP.

References

[1] Ministry of Cooperatives and SMEs RI, 2013. Local Product Development through OVOP (One Village One Product). Paparan Rakornas
[2] Tarigan, R (2005); Regional Development Planning, Bumi Aksara, Jakarta
[3] Triharini, M. Larasati, D dan Susanto, R.2014. One Village One Product (OVOP) Approach to Develop Potential of Regional Handicraft Case Study: Pottery Craft in District Plered, Purwakarta Regency. Journal of ITB J.Vis Art & Des, Vol.6, No. 1, 2014, 29-42
[4] Wangke, Humphrey. 2014. Indonesia Opportunities in ASEAN Economic Community 2015. Brief Information on International Relations Vol. VI, No.10/II/P3DI/Mei/2014

IOP Conf. Series: Earth and Environmental Science **79** (2017) 012031 doi:10.1088/1755-1315/79/1/012031

Dominant Factors for Kampung Laut Bontang Kuala Development through Ecotourism Concept

Hertiari Idajati, Fonita Andastry

Urban and Regional Planning Department,
Civil Engineering and Planning Faculty,
Sepuluh Nopember Institute of Technology,
Surabaya 60111, Indonesia

email: ide_archits@yahoo.com

Abstract. Bontang is one of the cities in East Kalimantan Province which is known oil and gas industry and condensate. Besides that, Bontang also has potential tourism that can be developed in the form of Kampung Laut Bontang Kuala. This tourism area can be a leading tourism spot based on ecotourism in Bontang. Based on these fact, conducted a study to find out how Kampung Laut Bontang Kuala can be developed. This study aims to determine the dominant factors that affect Kampung Laut Bontang Kuala development through ecotourism. This research stage begins with interviewing stakeholders related to Kampung Laut Bontang Kuala development through ecotourism. The interview results were analysed using quantitative content analysis which showed the factors that affect Kampung Laut Bontang Kuala development through ecotourism in accordance with the assessment of stakeholders. The research result showed there are 9 of dominant factors for Kampung Laut Bontang Kuala Development ecotourism development.

1. Introduction

The tourism sector is one of the development sectors that is constantly being promoted by the government. This is because tourism has a very important role in the economy of the local area. The tourism sector can be categorized into one of the largest industrial groups, where 8 % of its exports of goods and services derived from tourism [1]. In Indonesia, tourism has a strategic position in economic development. The role of tourism is increasingly important to the development of trade and foreign investment. Hence the need for the development and marketing of tourism destinations in Indonesia in order to increase foreign exchange earnings in each region in Indonesia. One of them is East Kalimantan Province that has a lot of potential and natural resources that have not been fully developed. Culture and tourism is one way in protecting the environment and increase the income of local communities [2].

East Kalimantan Province is one of the tourist destinations with cultural and tourism potential that is not inferior to another tourist area in Indonesia. Almost 90% of the existing tourist area in East Kalimantan in the form of nature and the other 10% is a tourist area made in support of tourism there. Some tourist area in East Kalimantan owned attractions including Balikpapan, Bontang, Tarakan, Kutai, Kutai Barat, East Kutai, and Paser [3].

According to Department of Culture and Tourism Bontang, the potential and attractions in Bontang include Beras Basah Island, Segajah Island, Gusung Island, mangrove forest, Kutai National Park, Bontang Kuala Sea Settlement, Selangan Island, Thitik-thitik Island, Melahing Island, Selambai Island, Guntung Village, Kanaan Lake and Tomb of Toraja, Berbas Pantai area, city forest, Cafe of Singapore tourist areas, Cibodas Park, Wanatirta City Forest, Marina Badak Beach, Pupuk Kaltim Industrial Area, and Badak Company Industrial Area. Each tourist region has its own characteristics. Most of the potential and appeal offered in Bontang is a natural attraction, such as Bersa Basah Island. Besides, there is a tourist area that offers cultural tourism that is Kampung Laut Bontang Kuala. It is also in accordance with the directives contained in the documents of spatial plans Bontang [4].

Kampung Laut Bontang Kuala is located in the eastern region of Bontang city in the area around the west coast of Makassar Strait. With a population of 4,823 inhabitants with a total area of 585 hectares and the total built area of 64 hectares. Kampung Laut Bontang Kuala is one of the local and foreign tourist destination. This tourist area was originally inhabited by fishermen and is also a mangrove conservation area [5].

In terms of fulfilment infrastructure, Kampung Laut Bontang Kuala can be considered adequate and reasonably orderly. The number of visitors annually is quite a lot. Recorded in 2015 the total visitors to the region in a year as many as 2,377 visitors (Bontang City Tourism Office, 2015). Unfortunately, there is no program or clear directives relating to the development of this tourist area, whereas the value of the region selling quite well [6].

In the tourist area of Kampung Laut Bontang Kuala there are two important activities such as activity above the sea settlement itself and also the other side in areas of mangrove conservation. However, in the development of Kampung Laut Bontang Kuala area is not enough to just maintain mangrove forests due to unfavourable directly to local communities, governments, and investors, but rather how to accommodate the interests of conservation with the interests of the population, the government and investors. The appropriate development for Kampung Laut Bontang Kuala area is the development of ecotourism [7].

According to the Ministry of Tourism of the Republic of Indonesia, ecotourism is a concept of sustainable tourism development that aims to support the efforts of environmental preservation (natural and cultural) and increase public participation in the management so as to provide economic benefits to local communities and governments. Ecotourism can support nature conservation activities in the area of Kampung Laut Bontang Kuala accordance with the potential and carry capacity of the region. Ecotourism can be regarded as environmental tourism activities with an emphasis on nature conservation activities, educational aspects, and empowerment of local communities in social, cultural and economic [8]. Yet, there is no directive for the development of tourism development in Kampung Laut Bontang Kuala as ecotourism. Therefore, there is the need for research to find the direction of development of the tourist area of Kampung Laut Bontang Kuala through ecotourism.

2. Research Methods

2.1. Population and Sample

The approach used in this study is rationalistic approach [9]. The population is the generalization region, consisting of the object or subject in the quantity and specific characteristics defined by the researchers to learn and then drawn conclusions. Based on such understanding, the population in this study is divided into two: the study area itself and the stakeholders associated with the development of Kampung Laut Bontang Kuala through ecotourism, such as Bontang City Government, communities, travellers and business around that tourist area. The population used to analyse each goal for the sample, this study using purposive sampling and accidental sampling. By purposive sampling requires the sample to choose subjects based on specific criteria set forth by the researcher or on certain considerations such as population characteristics or traits are predetermined. While accidental sampling is a method of determining samples from respondents who met incidentally without a specific consideration. However, the technique of sampling was not established earlier. Researchers directly collect data from the sampling unit encountered. For example, research on public opinion about the elections by using every adult citizen as the sampling unit. Here is shown a number of stakeholders in Table 1 as respondents in this study.

Table 1. Stakeholders Mapping

No.	Stakeholder	Institution
1.	Government	1. Bontang City Department of Culture and Tourism 2. Bontang City Government (BAPPEDA)
2.	Communities	1. Communities around Kampung Laut Bontang Kuala 2. Visitors of Kampung Laut Bontang Kuala
3.	Private Sector	Business actors

Source: Analysis result, 2016

2.2. Methods of Analysis

The analysis used in analysing the factors that influence the development of Kampung Laut Kuala Bontang is a quantitative content analysis. Methods of data collection in this research is divided into two primary and secondary. Primary through interviews relevant stakeholders with Kampung Laut Bontang Kuala. While secondary seen from the analysis results of each stakeholder interviews using quantitative content analysis method.

3. Result and Discussion

3.1. Determination of Factors Affecting the Floating Sea Area Tourism Bontang Kampung Kuala Based Ecotourism

This study used quantitative content analysis. The quantitative content analysis focuses on the recording of the values of numbers or frequencies to describe different types of content are defined. This analysis is used in communication text with the real content. At this stage, the factors that influence the development of Kampung Laut Bontang Kuala through ecotourism produced from the synthesis review of the literature and the literature include studies, research, and theories related to tourism and ecotourism. The results confirmed the synthesis of the literature review on stakeholders were selected through purposive sampling technique. Here are the results of content analysis was done to determine the factors that influence the development of Kampung Laut Bontang Kuala.

1. ***Results of In-Depth Interview: Government Group (Stakeholders 1)***
   Stakeholder 1 derived from Bontang City Department of Tourism and Culture. In response ecotourism variables proposed by researchers, stakeholder 1 indicates all variables affect the development of Kampung Laut Bontang Kuala. In table 2 is shown tabulation unit of analysis based on the intended purpose in the form of an influence indication of an ecotourism variable which also serves to see the consistency of the stakeholders to the effect of ecotourism variables.

Table 2. Frequency of Analysis Unit with the Intended Purpose in Transcript 1

Variable Code	Ecotourism Variables	Affected Indication (Xn)	Unaffected Indication (Yn)
F.1	Natural	3	1
F.2	Conservation	1	-
F.3	Sustainability	2	2
F.4	Local Community Quality	1	1
F.5	Education	2	-
F.6	Local Community Empowerment	2	1
F.7	Management	1	-
F.8	Promotion	6	4
F.9	Local Economic Level	2	-
F.10	Local Community Income Level	2	-
F.11	Job Opportunities for Local Community	1	1
F.12	Other variables	1	-

Source: Analysis result, 2016

In Table 2 is known that from those ecotourism variables that has an effect to the development of Kampung Laut Bontang Kuala in accordance with the repetition unit of analysis for the intended purpose in the transcript 1 is the whole variables. But there are some variables are disclosed by only one such variable of conservation, local community quality, management, job opportunities for local communities, and local community involvement. The most influential ecotourism variable according to Stakeholder 1 is promotion variable followed by the natural variable, sustainability variable, local people's income level variable and job opportunities for local community variable.

2. ***Results of In-Depth Interview: Government Group (Stakeholders 2)***
Stakeholder 2 is from BAPPEDA Bontang. In response ecotourism variables proposed by researchers, stakeholder 2 indicates some of the variables that influence the development of Kampung Laut Bontang Kuala. Table 3 is shown the tabulation based analysis unit with the intended purpose in the form of an influence indication of an ecotourism variable which also serves to see the consistency of the stakeholders to the effect of ecotourism variables.

Table 3. Frequency of Analysis Unit with the Intended Purpose in Transcript 2

Variable Code	Ecotourism Variables	Affected Indication (Xn)	Unaffected Indication (Yn)
F.1	Natural	4	1
F.2	Conservation	3	1
F.3	Sustainability	1	1
F.4	Local Community Quality	-	2
F.5	Education	1	3
F.6	Local Community Empowerment	1	2
F.7	Management	1	1
F.8	Promotion	1	1
F.9	Local Economic Level	1	1
F.10	Local Community Income Level	2	2
F.11	Job Opportunities for Local Community	-	2
F.12	Other variable	1	-

Source: Analysis result, 2016

Based on Table 3, it is known that from those ecotourism variables that has affect to the development of Kampung Laut Bontang Kuala in accordance with the repetition unit of analysis for the intended purpose in the transcript 2 are nature variable and conservation variable. There are several variables are disclosed only one such variable of sustainability, education, local community empowerment, management, promotion, local economic level, local community income level, and local community involvement—other variable. While local community quality variable and job opportunities for local community variable indicated no effect. The most influential ecotourism variable by Stakeholder 2 is variable nature and variable conservation.

3. ***Results of In-Depth Interview: Civil Society (Stakeholders 3)***

Stakeholder 3 derived from civil society around Kampung Laut Bontang Kuala as local communities. In response ecotourism variables proposed by researchers, stakeholder 3 indicates several variables affect the development of Kampung Laut Bontang Kuala. Here below (table 4) is a tabulation unit of analysis based on tabulation based analysis unit with the intended purpose in the form of an influence indication of an ecotourism variable which also serves to see the consistency of the stakeholders to the effect of ecotourism variables.

Table 4. Frequency of Analysis Unit with the Intended Purpose in Transcript 3

Variable Code	Ecotourism Variables	Affected Indication (Xn)	Unaffected Indication (Yn)
F.1	Natural	4	-
F.2	Conservation	5	-
F.3	Sustainability	1	1
F.4	Local Community Quality	2	1
F.5	Education	2	1
F.6	Local Community Empowerment	1	1
F.7	Management	2	1
F.8	Promotion	1	1
F.9	Local Economic Level	1	1
F.10	Local Community Income Level	1	1
F.11	Job Opportunities for Local Community	1	1
F.12	Other variable	-	-

Source: Analysis result, 2016

Based on the table above, known from that ecotourism variable that has affect to the development of Kampung Laut Bontang Kuala in accordance with the repetition unit of analysis for the intended purpose in the transcript 3 is the conservation variable that that has repetition 5 times, followed by the natural variables that has 4 times. The rest several variables such as sustainability, local economic level, local community income level revealed only one. While variable of job opportunities for local community indicated no effect. The most influential variable according to Stakeholders 3 is variable of local community income level.

4. ***Results of In-Depth Interview: Civil Society (Stakeholders 4)***

Stakeholder 4 is from civil society as visitor in Kampung Laut Bontang Kuala. In response ecotourism variables proposed by researchers, stakeholder 4 indicates several variables affect the development of Kampung Laut Bontang Kuala. Here below (table 5) is a tabulation unit of analysis based on tabulation based analysis unit with the intended purpose in the form of an influence indication of an ecotourism variable which also serves to see the consistency of the stakeholders to the effect of ecotourism variables.

Table 5. Frequency of Analysis Unit with the Intended Purpose in Transcript 4

Variable Code	Ecotourism Variable	Affected Indication (Xn)	Unaffected Indication (Yn)
F.1	Natural	4	-
F.2	Conservation	3	1
F.3	Sustainability	1	1
F.4	Local Community Quality	2	-
F.5	Education	-	2
F.6	Local Community Empowerment	1	1
F.7	Management	1	1
F.8	Promotion	1	1
F.9	Local Economic Level	1	1
F.10	Local Community Income Level	1	1
F.11	Job Opportunities for Local Community	-	1
F.12	Other variable	-	-

Source: Analysis result, 2016

Based on the table above, known from that ecotourism variable that has affect to the development of Kampung Laut Bontang Kuala in accordance with the repetition unit of analysis for the intended purpose in the transcript 4 is natural variable that goes into 4 times, followed by variable of local community quality and conservation. The rest several variables such as sustainability, local economic level, local community income level revealed only one. While for variable of education, job opportunities for local community indicated no effect. The most influential ecotourism variable according to Stakeholders 4 is variable of local community.

5. ***Results of In-Depth Interview: Private Sector (Stakeholders 5)***
Stakeholder 5 derived from the private sector as business actor who runs their business around Kampung Laut Bontang Kuala. In response ecotourism variables proposed by researchers, stakeholder 5 indicates several variables affect the development of Kampung Laut Bontang Kuala. Here below (table 6) is a tabulation unit of analysis based on tabulation based analysis unit with the intended purpose in the form of an influence indication of an ecotourism variable which also serves to see the consistency of the stakeholders to the effect of ecotourism variables.

Table 6. Frequency of Analysis Unit with the Intended Purpose in Transcript 5

Variable Code	Ecotourism Variable	Affected Indication (Xn)	Unaffected Indication (Yn)
F.1	Natural	2	1
F.2	Conservation	2	2
F.3	Sustainability	1	1
F.4	Local Community Quality	-	2
F.5	Education	1	3
F.6	Local Community Empowerment	1	1
F.7	Management	2	2
F.8	Promotion	2	1
F.9	Local Economic Level	1	1
F.10	Local Community Income Level	3	1

Variable Code	Ecotourism Variable	Affected Indication (Xn)	Unaffected Indication (Yn)
F.11	Job Opportunities for Local Community	-	2
F.12	Other variables	-	-

Source: Analysis result, 2016

Based on the table above, it is known that from those ecotourism variables that has affect to the development of Kampung Laut Bontang Kuala in accordance with repetition unit of analysis for the same purpose at the transcript 3 is local community income level variable which has repetition 3 times, followed by the variable of nature, conservation, management, and promotion. The rest several variables such as the sustainability of ecotourism, education, local community empowerment and local economic level revealed only one. While for the variable of local community quality and job opportunities for local community indicated no effect. The most influential variable according to stakeholders 5 is variable local community income level.

6. ***Factors Affecting the Development of Kampung Laut Bontang Kuala through Ecotourism***
The final step of content analysis process is narrating. This stage will be generated the answer from research question namely the factors affecting the development of Kampung Laut Bontang Kuala through ecotourism. Those results will show the significance effect of an ecotourism variables for tourism development. In its determination, the researchers use a frequency analysis unit with the intended purpose and the same style of speech originating from the preceding step that is understanding data. Those results will illustrate the significance effect of each variable thereafter combined in order to show the influence distribution of each stakeholder. Influence variable that agreed from each stakeholder in whole or dominant will absolutely be the affected ecotourism variable and vice versa. While for the half-agreed variable will be defined by the strength of the statement from stakeholders who considered it has an affect thereafter will be eliminated if has no significant.
From all ecotourism variables that has an effect, they will be reduced in the factors of tourism development in which those factors have the same ecotourism characteristics. The sort of ecotourism variables is looked at the effect of the distribution and the number of iterations. The more agree on and repeat in a transcript is getting on the list. Here below is a matrix combination of each stakeholder.

Table 7. Determining the Dominant Affected Ecotourism Variable for Kampung Laut Bontang Kuala Development

No	Variable Code	Stakeholder				
		A.1	A.2	A.3	A.4	A.5
F.1	Natural	3	4	2	4	4
		Very Affected	Very Affected	Affected	Very Affected	Very Affected
F.2	Conservation	1	3	2	3	5
		Affected	Very Affected	Affected	Affected	Very Affected
F.3	Sustainability	2	1	1	1	2
		Affected	Very Affected	Affected	Affected	Affected
F.4	Local Community Quality	1	-	-	1	1
		Very Affected	Not Affected	Not Affected	Not so Affected	Not so Affected
F.5	Education	2	1	1	-	1
		Affected	Affected	Affected	Not so Affected	Not so Affected
F.6	Local Community Empowerment	1	1	1	1	1
		Affected	Affected	Affected	Not so Affected	Not so Affected
F.7	Management	1	1	2	2	1
		Affected	Affected	Affected	Not so Affected	Not so Affected
F.8	Promotion	6	1	2	1	1

No	Variable Code	Stakeholder				
		A.1	A.2	A.3	A.4	A.5
		Very Affected	Affected	Affected	Not so Affected	Not so Affected
F.9	Local Economic Level	2	1	1	1	1
		Very Affected	Affected	Affected	Not so Affected	Not so Affected
F.10	Local Community Income Level	2	2	3	1	1
		Very Affected	Affected	Very Affected	Not so Affected	Not so Affected
F.11	Job Opportunities for Local Community	1	-	-	-	1
		Affected	Not Affected	Not Affected	Not so Affected	Not so Affected

Notes:

Stakeholder	Institution
A1	Bontang City Department of Culture and Tourism
A2	Bontang City Goverment (BAPPEDA)
A3	Commmunities around Kampung Laut Bontang Kuala
A4	Visitors of Kampung Laut Bontang Kuala
A5	Bussiness actors

Based on table above, all variables can be concluded that the affected ecotourism factors for Kampung Laut Bontang Kuala development are: (1) Natural Factor, (2) Conservation Factor, (3) Sustainability Factor, (4) Factor of Local Community Quality, (5) Education Factors, (6) Local Community Empowerment Factor, (7) Management Factor, (8) Promotion Factor, (9) Factor of Local Economic Level, (10) Factor of Local Communities Income Level, (11) Job Opportunities in Local Communities.

4. Conclusion

This article aims to determine what factors may affect the ecotourism development in kampong laut Bontang Kuala. Conducted various stages, which began by making a list of ecotourism factors, then run the in-depth interviews to several selected stakeholders related to the development of ecotourism. Results of in-depth interviews will be analysed using content analysis which would then be translated what the purpose of each transcript of the interview. The research result showed that the affected factor to ecotourism development are: (1) Natural Factor, (2) Conservation Factor, (3) Sustainability Factor, (4) Factor of Local Community Quality, (5) Education Factors, (6) Local Community Empowerment Factor, (7) Management Factor, (8) Promotion Factor, (9) Factor of Local Economic Level, (10) Factor of Local Communities Income Level, (11) Job Opportunities in Local Communities

5. References

[1] Bontang tourism and culture office, *Bontang Tourism Data Document* ,2014
[2] Idajati, Hertiari. 2013.*Cultural And Tourism Planning As Tool For City Revitalization.* Surabaya : Urban and Regional Planning ITS, 2013.
[3] www.kaltimprov.go.id (access at Oktober , 11 2015)
[4] Bontang Ciy Profile, 2014
[5] Medium Term Regional Development Plan of Bonntang City 2011 – 2016
[6] Bontang tourismand culture office, *Bontang Tourism Data Document* ,2014
[7] www.pariwisatakaltim.com (access at Oktober, 13 2015)

[8] Idajati, Hertiari. 2015. *Final Report of Participatory Planning Approach in Optimizing Mangrove Ecotourism.* Surabaya : LPPM ITS, 2015.
[9] Imam Asy'ari, Sapari. 1993. *Sociology of City and Village* , Usaha Nasional Surabaya
[10] Bontang Statistic Agency, Monograph of Bontang Kuala Subdistrict year 2014/2015
[11] Bontang Statistic Agency, Monograph of Bontang Kuala Subdistrict year 2014/2015
[12] Bontang Statistic Agency, *North Bontang in Numbers*, 2014

Changes in Managerial Decision on Pond Management to Adapt to Climate Anomalies in the Coastal Area of Pare-Pare Gulf, District of Pinrang

M H Fattah[1], S R Busaeri[2], M Kasnir[1] and Siswanto[1]
[1]Department of Coastal Management and Marine Technology, Universitas Muslim Indonesia (UMI) Makassar, 90231, Indonesia
[2]Department of Agriculture Sosio-economic, Universitas Muslim Indonesia (UMI) Makassar, 90231, Indonesia

Phone : +62811447008
E-Mail : emhattahfattah@yahoo.com

Abstract. The climate anomaly was adapted through the adjustment of tiger shrimp stocking patterns and optimum use of locally endemic *Phronima Suppa* (PS) to suit the season. Thus, the batches period determined was adjusted to suit climate change dynamics, reducing the shrimp stocking density down to 10.000 – 15.000 per ha and applying PS so the shrimp became tolerant of any environmental stress and pathogen infection. The knowledge of adaptation to climate anomalies, a learning outcome of the field school, managed to increase the average production of tiger shrimp to 217 kg/ha with an average receipt of IDR 22.60 million/ha.

1. Introduction

Nowadays, climate anomaly phenomenon has been a global issue. Climate change occurs as a result of increasing accumulation of gas emission causing temperature rise through the Earth's atmosphere. The main indicator of climate change consists of changes in pattern and intensity of various climate parameters such as temperature, precipitation, humidity, cloud cover, and evaporation. This change causes extreme changes in rainfall and season patterns known as climate anomaly [1,2]. The temperature change happens in both local and global scale. In local scale, it affects the microclimate. The climate anomaly gives various effects locally [3,4,5]. It brings some consequences dealing with seasons timing. Thus, the seasons in Indonesia have been experiencing a shift either in commencement or in length.

Climate anomaly has such a significant impact on the existence of coastal community's resources and welfare. The abnormal climate changes lead to coastal environment degradation, pathogens outbreak, phytoplankton instability, and biota growth delay [6,5,7,8]. This greatly impacts the tiger shrimp farming success. Since any harvest failure will lead to an increase of poor population in coastal areas [1,9,4], there should be an innovation to anticipate and adapt to climate anomaly.

Since 1998, Pinrang, the district which has been determined as a shrimp farming center in Indonesia, has been experiencing harvest failures as a result of environmental stress and pathogen infection, particularly white spot syndrome virus (WSSV) and *Vibrio harvey*. Climate anomaly has contributed to the failures as well as the increase of poor population in coastal areas in Indonesia [1,9,4,10]. Based on the data from the Badan Pusat Statistik Provinsi Sulawesi Selatan (2015), shrimp ponds area in the province of South Sulawesi has a total width of 85,500 ha and employs 156,000 workers [11]. Meanwhile, it is predicted that there are about 15 million people involved, directly and indirectly, in the shrimp farming and processing business in the country. The shrimp disease attack has killed productivity of around 39,022 ha in South Sulawesi. It is also predicted that the harvest failures have caused an annual loss of USD 33.4 million in the province and more than USD 300 million in the country [12].

The climate anomaly issue needs to be anticipated to increase the tiger shrimp production in Indonesia. The anomaly effect can be adapted through environmental quality improvement and pond managerial innovation. *Tasiwalie* and *Wiringtasi* are two coastal villages in the district of Pinrang having shrimp ponds with a total width of 634.45 ha impacted by the climate anomaly [13]. A field school, a systematic informal learning method by, from, and for farmers at field level [14], has been organized here as an attempt to adapt to the climate anomaly. This school was designed and conducted based on farmers' needs to result in changes in farmers' managerial behaviors.

2. Methods

The research was conducted in *Tasiwalie* and *Wiringtasi* village in Pare-Pare gulf coast, sub district of Suppa, district of Pinrang, South Sulawesi province, from March 2014 to June 2016. The data was collected through observations, interviews, and secondary data investigation. The parameters observed were climate, water quality, plankton, pathogen and the community's income. The farming business and income of the field school participants (A) was to be compared with that of those who did not attend the field school (B).

2.1. El-Nino and La-Nina Analysis

Southern Oscillation Index (SOI) is an index value that indicates the difference in Sea Level Pressure (SLP) between Tahiti and Darwin, Australia [16]:

$$SOI = 10\left[\frac{P_{diff}-P_{diffay}}{SD\,(P_{diff})}\right] \tag{1}$$

Where : P_{diff} = the difference between Tahiti and Darwin in average monthly SLP
P_{diffav} = the average long-term P_{diff} in a given month
$SD\,(P_{diff})$ = the long-term deviation standard of P_{diff} in a given month

La Nina phenomenon is characterized by the SOI index value of more than +10 (minimum 3 months), while the El Nino phenomenon is characterized by the SOI index value of less than -10 (minimum 3 months).

2.2. Plankton

The observation on plankton parameter included density, uniformity, and domination as described as follows;

2.2.1. Density. Plankton density was calculated through the modification of Lackley Drop Microtransect Counting (LDMC) formula based on APHA (1998) using microscope [17];

$$\sum\frac{ind}{l} = \frac{T}{L}\times\frac{P}{p}\times\frac{V}{v}\times\frac{1}{W} \tag{2}$$

Where : T = number of SRC boxes = 1,000
L = number of SRC boxes in one viewed field = 1
P = number of visible plankton individuals
p = number of observed SRC boxes = 100
V = concentrate volume in sample bottle = 100 ml
v = konentrat volume in SRC = 1ml

W = iltered water volume = 100 l

2.2.2. Uniformity. The plankton uniformity value was calculated based on the Indeks Shannon [18] formula;

$$H' = -\sum \left|\frac{ni}{N}\right| \ln \left|\frac{ni}{N}\right| \quad (3)$$

$$H_{maks} = \ln S \quad (4)$$

$$E = \frac{H'}{H_{maks}} \quad (5)$$

Where : E = uniformity index of plankton
H[1] = Shannon index
ni = number of individuals of the 1st species
N = number of individuals of the whole species
S = number of the whole species

2.2.3. Domination. The domination value was calculated through the Simpson domination index [18];

$$d = (1 - c) \quad (6)$$

$$c = \frac{ni}{N} \quad (7)$$

Where : d = plankton domination index
c = Simpson index
ni = number of individuals of the 1st species
N = number of individuals of the whole species

2.3. Production

The cultivation performance was determined based on the tiger shrimp production employing the following formula;

$$P = \frac{W_t}{A} \quad (8)$$

Where : P = tiger shrimp production (kg/ha)
W_t = shrimp weight in farming plot (kg)
A = farming plot width (ha)

2.4. Income Analysis

Net income was calculated based on formula;

$$\prod = TR - TC \quad (9)$$

Where : $\prod$ = net income (Rp)
TR = total revenue (Rp)
TC = total cost (Rp)
$\prod > 0$ means profitable business

2.5. R/C-ratio Analysis

The R/C-ratio was conducted to determine feasibility level of the following formula;

$$\frac{R}{C} - ratio = \frac{TR}{TC} \quad (10)$$

Where : TR = total revenue
TC = total cost
R/C-ratio > 1 means profitable business deserving to be developed

3. Result and Discussion

The climate anomaly has impacted on water quality, plankton dynamics, shrimp production, and coastal community's welfare. These parameters are related to each other, influencing society's rationality in making a decision dealing with adaptation to the climate anomaly.

3.1. Climate Anomaly

Pinrang is a district with *Monsun* type of rain. It means that there is an obvious difference between wet season and dry season in this district. Normally, the wet season lasts from November to April and reaches its peak between December and February. Meanwhile, the dry season occurs from May to October and reaches its peak between July and September [19]. Table 1 shows the calculation result of normal rain in Suppa, Pinrang.

Table 1. Calculation of Normal Rain in Suppa, Pinrang (mm)

	Month											
Normal	Jan	Feb	Mar	Apr	May	Jun	Jul	Aug	Sep	Oct	Nov	Dec
Rainfall	230	184	274	253	150	118	102	34	81	110	238	342

Source : Result of BMKG data (2009 – 2015) processing

Based on the result of SOI value analysis, El Nino and La Nina occurred for three years, causing some climate anomaly in Pinrang (Table 2).

Table 2. Southern Oscillation Index Values in 2009 – 2015

	Month											
Year	Jan	Feb	Mar	Apr	May	Jun	Jul	Aug	Sep	Oct	Nov	Dec
2009	09.04	14.08	00.02	08.06	-7.4	-2.3	01.06	-5.0	03.09	-14.7	-6.0	-7.0
2010	-10.1	-14.5	-10.6	15.02	10.00	01.08	20.05	18.08	24.09.00	18.03	16.04	27.01.00
2011	19.09	22.03	21.04	25.01.00	02.01	00.02	10.07	02.01	11.07	07.03	13.08	23.00
2012	09.04	02.05	02.09	-7.1	-2.7	-10.4	-1.7	-5.0	02.06	02.04	03.09	-6.0
2013	-1.1	-3.6	10.05	00.03	08.04	13.09	08.01	-0.5	03.09	-1.9	09.02	00.06
2014	12.02	-1.3	-13.3	08.06	04.04	-1.5	-3.0	-11.4	-7.6	-8.0	-10.0	-5.5
2015	-7.8	00.06	-11.2	-3.8	-13.7	-12.0	-14.7	-19.8	-17.8	-20.2	-5.3	-9.1

Source : Result of BMKG Data (2008 – 2015) Processing (El Nino = Red, La Nina = Blue)

Both El Nino and La Nina influence rainfall amount and cause season anomaly. A quite long El Nino occurred in 2010 and 2015. Meanwhile, La Nina took place in 2010 and 2011. The El Nino was marked by a dry season lasting for minimum three months. The longest El Nino happened for six months in 2015. The longest La Nina occurred in 2010, marked by a six-month rainy season, four months longer than usual. This climate anomaly has been occurring in Pinrang for three years. The most extreme climate anomalies occurred in 2010 as El Nino and La Nina came up simultaneously. The first three months of the El Nino phenomenon lasted from July to the end of the year, followed by the phenomenon of La Nina. Climatic conditions in 2009, 2012 and 2014 tended to be normal although in certain months symptoms of El Nino and La Nina occurred. Symptoms of El Nino occurred in 2012 and 2014, and in 2013 those of La Nina did. In 2009, symptoms of both came up.

La Nina has increased rainfall in the coastal areas of Pinrang. The highest increase of rainfall (431 mm) occured in January 1999, while the highest increase from normal rainfall took place in October

2010, reaching the level of 342,5% [19]. This has made rainy season in the coastal areas of Pinrang one until four months longer. The longest rainy season occured in 2010. This rainy season lasted in all months of the year but August.

3.2. Water Quality

The parameters influencing dynamics in coastal areas are salinity, dissolved oxygen, temperature, ammonia, and nitrite. The measurement results of each parameter in the research location were distinguished according to the seasons (Table 3);

Table 3. Average Value of Water Quality Parameters

Parameter	Wiringtasi Village		Tasiwalie Village		Eligible Range of Water Quality according to literature
	Dry	Wet	Dry	Wet	
Salinity (ppt)	32.7	18.9	24.7	7.9	15.0 – 25.0 (Poernomo, 1979)
Dissolved oxygen (ppm)	6.14	4.86	5.61	6.61	$\geq$ 5.00 (ManikdanMintardjo, 1980)
Temperature (° C)	34.80	33.98	34.48	33.30	26.00 – 32.00 (Tiensongrusmee, 1980)
Degree of acidity (pH)	8.10	7.96	7.71	7.75	7.00 – 8.50 (Tiensongrusmee, 1980)
Ammonia (ppm)	-	0.44	-	0.39	$\leq$ 0.10 (Poernomo, 1979)
Nitrite (ppm)	-	0.09	-	0.40	$\leq$ 0.01 (Poernomo, 1979)

The main aspect to notice in managing tiger shrimp ponds is water quality parameter [22] since it highly influences shrimp health. The parameters of salinity, temperature, and ammonia in the coastal areas of Pare-Pare gulf are beyond the eligible range according to literature (Table 3). This condition has caused low-level growth, survival, molting frequency, and pathogen organism population increase. Change in water quality along this gulf coast was influenced by the climate anomaly. The water quality sampling described in Table 3 was conducted in the middle of a quite long El Nino and La Nina.

The salinity level influences the levels of dissolved oxygen and phosphate impacting the lives of waters organisms. Black tiger shrimp grow well at salinity level of <15.0 ppt with no sudden salinity change. A good salinity range for shrimp growth is 10.0 – 35.0 ppt with optimum range of 15.0 – 25.0 ppt [23]. The waters osmosis stress on optimum salinity should be in the same condition or close on body cell osmosis stress so there is no energy wasted for salt regulating in shrimp cells. That salinity in Tasiwalie village in dry season and the lowest tide period reaches the level of 42.3 ppt has inhibited black tiger shrimp growth.

Adequacy of dissolved oxygen ensures shrimp survival in ponds. Low level of oxygen leads to balance loss and death. Dissolved oxygen can be used as indicator of organic materials contamination in waters. Most of waters organisms cannot directly utilize free oxygen from the air. The average level of dissolved oxygen in Pare-Pare gulf waters, particularly in Wiringtasi village, in wet season is beyond the tolerance limit of shrimp growth. Black tiger shrimp grows normally at oxygen level of $\geq$ 5.00 ppm [6, 24].

Almost all waters organisms, including shrimp, is so sensitive to environmental temperature changes. A temperature change of 5.00 °C occurring suddenly causes stress and death in ponds [25,26]. The average temperature of waters in Wiringtasi village is 34.80°C in dry season and 33.98°C in wet season, while that of waters in Tasiwalie village is 33.30°C in rainy season and 3.48°C in dry season. The optimum temperature for black tiger shrimp growth ranges between 29.00°C and 31.00 °C [27,23]. The rise of temperature in Pare-Pare gulf is an impact of global climate change.

The average level of ammonia in the coastal areas of the two villages in low tide condition in wet season is 0.30 ppm and 0.28 ppm. In high tide condition, the average level is 0.44 ppm in Wiringtasie and 0.39 in Tasiwalie. The free ammonia content is toxic and dependent on pH, temperature, and salinity of waters. The pH factor plays the most important role in expressing ammonia's toxic nature. Rise of pH will increase free ammonia content. At critical level, ammonia influence inhibits growth, decreases fecundity, and decreases biota resistance to disease [28,29,7]. The toxic level of free ammonia usually occurs at a concentration of 0.60 – 2.00 ppm [25, 30, 24].

Excess nitrite compounds will decrease shrimp blood's ability to bind oxygen. Nitrite will react more strongly to the lymphocytes, increasing level of mortality. High concentration of ammonia and nitrite compounds inhibits excretion process of ammonia and nitrite, accumulating in shrimp body. The concentration of nitrite permitted in the pond waters should be less than 0.01 ppm [22].

3.3. Plankton

The zooplankton found either in the waters of Wiringtasi and Tasiwalie is relatively low with only nine genera as described in Table 4. In the coastal areas of Pare-Pare gulf in the two villages, nine plankton genera are available, while *nauplius,zoea,* and fish larvae are not. The *Onychocamptus, Echinochamptus, Tortarus, Apocyclops, Acartia*, and *Temora* types can be found in Wiringtasi with *Acartia* sp type as dominant plankton. The plankton in Tasiwaliehas the same type as that found in Wiringtasi. *Schmackeria* sp is the dominant type in Tasiwalie.

In general, black tiger shrimp hatcheries in district of Pinrang use *Artemia salina*, an imported feed [13]. The plankton used as natural feed for this kind of shrimp consists of Diatom (*Chaetoceros sp* and *Skeletonema sp*), Branchipoda (A*rtemia salina*), Cladocera (*Moina sp* and Daphnia *sp*) and Rotifer (*Brachionus sp*) types [31,32,33,34]. These types of plankton were not discovered in this research. The zooplankton composition in Pare-Pare gulf waters is relatively low. Plankton diversity index is influenced by climate and waters management factors. The climate anomaly and cultivation management influence the progress of plankton in type, abundance, and stability in black tiger shrimp.

Table 4. Abundance, Diversity, Uniformity, and Domination of Plankton

Sampling Location	Plankton Type	Number of Planktons (ind/L)	Diversity Index	Biological Index Uniformity Index	Domination Index
Wiringtasi Village	*Onychocamptus* sp	4	0.5363	0.7740	0.5555
	Echinocamptus sp	16	2.1452	3.0960	2.2220
	Tortarus sp	4	0.5363	0.7740	0.5555
	Apocyclops sp	20	1.7539	1.9656	0.8579
	*Acartia*sp	44	1.3194	1.9035	1.1101
	Temora sp	8	0.6365	0.9183	0.5555
Tasiwalie Village	*Onychocamptus* sp	8	1.8469	0.9491	0.1734
	Schmackeria sp	44	2.6895	2.9181	1.3055
	Acartia sp	16	1.2135	1.7507	1.3088
	Apocyclops sp	8	0.6931	1.0000	0.5000
	*Tortarus*sp	4	0.3218	0.4642	0.7221

3.4. Pathogen

The pathogen type infecting black tiger shrimp and vaname cultivated in Pare-Pare gulf coast consists of *Epistylis* sp, *Vibrio* sp, *Vibrio damsella*, *Zoothamnium* sp, and WSSV types. Table 5 describes relationship among aquaculture biota, water quality, and pathogen prevalence.

Table 5. Aquaculture Biota, Water Quality, and Pathogen Prevalence

Year	Shrimp Type	Water Quality						Pathogen
		Temperature (°C)	Salinity (ppt)	DO (ppm)	pH	Brightness (%)	Water depth (cm)	
2010	Black Tiger	33.00.00	23.00	05.00	07.00	70	-	*Epistylis* sp *Vibrio* sp *Vibrio damsella*
2011	Black Tiger	-	-	-	-	-	-	*Zoothamnium* sp
2013	Black Tiger	32.00.00	0,2569444	07.32	07.50	30	60	*Zoothamnium* sp
2015	Vaname	26.00.00	26.50.00	0,3604167	0,3819444	30	70	WSSV

Source :Hasanuddin Fish Quarantine, Makassar (BBKIHM) (2016)

Table 3 and table 5 indicate that water quality in the ponds in the research location is not optimum and tends to be at the upper threshold of tolerated limit of black tiger. This condition has triggered pathogen outbreak and infection in cultivated shrimp. Based on the result of research by Fattah *et al.* (2014a); Kou *et al.* (1998), WSSV infection has occurred in black tiger cultivated in the coast of Pare-Pare gulf [35,36]. Table 5 shows an assumption that the VSSW transmission occurs horizontally between black tiger and vaname.

The water quality change in the coast of Pare-Pare gulf is influenced by the climate anomaly (Table 1, 2, and 3). The impact of the climate change on pathogen prevalence is described in Table 6.

Table 6. Climate Anomaly and Pathogen Prevalence

Pathogen and Climate Anomaly (Year)						
2009	**2010**	**2011**	**2012**	**2013**	**2014**	**2015**
Normal	El Nino and La Nina	La Nina	Normal	Normal	Normal	El Nino
Epistylis sp	*Epistylis* sp			*Epistylis* sp		
Vibrio sp	*Vibrio* sp		*Vibrio* sp	*Vibrio* sp		
	Vibrio damsella					
Zoothamnium sp		*Zoothamnium* sp		*Zoothamnium* sp	*Zoothamnium* sp	*Zoothamnium* sp WSSV

Source : Result of BMKG (2016) and BBKIHM (2016) Data Compilation

The climate anomalies over these seven years has been followed by the pathogen prevalence in tiger shrimp and vaname. The longest La Nina in 2010 triggered the highest pathogen outbreak epidemics of three pathogen types (Table 6). The La Nina phenomenon, which began with the longest El Nino in the early years, led to explosive growth of three kinds of pathogens; *Epistylis* sp, *Vibrio* sp, and *Vibrio damsella*. The El Nino phenomenon triggered the outbreak of *Zoothamnium* sp and WSSV pathogenic types. *Zoothamnium* sp phenomenon grows in any climate condition. The symptoms of El Nino occurring in 2012 and 2014 triggered Zoothamnium sp and *Vibrio* sp, and in 2013 the symptoms of La Nina triggered *Epistylis* sp, *Vibrio* sp, and *Zoothamnium* sp. In 2009, the combination of El Nino and La Nina symptoms triggered *Epistylis* sp, *Vibrio* sp, and *Zoothamnium* sp. The findings of *Zoothamnium* sp in any climate condition indicate that pond management at the level of farmers has not been fully in accordance with the instructions of best management practice. The symptoms of La Nina in 2013 mainly affected the water quality parameters of salinity waters of the Pare-Pare Gulf in the range of 7.75 - 7.90 ppt (Table 3) and 5.7 ppt (Table 5) in wet season, triggering the *Zoothanmiums* p development.

3.5. Adaptation to Climate Anomalies

The climate anomaly impact should be anticipated by upgrading community's knowledge of the climate changes, improving environmental management, applying environmentally-friendly shrimp cultivation technology, and applying the *Phronima Suppa* endemic organism. The development of community's capacity is done through the organizing of the "Phronima Suppa" field school. The managerial decision improvement has been committed in the aspect of management target function, while the functions of planning, decision making, and observation have not made any significant changes [37,39]. In this research, the production of pond with the phronima (B) was compared with that of pond without the phronima (A) as described in Table 6.

Table 6. Survival and Production in Pond with PhronimaSuppa

Application	Pond Width (ha)	Cultivation Period (days)	Stocking Density (seeds) Shrimp	Stocking Density (seeds) Milk Fish	Shrimp Survival (%)	Production (kg) Shrimp	Production (kg) Milk Fish
Without Phronima Suppa Application (A)							
Average	1.25±0.52	112.50 ± 21.21	14,687.50 ± 6,870.94	1,962.50 ± 1,047.36	17.20 ± 7.34	50.63 ± 24.27	337,50± 180.77
With Phronima Suppa Application (B)							
Average	1.31±0.37	47.00 ± 2.12	18.555,56± 247.60	-	61,54 ± 10,36	285.44 ± 88.02	-

The implementation of the field school learning outcomes with the application of phronima suppa (B) has increased black tiger shrimp production to 285.44±88.02 kg/1.31 ha or 217.89 kg/ha for 47 days growing season, higher than the production of pond without phronima suppa (A) which is only 50,63±24,27 kg/1,25 ha or 40,50 kg/ha for 112.50 days growing season. The application of phronima suppa employed an operational standard which was different from the conventional method used by the control group (A) consisting of eight farmers. The nine farmers applying the phronima (B) posted PL-12 black tiger shrimp in the transition pond for 30 days before the stocking into the growing pond and reduced the shrimp stocking density down to 10.000 – 15.000 per ha. The posting by group B produced black tiger shrimp adaptive to environmental stress caused by the climate anomaly.

The application of phronima suppa in ponds areas infected by WSSV and *V. Harvey* managed to result in a harvest of black tiger shrimp with a survival level of 65.00 %, while the pond which did not use the phronimaonly resulted in a survival level of 10.00 % [38]. Phronima suppa can provide nutrients needed by black tiger shrimp, form immune stimulant, and improve environmental quality of aquaculture [33]. The application of phronima suppa is predicted to result in immune stimulant towards environmental stress so shrimp become more tolerant of extreme climates. The survival level of shrimp provided with phronima suppa is higher although the ponds area is not free of WSSV and *V. harvey* yet. The development of modular aquaculture method and the continuous supply of phronima suppa will increase production.

The receipt value through the application of phronima suppa (B) is higher. The result of R/C-ratio analysis also indicates that B (8.48) is higher than A (2.35) as described in Table 7.

Table 7. Cost and Receipt through the Application of Phronima Suppa

		Receipt (IDR million)		
Application	Cost (IDR million)	Shrimp	Milk Fish	Amount
Without Phronima Suppa (A)				
Average	2.78	4.92	3.07	7.99
With Phronima Suppa (B)				
Average	3.12	22.60	-	22.60

The pond applying the phronima (B) is better at survival level, production, and size of shrimp. These three aspects absolutely influence farmers' income and profit. Shrimp price varies according to size. The phronima application results in relatively bigger black tiger shrimp (39.67±0.71/kg) in relatively shorter time (47.00±2.12 days). The application has big potential to empower farmers of marginal ponds along Pare-Pare gulf coast. The R/C-ratio value of the application, which is 8.48, is considered high and means that each unit of spending will result in eight units of income so the application has big potential to increase farmers' income and welfare. The result of research by Burhan (2014) indicates that the average R/C-ratio value of ponds in Lowita, Pinrang [40], is 1.83± 0.33, lower than that of ponds without phronima suppa (2.35). The difference in value indicates a gap between farmers guidance and cultivation management in the coast of Pare-Pare gulf.

Contribution of phronima suppa to farmers' income and welfare improvement goes with ponds managing success, resulting in additional value for black tiger shrimp cultivation, particularly in marginal ponds in the coast of Pare-Pare gulf. What happens in these ponds needs to be handled in integrated way in order to normalize environmental support capacity. Continuous application of phronima suppa determines farmers' success in the cultivation in the coastal area impacted by the climate anomaly. Ridwan, one of the farmers using phronima suppa, now has a 3-ha pond and hires 12 workers. Previously, this man was just a tenant of a 1-ha pond.

Good knowledge of climate anomaly and adaptation process has managed to increase black tiger shrimp production to 217.89 kg/ha with a receipt of IDR 22.60 million/ha. The community's production and income increase significantly through the improvement of environmental management and the application of phronima suppa.

4. Conclusions

The climate anomaly has been occurring in the coast of Pare-Pare gulf for three years, impacting water quality, plankton availability, and pathogen explosion.The field school learning has resulted in changes in managerial decision and efforts to adapt to the climate anomaly through the adjustment of black tiger shrimp stocking time to October until November and the optimizing of phronima suppa application. The extreme climate anticipation has succeeded in increasing the black tiger shrimp production to 217.89 kg/ha for 47 days growing season with a receipt of IDR 22.60 million/ha in the coast of Pare-Pare gulf. The climate anomaly learning and managerial decision quality improvement need to be developed to be a part of community's adaptation to the coastal climate anomaly.

References

[1] Badjeck M, Allison E H, Halss A S and Dulvy N K 2010 *Impact of climate vulnerability are change on fishery-based livelihood* (Marine Policy) p 34
[2] IPCC 2007 *A Report of the working group of the intergovernmental panel on climate change summary for policymaker* (Geneva : IPCC) p 18
[3] Rizaldi B 2009 *Sekolah lapang iklim : Antisipasi risiko perubahan iklim* (Bogor : IPB) p 8 -10
[4] Safril A and Marjuki 2009 Desain sistem informasi cuaca dan iklim maritim untuk pemberdayaan ekonomi masyarakat wilayah pesisir *Jurnal BMKG*. Jakarta. 13
[5] Sagala S, Argo T, Asirin, Adhitama, P and Dodon 2014 Strategi adaptasi nelayan terhadap dampak perubahan lingkungan *Jurnal Penataan Ruang.ITS* Surabaya 12

[6] Boyd C E and Clay J W 1998 Shrimp aquaculture and the environment *Sci. Am.* **278** p 58–65
[7] Tookwinas S 2003 *Country Papers: Thailand (2). In* Cruz DA (ed): Aquaculture Management. APO Seminar on Aquaculture Management held in the Republic of China, 3–8 December 2003. Asian Productivity Organization, Tokyo, and Taiwan Fisheries Research Institute, Keelung, pp. 239–254, 2003.
[8] Zhao G X, Liu L P, Li Y and Hu Z X 2012 *Comparisons on water quality and production performance between integrated culture system of Litopenaeus vanammei and monoculture tank* (Jiangsu) *Agric Sci.* **40** p 202–06
[9] Kurniawan T 2009 *Bentuk bentuk respon nelayan terhadap penurunan pendapatan sebagai akibat perubahan iklim Seminar Nasional Perikanan Indonesia* (Jakarta : Sekolah Tinggi Perikanan) p 567-77
[10] Fattah M H and Busaeri S R 2003 Perubahan perilaku manajerial petani penerima bantuan saluran tambak *Jurnal Ilmiah Mutiara Perikanan dan Ilmu Kelautan* **2** p 12 – 23
[11] Badan Pusat Statistik Provinsi Sulawesi Selatan 2015 *Provinsi Sulawesi Selatan Dalam Angka* (BPS Statistik Provinsi Sulawesi Selatan) p 245
[12] Rukyani A 2003 *Strategi penanggulangan penyakit udang. Rapat Kerja dalam rangka Pengaturan Lalu-lintas Udang Hidup* (Jakarta : Direktorat Jenderal Perikanan Budidaya DKP) p 30
[13] Fattah M H and Saenong M 2012 *Profil Pengembangan Kawasan Minapolitan Percontohan Kabupaten Pinrang* (Dinas Kelautan dan Perikanan Kabupaten Pinrang dan Fakultas Perikanan dan Ilmu Kelautan Universitas Muslim Indonesia) p 65
[14] Fattah M H, Busaeri S R and Atjo S 2015 *Sekolah Lapang Phronima Suppa pada Kawasan Minapolitan Kabupaten Pinrang* p 178
[15] Fattah M H, Asbar and Busaeri S R 2015 Dynamics of endemic habitats microcrastacea phronima suppa (*Phronima* sp.) as determinants of artificial production development. *Journal of life science and Teknologies* **3** No. 1 and 2
[16] Troup A J 1965 The Southern Oscillation Index *Quarterly Journal of Royal Meteorological Society* (http:// www.longpaddock.qld.gov.au. Accessed on September 1st 2015)
[17] APHA (American Publich Health Association) 2005 *Standard Methods for the examination of water and wastewater* (Washington : Twentieth edition APHA-AWWA-WEF)
[18] Odum E P 1971 *Fundamental of ecology* 3rd (Philadelphia : WB.Saunders Company) p 574
[19] BMKG 2016 *Buku informasi perubahan iklim dan kualitas udara di Indonesia* (Jakarta : BMKG Pusat p 10-30.
[20] Manik R and Mintarjo K 1980 Kolam Ipukan *Dalam* Pedoman Pembenihan Udang Penaeid (Jakarta : Direktorat Jenderal Perikanan, Departemen Pertanian) p 117-24.
[21] Tiensongrusmee B 1980 Shrimp Culture and It's Improvement in Indonesia (Bogor: Research Institute for Inland Fisheries) p 27
[22] Poernomo A 1979 Budidaya Udang di Tambak *Dalam* Udang, Biologi, Potensi, Budidaya, Produksi, dan Udang sebagai Bahan Makanan di Indonesia Soegiarto K, Soegiarto A, and Toro V (Ed) (Jakarta.: LON-LIPI) p 77-170
[23] Poernomo A 1992 *Pemilihan Lokasi Tambak Berwawasan Lingkungan* (Jakarta : Pusat Penelitian dan Pengembangan Perikanan. Badan Penelitian dan Pengembangan Pertanian) p 40
[24] Widigdo B 2000 *Pembakuan Kriteria Eko-Biologis untuk Menentukan Potensi Alami Kawasan Pesisir untuk Budidaya Udang* (Bogor : Prosiding Pelatihan Pengelolaan Wilayah Pesisir Terpadu. PKSPL IPB) p 5
[25] Boyd C E 1990 *Water quality in ponds for aquaculture* (Alabama : Alabama Agricultural Experiment Station, Auburn University) p 482

[26] Harwati 2014 *Analisis aspek biofisik perairan untuk pengembangan budidaya tambak marjinal di Kawasan Wiringtasi, Tasiwalie, dan Lotangsalo Kabupaten Pinrang* (Makassar : Program Pascasarjana Universitas Muslim Indonesia) p 69
[27] Environment Ministry Environment Ministry 2004 Decision No.51 in 2004 about sea water quality standard (Jakarta) 56 pages.
[28] Chen D and Hanna P J 1994 Immunotection of specific vibrio bacteria attaching tissue of aquatic organism. **20** p 159 – 62
[29] Gauger E, Slomowitz R., Uhlinger K, Casey J and Gomez-Chiarri M 2006 *Vibrio harveyi* and other bacterial pathogens in cultured summer flounder, *Paralichthys dentatus. Aquaculture* **260** p 10-20
[30] Jackson C, Preston N, Thompson PJ and Burford M 2003 Nitrogen budget and effluent nitrogen components at an intensive shrimp farm *Aquaculture* **218** p 397-411
[31] Amin A and Jompa H 2006 Komposisi dan kelimpahan plankton di sepanjang aliran Sungai Maranak, Sulawesi Selatan *Dalam* T Ahmad T, Rachman Syah, Mustafa A and Priono B (eds) Kajian Keragaan dan pemanfaatan lingkungan perikanan budidaya ISBN 979-786-014-0 : 57-64
[32] Reinolds C S, Tundisi J G and Hino K 1984 Observation on a metalimnetic phytoplankton population in stably stratified tropical lake *Arch Hydrobyol* Argentina **97** p 7-17
[33] Fattah M H and Busaeri S R 2014 *Produksi mikro crustacea endemic Phronima suppa* (Phronima sp) *secara terkendali dengan pemberian zoopplankton* (Makassar : Prosiding Simposium Nasional I Kelautan dan Perikanan ISBN 976-602-71759-0-7) p 1-8
[34] Sutomo 2005 Kultur tiga jenis mikroalga *(Tetraselmis* sp., *Chlorella* sp. dan *Chaetoceros gracilis)* dan pengaruh kepadatan awal terhadap pertumbuhan *C. Gracilis* di laboratorium *Oseanologi dan Limnologi* **37** p 43-58
[35] Fattah M H, Saenong M, Asbar and Busaeri S R 2014a Production of endemic microcrustacean Phronima suppa (*Phronima* sp) to Subtitute *Artemia salina* in tiger prawn cultivation *J. Aquac. Res Development* **5** p 257
[36] Kou G H Peng, S E Chiu, Y L and Lo C F 1998 Tissue distribution of white spot syndrome virus (WSSV) in shrimp and crabs *In* Flegel TW (ed): Advance in shrimp biotechnology (Bangkok : National Center for Genetic Engineering and Biotechnology) p 268 – 71
[37] Fattah M H, Saenong M, Haryati, Asbar and Busaeri S R 2013 *Using of substrate for a micro-crustaceans endemic production Phronima suppa (*Phronima sp*) in controlled tank* (Surabaya : Proceeding of International Seminar University of Hangtuah) ISBN 978-979-3153-80-3 p 107 – 14
[38] Fattah M H and Saenong M 2008 Uji pendahuluan kultur Phronima suppa (*Phronima* sp). (Makassar : Laboratorium Lapang Akuakultur, Fakultas Perikanan dan Ilmu Kelautan, Universitas Muslim Indonesia (UMI)) 45 pages
[39] Fattah M H, Busaeri S R, Sabir T and Atjo A S 2014b Pond marginal farmers empowerment through development phronima suppa (*Phronima* sp) (Jakarta : Indonesian Aquaculture (Indoaqua) Seminar, August 26-29, 2014) p 14
[40] Burhan A 2014 *Analisis aspek sosial ekonomi dan budaya untuk pengembangan tambak di desa wiringtasi, tasiwalie, dan lotangsalo kabupaten pinrang* (Makassar : Pascasarjana UMI) 75 pages

Acknowledgements
We would like to thank the society and government of Pinrang and the Research and Public Service Directorate (DRM) – Ministry of Research, Technology and High Education for the financial support and all kinds of assistance supporting the research. Our thanks also go to the head of Meteorology, Climatology and Geophysics Office (BMKG) Makassar and Hasanuddin Fish Quarantine (BKIH) Makassar for the helpful data.

IOP Conf. Series: Earth and Environmental Science **79** (2017) 012033 doi:10.1088/1755-1315/79/1/012033

Coastal Planning for Sustainable Maritime Management

F Hakim[1], E B Santoso[1,2] and R Supriharjo
[1] Graduated Student of the Urban Development Management, Departement of Architecture, Institut Technologi Sepuluh Nopember
[2] Departement of Urban and Regional Planning, Institut Teknologi Sepuluh Nopember

Tel. : +6285241610906;
E-mail : fiha13hakim@gmail.com

Abstract. The Kendari Bay has a unique asset as a tourist attraction for the residents of the city of Kendari. The coastal area with all its potential like as a green open space, mangrove forests, the play area, is still a main destination to attract visitors. The function of Kendari Bay area as a tourist attraction makes this area as a place that has potential as a center of the economic vibrant and social interaction. Unfortunately, the arrangement of the area has not been done so that the integrated development of the region is not optimal. Therefore, it is important to promote a concept of area development as a tourist destination of coastal areas in order to improve function of the area. The concept of the coastal development area of Kendari Bay as tourist areas is formulated by the development criteria that influence to capable of attracting tourists. The criteria is formulated by the factors that play a role in the development of tourist areas, further exploration by qualitative descriptive analysis based on the information respondents. Fixation of the results of the criteria development was done with descriptive analysis assessed based on theoretically references through literature and regulations regarding the criteria for the development of tourism. To formulating the concept of tourism development used qualitative descriptive analysis technique with validation using triangulation techniques. The concept of tourism development based on the potential of the region is divided into three zones, namely area development of the core zone, direct supporting zone and indirect supporting zone. The macro spatial concept is necessary for the development of the area through the improvement of accessibility to tourist attraction, while the micro spatial concept includes improvements and additions to the activity in each zone to provide the convenience facilities for the tourists.

1. Introduction

Kendari city is geographically located in the southeast of the island of Sulawesi, which most areas of the city consist of waters of the bay and mangrove vegetation, has a potential tourist attraction and encourage social and economic activities of the urban [1]. The main landmark is the Kendari City Kendari Bay which is one of the most important tourist destinations in the city of Kendari. One of the uniqueness that stands out and makes this region different from other regions is the physical condition of the region that resembles an estuary. Kendari Bay Area is also the gateway city of Kendari from the sea. Potential Kendari bay are coastal tourism, paddle travel, culinary tourism, agro-tourism area of mangrove plants are found in the waters of the bay, fishing waters as a fishing area (fishing ground, as well as a training tool rowing).

But in its development, there are problems in the Gulf region Kendari, such as the spread of street vendors in the form of cafe that uses the shoulders as a space to sell the result is not well-organized area of Kendari Bay. The lack of supporting facilities such as bins, lighting, pedestrian ways, as well as a shade plant in this area is also one of the factors that make this region not optimal as a means of travel and recreation community [2]. The north side of the Gulf region are the location Kendari Bay Tourism Parks whose condition is less maintenance and less maintained, so that by the Southeast Sulawesi provincial government needs to do the arrangement [1].

Of exposure potential and the matter can be concluded that the Kendari Bay area has the potential to become a tourist area but still has not been developed until now. In connection with this it takes the concept of the development of tourism in the Gulf city of Kendari.

2. Methods

In formulating the concept of tourism development used qualitative descriptive analysis technique with validation using triangulation techniques. Descriptive and qualitative analysis in this study aimed to define the concept in accordance with the development of tourism in the city of Kendari then using validation using triangulation techniques. Analysis triangulation basically use three sources of data which will be used as a consideration in determining the direction of development of tourism in the city of Kendari implementation.

Table 1. Triangulation Analysis

No.	Aspects	Triangulation Analysis
1	Resources	• Expert competent • Research result • Interviews empiric
2	Goals	Looking priorities, interventions and egress from all sides
3	Conflict	Formulate together to achieve the best choice for this analysis departs from participatory techniques
4	analyzer	Questionnaire. Interview, and literature study of empirical experience elsewhere
5	Validation	Third accommodated resources into solving the problem of the best according to researchers (triangulation analysis itself)

Source : Sugiyono, 2009

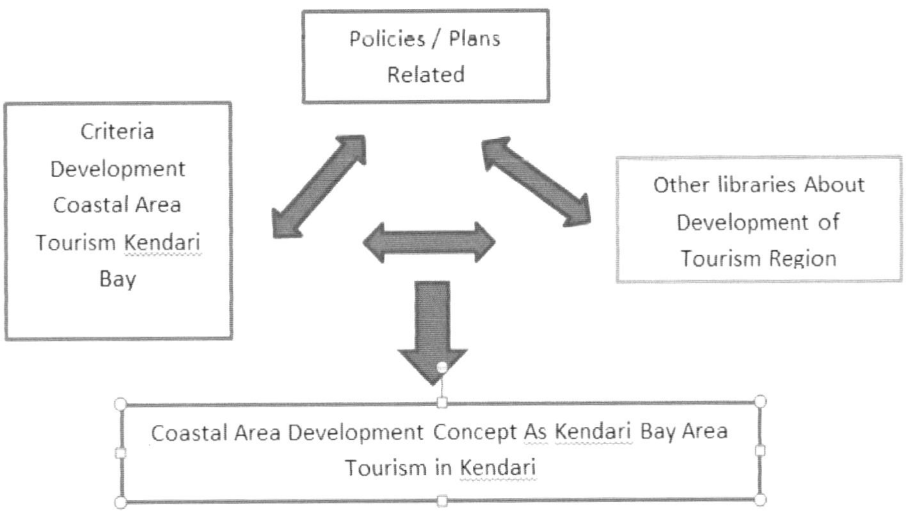

Figure 1. The Process of Tourism Development Concept Analysis

3. Result and Discussions

In the tourist area of the Gulf Kendari divided into three regions, namely coastal resorts travel, culinary tourism area and the tourist area of mangrove forests. Each region have the potential to have a tourist attraction in the city of Kendari bay.

In a tourist area, the relationship between the community and tourism activities will be very tight, because of the attractiveness of town travel is actually the people themselves, so that the tourism activity will not continue if there is no support from the community around. In development of a tourist area consists of spatial development and non-spatial. Therefore, Smith (1980) describes the spatial relationship between tourist activities with the activities of the locals into zones as follows [3]:

a. The core zone, containing the main tourist attraction that makes an area as a tourist destination;
b. Zona direct support, is central to a care facility that is already available by the community as well as tourists like shopping / perbelanjaan center, hotel / inn /, as well as other tourism industries supporting services;
c. Indirect supporting zone, an area around which is still affected indirectly tourist areas.

In the area of coastal tourism, culinary tours and mangrove forest tourism development zone can be formulated to clearly divide the functions of each zone in the region, in a system spasial. In development of each region is divided into zones, tourism, following a regional division Beaches, culinary and mangrove forests that adopt the model development zone Smith (1980), which divides the region into three regional development zones [3], namely:

1. Coast Tourism Region
   a. Top Regional Tourism Activity
   In accordance with the criteria of the model development zone Smith (1980), this area is where the concentration of owned tourist attraction tourist area [3]. In other words, this region is a zone that is a center of tourist offerings in the region. The main attraction of the tourist area of the beach is the condition existing which is the main form of travel offerings as a major attraction in this coastal region.
   b. Direct Support Region Tourism Region
   Based on the model development zone Smith (1980), this region is an area that directly supports the activities of tourist cities in the region [3]. Regions that functions to provide services for tourists during the tourist activities taking place in the region. This area is a place of concentrated trade and services in support of tourism activities, such as trade and services, accommodation services and facilities, travel, shopping, parking, and various other facilities that support tourism activities. By looking at the role of the region, it can be deduced that this region has a very important role in the smooth and comfort for travelers while in the region.
   c. Indirect Support Region Tourism Region
   That is an area that is still affected by tourist activities of coastal resorts. In the development of this area of indirect supporting the intended form of activity that is outside and around the tourist areas, either in the form of trading activities and community activities as well as in the form of tourist attraction another which can be used as a support and a tourist destination in addition to a visit to the coastal tourist city of Kendari. Surely this area are directed outside the tourist area is still affected by activities in tourist areas. Region including in the area of indirect support in the western and northern borders of the physical network of roads. For more details, zoning of the development of the tourist area of the Gulf Coast.

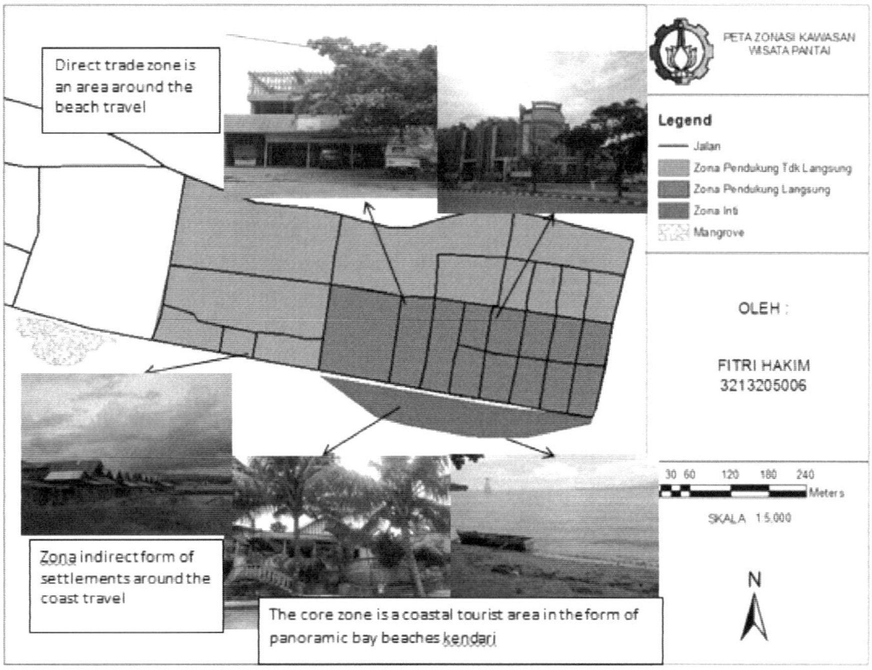

Figure 2. Zone coastal resorts

2. Culinary Tourism Region
   a. Top Regional Tourism Activity
   The main attraction of regional culinary tourism that is their tents along Kendari Bay as well as cafes that sell typical food Sulawesi. Eksisiting described conditions is a major tourist offerings form a major attraction for tourists to visit the Culinary Tourism Region Kendari.
   b. Direct Support Region Tourism Region
   Based on the model development zone Smith (1980), this region is an area that directly supports the activities of tourist cities in the region [3]. Regions that functions to provide services for tourists during the tourist activities taking place in the region. This area is a place of concentrated trade and services in support of tourism activities, such as trading and service operations in the form of shops, and a variety of other facilities that support tourism activities and the community. By looking at the role of the region, it can be deduced that this region has a very important role in the smooth and comfort for travelers while in culinary tourism Kendari. Existing conditions of trade and accommodation is in the neighborhood around the area of culinary tourism.
   c. Indirect Support Region Tourism Region
   That is an area that is still affected by the activities of regional culinary tourism travel. In the development of this area of indirect supporting the intended form of activity that is outside and around the tourist areas, either in the form of trading activities and community activities as well as in the form of tourist attraction another which can be used as a support and a tourist destination in addition to a visit to the tourist area of culinary. Surely this area are directed outside the tourist area is still affected by activities in tourist areas. Region including in the area of indirect support in the western part of the north and east.

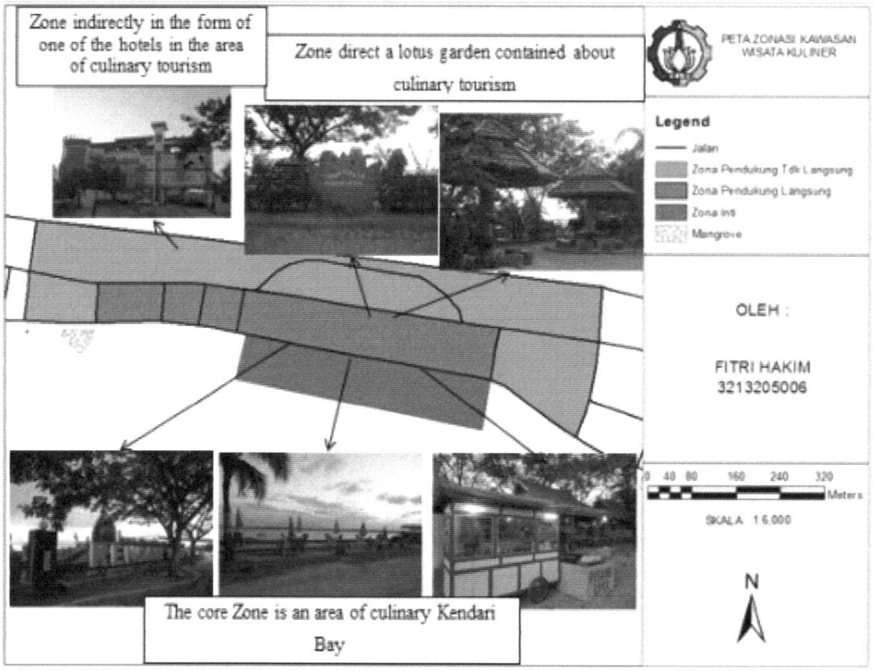

Figure 3. Zone Culinary Tourism Area

3. Tourism Area Mangrove Forests
   a. Top Regional Tourism Activity
      The main attraction of the tourist area of mangrove forests that mangrove forest area into a natural attraction for visitors both tracking travel and coastal cruising showing mangrove plant species contained in Kendari.
   b. Direct Support Region Tourism Region
      Based on the model development zone Smith (1980), this region is an area that directly supports the activities of tourist cities in the region [3]. Regions that functions to provide services for tourists during the tourist activities taking place in the area of the region is a good trade area in the form of restaurant and mini market located just outside the tourist area of mangrove forests.
      By looking at the immediate support zone criteria based on the model Smith (1980), then the support zone immediately directed to be around the main area or directly adjacent to the main regional tourist activities, it also aims to support the main attraction of mangrove forest area [3].
   c. Indirect Support Region Tourism Region
      That is an area that is still affected by the travel activities of the mangrove forest. In the development of this region do not directly support the intended form of activity that is outside and around the tourist areas, community activities or any other form of tourist attraction that can be used as a supporting and tourist destination in addition to a visit to the mangrove forest. Surely this area are directed outside the tourist area is still affected by activities in tourist areas. Region including in the area of indirect support in the north.

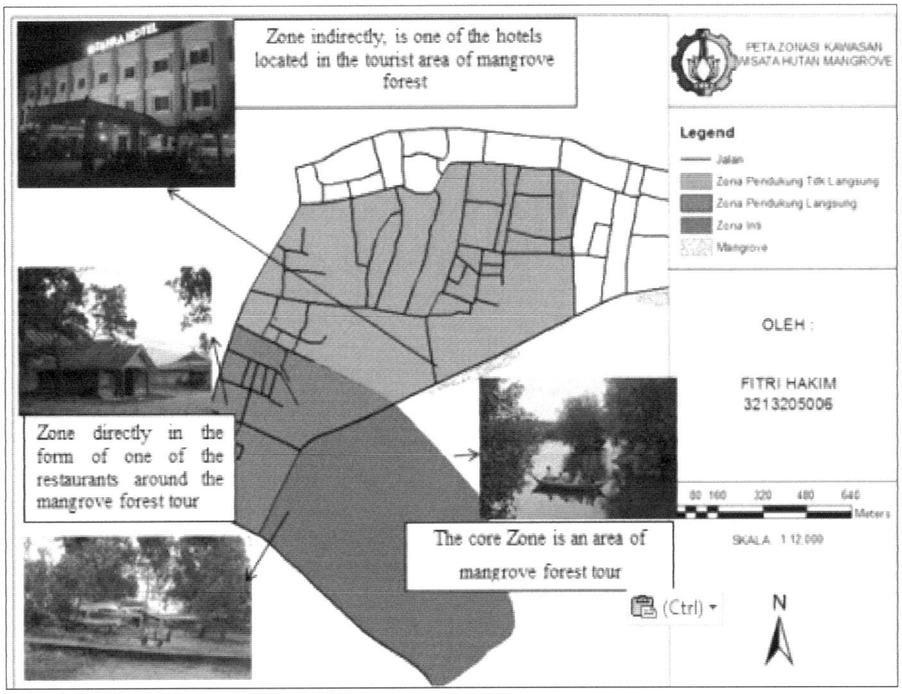

Figure 4. Zone tourist area of mangrove forest

Concept that resulted from previous analysis, will be grouped into macro and micro development concept tailored to the regional scope of application of each concept. The concept applies only to the internal one area only and refer to an object or aspect, will be classified into micro concept. While the concept does not only affect the scope of its application for a regional internal only, but also for the development of tourism areas in Kendari in general, so the concepts included in the scope of this application will be used as a macro-region concept. On the basis of these groupings, it can be concluded that the concept of a macro is a concept that is general and complete of all the tourist areas of Kendari as a tourist town, while the concept of micro is a concept that refers to the development of an object or a certain aspect in one area of tourism in the city Kendari and the nature of the concept is only an internal concept of an area.

4. Conclusions

In developing regions need to know the activity in each region so as to determine the allotment of the core zone, the zone directly or indirectly in the zone of drafting the development of tourism in the city of Kendari Bay.

References

[1] Rachim M D 2009 *Krisis Teluk Kendari* (Kendari: Universitas Haluoleo Kendari)

[2] Jaya F S 2013 *Konsep Penataan Ruang Terbuka Publik pada Kawasan Teluk Kendari Sebagai Kawasan Rekreatif* (Makassar : Fakultas Teknik Universitas Hasanuddin)

[3] Smith 1986 *Finding Lost Space : Theories Urban Design* (New York : Van Nostrand Reinhold Company)

[4] Sugiyono *Metode Penelitian Kuantitatif Kualitatif dan R & D* (Bandung : Alfabeta)

[5] Undang-Undang Republik Indonesia Nomor 10 Tahun 2009 Tentang Kepariwisataan

IOP Conf. Series: Earth and Environmental Science **79** (2017) 012034 doi:10.1088/1755-1315/79/1/012034

Model of Coastal Transit Oriented Development (TOD) Based on the Potential of Local Port and marine Tourism Port, Case Study: Fort Rotterdam Makassar and the Surrounding Areas

A B Arief[1], A Yudono[2], A Akil[3], and I Ramli[4]
[1,2,3]Architecture Department, Hasanuddin University, Makassar, Indonesia
[4]Civil Engineering Department, Hasanuddin University, Makassar, Indonesia

E-Mail: andibachtiararief@gmail.com

Abstract. The lack of social and public facilties of seven small islands around Makassar, causing the commuters to experience inefficiency in fulfilling their basic needs in the mainland of Makassar city. The purpose of this study is finding the location of coastal TOD in accordance with the principles of development model of coastal TOD. The result showed that inefficiency of time, cost and distance could be eliminated by applying vertical, united and integrated development model of coastal TOD. Using survey, interview and literature study through expert system analysis based on GIS deliniates coastal TOD.

1. Introduction

Research about the principles of transit oriented development (TOD) spatial plan for coasts based on the potency of local and marine tourism port, case study: Fort Rotterdam and the surrounding areas in Makassar. Makassar is the capital city of South Sulawesi. Makassar is situated at coordinate 119° 18' 27.79" east longitude and 5° 3' 30.81" south latitude with boundaries; in the north is Maros city in the east is Maros regency, in the south is with Gowa regency, in the west is with Makassar straits. Makassar city has several ports, some of them are ports around Fort Rotterdam (Kayu Bangkoa local port, Kayangan and Association of Boat Sport water ski marine tourism port), Paotere commerce port, Untia fishery port, and Soekarno-Hatta national port, as shown in this following figure 1 (b).

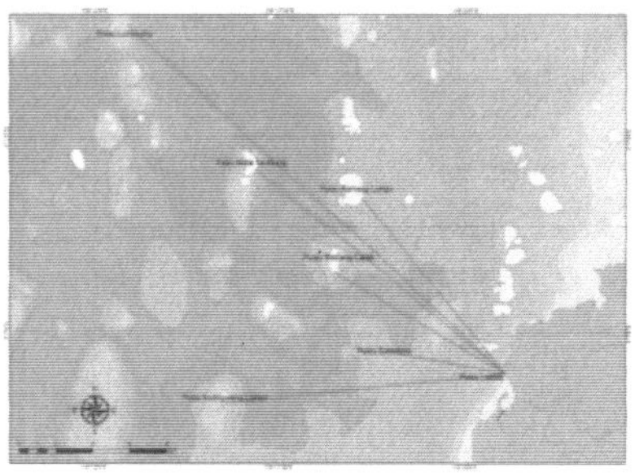

(a) (b)

Figure 1. (a) position of smal islands toward Makassar city; (b) Position of ports around Fort Rotterdam

The trip of goods and people in urban area in transit location around Fort Rotterdam will be more increased along with the increase of needs and activities of commuters. Commuters from mainland of Makassar to small islands are dominated by the purpose to have water sport and tourism or picnic such as boat sport, swimming, diving, dan snorkeling. Generally, the vehicle used from their residence to transit location are motorcycle, public transportation (*pete-pete*) and private cars. Besides, commuters from small islands (Lae-lae, Samalona, Barrang Caddi, Barrang Lompo and Kodingareng) to ports around Fort Rotterdam use speed boat or motor boat. The trip will be continued to the main land of Makassar city by using pedicab, motor boat, *pete-pete*, taxi, bus or private cars.

The purpose of commuters are quite varied such as visiting relatives, traditional market, mall and shops to shop for their daily needs whether they are physical or non phisical [1] [2]. Physical and non physical needs are about shopping for food and drinks needs,clothes, building materials, joining the education to improve their intelegency ability, receiving the health treatment, enjoying the beauty of nature, doing sport, picnic or marine tourism, and also doing the worship according to their belief [3], [4], and [5].

Generally, in travelling commuters use motorcycle and that is not massive, only a few of them use bus, non motor vehicle (pedicab) or walking, hence the number of vehicles tend to be more increased as well as the use of fuel , there was redundancy in cost congestion and traffic accident [6] [7]. In overcoming several problems of development urban and traffic on the main land then it is applied some urban policies, they are development city which oriented to TOD. It is a kind of policy strategy using intergrated concept among land use and transportation, compact, diverse use, walking in 5-10 minutes from transit location or bus stops [8]. The lack of the willingness to walk within ½ mile radius is also affected by the development pattern of shopping facilities and residential, building density, and mixed used land planning [9]

The development of urban is strongly related with accessibility of citizen and job so the development of massive public transportation is concentrated around residential, working places, transit area or stop stations of massive public transportation[15]. Success key of development of coastal TOD is the integrated land use from various activities, massive public transportation, pedestrians for doing sport, recreation and many other services safely and friendly around port or transit location [11] . TOD principles and compact city are a reference to bring closer various services of goods and service around port where the commuters transit. The service of goods and service are supposed to be able to be accessed through walking comfortably, cycling safely and better quality of the using of massive public transportation[12]. It will be beneficial for commuters both from small islands to Makassar mainland and vice versa in order to increase the efficiency of distance and travel time, energy, trip cost and also prevent traffic congestion and traffic accident [13][14]. Besides, the application of TOD principles and compact city development are hoped to reduce goods and passangers mobility and number of private motor vehicles and to make massive public transportation effective, provide safe pedestrian route and also spatial urban palnning with transit oriented [16][17] and [11].

Vehicle oriented development around transit location is conducted by putting pedestrian, high density of residential building and the variety of buildings as priority [18] . It happens also for environment development around massive public transportation, it will be friendly to pedestrian, disability people, wheelchair users, and bycycle users by providing pedestrian route, crossroads which built in with garden and trees, they are comfortable and safe to be able to access shopping places which are closer than using motor vehicle[8]. The use of land in mixed and various ways with high density whether they are residential, jobs, shopping places and easy access to walk to transit location or transportation stops, reduction of parking area will boost citizen and commuters to travel by using public transportation facility instead motor vehicle. Therefore, it can decrease dependency for land [8][19]. This transportation problem happen due to the lack of transportation management and low quality of massive public transportation [20]. another problem is massive public transportation facility is not avaliable every time it is needed, it is not regularly avaliable and it can not arrive and depart on time[21].

This Model of city development is a kind of compact and sustainable city [22]. Compact city with integrated land planning and transportation supported by policy and politic related with territory development to have high quality and suistainable city, the increase of diverse function and high density of building and located in transportation center by prioritizing the mobility of non motor vehicle to reduce dependency on land use and cars [17]. Hence, it is important to improve the transportation management of land and sea transportation management to be better in integrated way and coordinated especially in the port area and its surroundings as transit location without traffic congestion and trafffic accident. The improvement here means certainty of departure and arrival schedule for motor boat, comfortable and safe pedestrian route, open public space and enough green open space, efficiency of transportation cost. Open public space and enough green open space from those two ports are utilized together to make it efficient[20] .

The mix of ddiverse land use planning happen and it complete each other among residential or housing with non-housing in the same block or closed block, society residential for low income can accesss the location of fresh food seller, less than ½ mile and it is completed with parking area [8]. Based on theory and existing condition of Kayu Bangkoa local port and coastal tourism port Fort Rotterdam, the writer assumed; spatial planning of coastal TOD based on port is very urgent to be developed to fasten the improvement of prosperity of citizen and commuter from small islands by applying principles; efficiency, accessibility, density, distance, and diversity.

Those negative effects can be reduced by encouraging growth of areas smartly, applying integrated spatial planning development model among land function, diverse land uses, high density of building, connectivity of pedestrian and people with disability and massive public transportation in friendly way around transit location such as port [23][25][24][11]. To improve spatial planning of coastal TOD to be better, to be able to eliminate redundance of cost, time and distance, it requires integrated overcoming and management of public infrastructure development as well as corporate patnership among few parties; goverment, private sectors and bussiness world, it is hoped it may impact on commuters to be able to create efficiency toward redundance of cost, time and distance, hence the price of goods and services will be under controlled [23] and [26]. This problem happen due to the lack of transportation management and low quality of masssive public transportation[20]. The other is inavaliability of masssive public transportation when it is needed by the commuters, fixed schedule and also it doesnt depart and arrrive on time. The objective of this research is to identify land function, the potency of physical problem, the pattern and purpose of commuters trip from and to small islands around Fort Rotterdam for coastal TOD development.

2. Data and Methods

2.1 Modelling methodology

The research location is at western of Makassar city at Losari beach, its exact location is around Fort Rotterdam, Makassar as shown in this following figure 3 within radius about ½ mile from transit location includes several local ports: Kayu Bagkoa, Kayangan and the Association of boat sport dan water ski - Persatuan Olah Raga Perahu dan Ski Air (POPSA) marine tourism port. Kayu Bangkoa local port is dominated by commuters of households from small islands to shop for their basic needs in the mainland of Makassar. Kayangan or Benteng Pannyua marine tourism port is dominated by commuters ranging from teenagers and juvenille who will head to islands like Lae-lae, Kayangan and another small islands to visit their relatives or recreational activities. POPSA marine tourism port is used by the commuters from low end to middle end to have water sport such as swimming, water ski, diving, and snorkeling.

a

b

Figure 2. (a) dan (b) The acivity of goods and passanger loading around Fort Rotterdam

The data obtained in this research is pattern of commuter trip from and to small islands; Lae-lae, Samalona, Barang Caddi, Barrang Lompo and Kodingareng. Big redundancy experienced by commuters can be decreased if all the facilities of public services of shopping, sports and recreational put closer to ports around fort Rotterdam. Existing condition, building diversity, and building density around Fort Rotterdam - Figure 3 (a), (b), and (c).

(a)
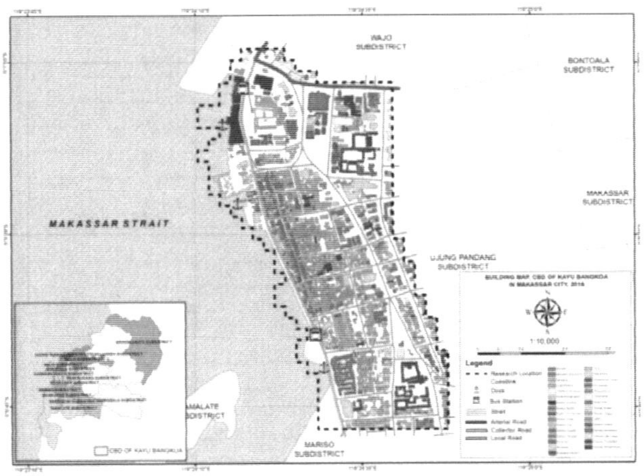

(b)
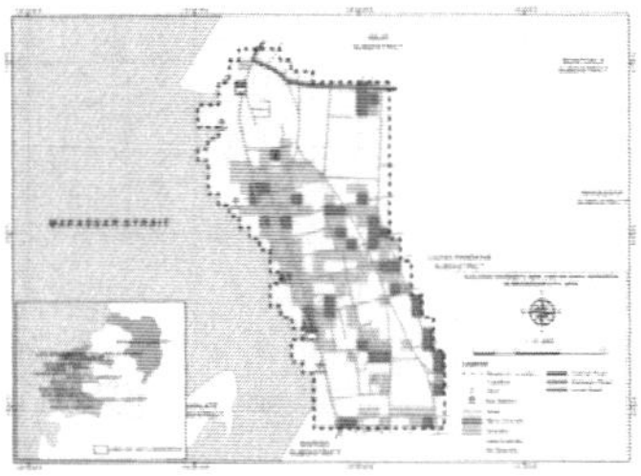

(c)
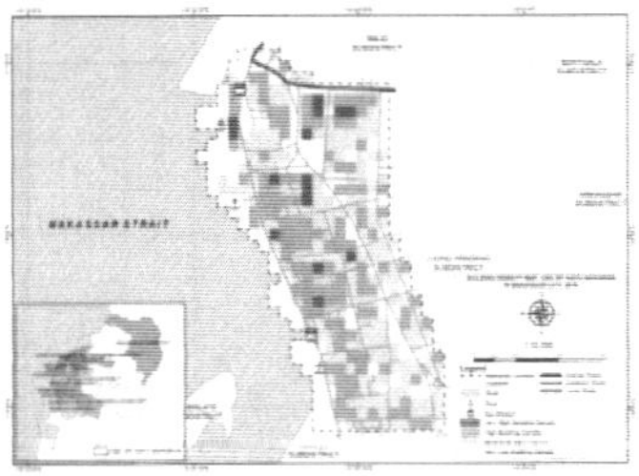

Figure 3. (a) Existing Building Function Map around Fort Rotterdam, (b) Building Diversity Map, (c) Building Density Map. Source: Makassar City Spatial Planning, observation, processed, 2016

2.2 Methods

The data is collected through survey method and direct and random interview about commuters trip pattern from small islands to mainland of Makassar and vice versa. Data of function building is conducted through literature study and field survey within walking radius around ½ mile from ports or quay as transit location for sea transportation. Analysis technique used is expert system in figure 5 and set up the certainty factor (CF) in figure 6 for judgement based on geographic information system (GIS) to delineate areas in accordance to development principles of coastal transit oriented development (TOD) vertically and integrated related with efficiency, density, diversity and distance.

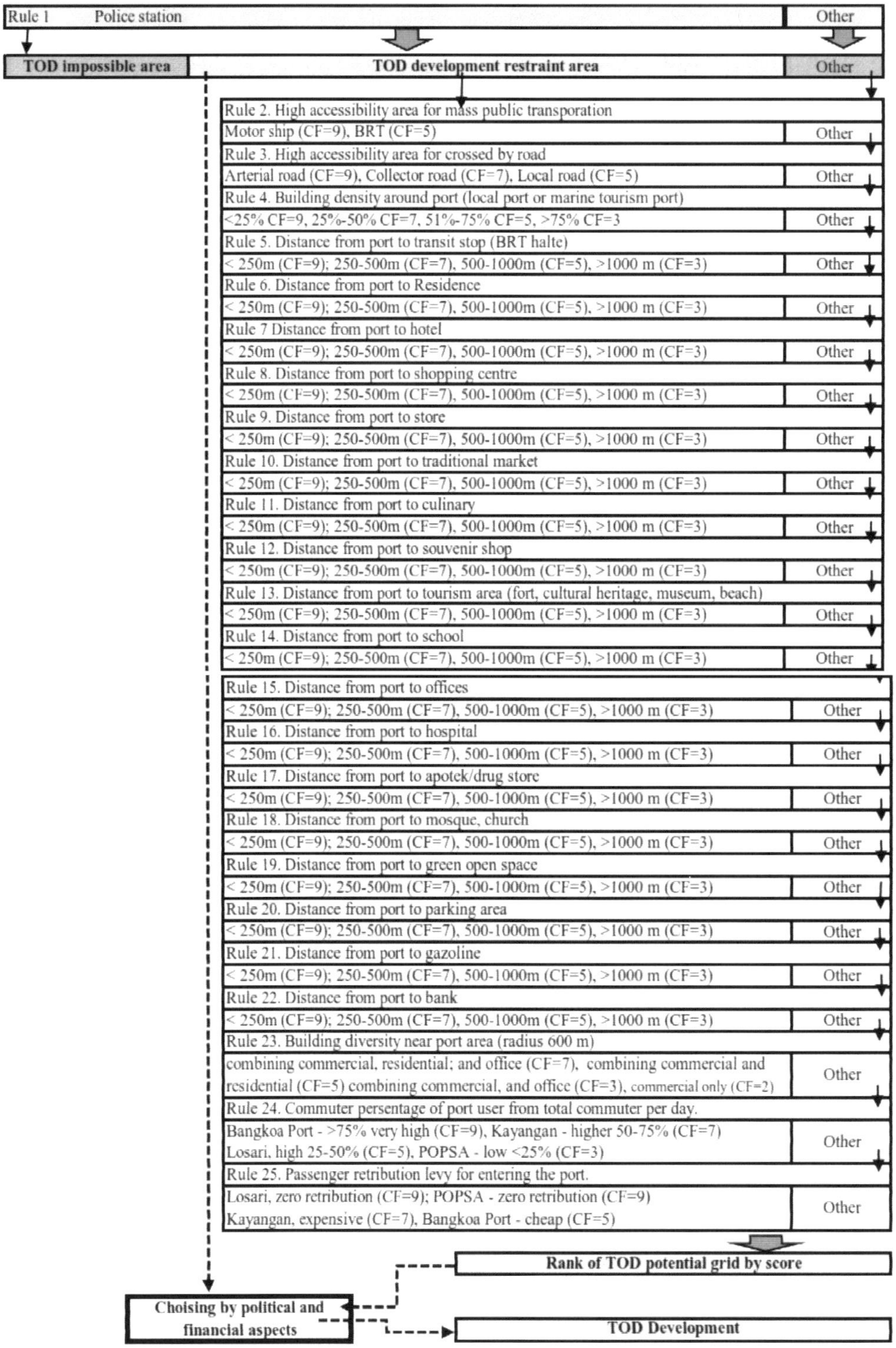

Figure 4. Export System

Tabel 1. The set up the certainty factor (CF)

FNo.	Factor	Category	CF	Code
1.	Police station	Police station	Y/N	PS
2.	High accessibility area for mass public transporation	Motor ship	9	MS
		BRT	5	BR
3.	High accessibility area for crossed by road	Arterial road	9	AR
		Collector road	7	CR
		Local road	5	LR
4.	Building density	<25% (very low)	9	BD
		25-50% (low)	7	
		51-75% (high)	5	
		>75% (very high)	3	
5.	Distance to transit stop (BRT halte)	< 250m (very near)	9	TS
		250-500m (near)	7	
		500-1000m (far)	5	
		>1000 m (very far)	3	
6.	Distance to residence	< 250m (very near)	9	RS
		250-500m (near)	7	
		500-1000m (far)	5	
		>1000 m (very far)	3	
7.	Distance to hotel	< 250m (very near)	9	HT
		250-500m (near)	7	
		500-1000m (far)	5	
		>1000 m (very far)	3	
8.	Distance to shopping centre	< 250m (very near)	9	SC
		250-500m (near)	7	
		500-1000m (far)	5	
		>1000 m (very far)	3	
9.	Distance to store	< 250m (very near)	9	ST
		250-500m (near)	7	
		500-1000m (far)	5	
		>1000 m (very far)	3	
10.	Distance to traditional market	< 250m (very near)	9	TM
		250-500m (near)	7	
		500-1000m (far)	5	
		>1000 m (very far)	3	
11.	Distance to culinary area	< 250m (very near)	9	CA
		250-500m (near)	7	
		500-1000m (far)	5	
		>1000 m (very far)	3	
12.	Distance to port to souvenir shop	< 250m (very near)	9	SS
		250-500m (near)	7	
		500-1000m (far)	5	
		>1000 m (very far)	3	
13.	Distance to port to tourism area (fort, cultural heritage, museum, beach)	< 250m (very near)	9	TA
		250-500m (near)	7	
		500-1000m (far)	5	
		>1000 m (very far)	3	
14.	Distance to School	< 250m (very near)	9	SH
		250-500m (near)	7	
		500-1000m (far)	5	
		>1000 m (very far)	3	
15.	Distance to offices	< 250m (very near)	9	OF
		250-500m (near)	7	
		500-1000m (far)	5	
		>1000 m (very far)	3	
16.	Distance to Hospital	< 250m (very near)	9	HO
		250-500m (near)	7	
		500-1000m (far)	5	
		>1000 m (very far)	3	
17.	Distance to Apotek/Drug Store	< 250m (very near)	9	DS
		250-500m (near)	7	
		500-1000m (far)	5	
		>1000 m (very far)	3	
18.	Distance to Mosque, Church	< 250m (very near)	9	MC
		250-500m (near)	7	
		500-1000m (far)	5	
		>1000 m (very far)	3	
19.	Distance to green open space	< 250m (very near)	9	OS
		250-500m (near)	7	
		500-1000m (far)	5	
		>1000 m (very far)	3	
20.	Distance to parking area	< 250m (very near)	9	PA
		250-500m (near)	7	
		500-1000m (far)	5	
		>1000 m (very far)	3	
21.	Distance to gazoline	< 250m (very near)	9	GZ
		250-500m (near)	7	
		500-1000m (far)	5	

3. Result and Discussions

The following table is the result of fortran to determine the location to develop coastal TOD. This following map shows the location of existing TOD around Fort Rotterdam Makassar, in sequence; from north to the south in Figure 7, they are POPSA marine tourism port, Kayu Bangkoa local port, and Losari beach.

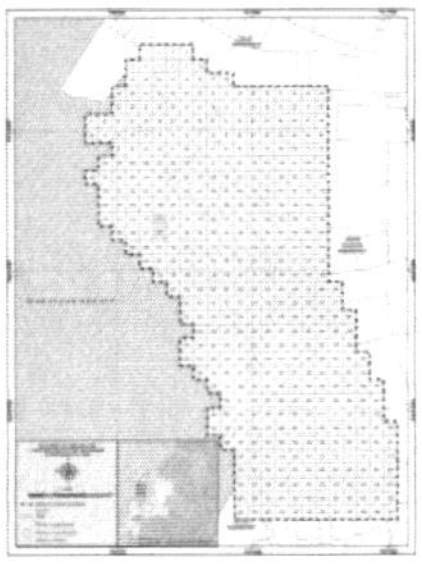

No. Grid	X axis	Y axis	Value	Ranking	Explanation
71	3	7	31	III	Around POPSA marine tourism port
89	5	8	40		
192	6	14	68	II	Around Kayu Bangkoa Port
235	7	17	72		
351	12	26	75	I	Around Losari Beach
364	12	27	77		

Figure 5. Grid Ranking of chosen TOD
(Source: result of data processing, 2016)

4. Conclusions

Land function around Fort Rotterdam is for shopping places, offices, residential, education, health, sports and recreational. Potency around Fort Rotterdam can support development of coastal TOD covers ports area, massive public transportation route of bus rapid transit (BRT) supported by transit stops and parking areas, and other area marine tourism, culture tourism, culinary tourism, Somba Opu shopping, hotels, restaurants, health facilities such as hospitals and drugstores. Physical problem found; generally, land and buildings around Fort Rotterdam are owned privately and individually. Trip pattern of commuters from small island to mainland of Makassar is generally done in the morning and the afternoon.

Development model of coastal TOD is recommended to build vertically, diverse, dense, integrated and bring closer the facilities of goods and services around TOD. Development principles of coastal TOD is to make the avaliability of land and buildings as the center of goods and services shopping which guarantee the price is same and under controlled, the avaliability of residential building, working places- education and health facility includes the avaliability of massive public transportation, fuel and clean water in the ports area around Fort Rotterdam effective. Based on distance and time, shows that TOD at three locations are needed to be developed. Commuters spend a lot of time and long distance to buy something. In order to increase the quality of TOD, some facilities have to be located near by TOD. POPSA local port around Fort Rotterdam used commuters from mainland of Makassar including tourist from hotels for having water sport activity and recreation at small islands around Makassar. Kayu Bangkoa local port is generally used by commuters from small islands to mainland of Makassar to fulfil their basic needs, yet the distance is quite far from TOD, hence it leads redundancy of time, energy and cost to happen. Losari beach is situated at collector road, around hotels, is used for beach recreation, culinary, and to small islands around mainland of Makassar (Lae-lae, Samalona, Kodingareng, Barrang Caddi, and Barrang Lompo).

References

[1] Maslow A H 1943 A Theory of Human Motivation *Psychological Review* **50**

[2] Saeednia Y and Nor M M D 2013 Measuring Hierarchy of Basic Needs among Adults. *Procedia - Social and Behavioral Sciences* **82** 417-420

[3] Rishi P Moghe S and Upadhyay B K 2008 Analysis of hierarchy of needs and motivational strategies for eco-development planning in two national parks of India *Resources, Conservation and Recycling* **52** 707-718

[4] Burhan N A S, Mohamad M R, Kurniawan Y and Sidek A H 2014 National intelligence, basic human needs, and their effect on economic growth *Intelligence* **44** 103-111

[5] Yudono A Akil A Ihsan and Arief A B 2016 Some City Planning Wisdoms since Pre until Colonial Era in Indonesia *Proceedings, Asian Urban Research Group 10th* 53-57

[6] Rahul T M and Ashish V 2013 Economic impact of non-motorized transportation in Indian cities *Elsevier* **38** 22–34

[7] Davies N J and Weston R 2015 Reducing car-use for leisure: Can organised walking groups

switch from car travel to bus and train walks? *Journal of Transport Geography* **48** 23-29
[8] Balachandran B R, Cevero R, Deakin B, Mehndiratta K M, Nadal S L and Suzuki H 2014 *TOD Standard 2.1* (New York: Despacio)
[9] Litman T 2014 Implementing Transit Oriented Development in Indian Cities *Learning and Challenges Victoria Transport Policy Intitute* 1-42
[10] Jun M J, Choi K., Jeong J E , Kwon K H and Kim H J 2015 Land use characteristics of subway catchment areas and their influence on subway ridership in Seoul *Journal of Transport Geography* **48** 30-40
[11] Vale D S 2015 Transit-oriented development, integration of land use and transport, and pedestrian accessibility: Combining node-place model with pedestrian shed ratio to evaluate and classify station areas in Lisbon *Journal of Transport Geography* **45** 70-80.
[12] Cervero R and Kockelman K 1997 Travel demand and the 3Ds: Density, diversity, and design. *Transportation Research Part D: Transport and Environment,* **2** 199-219
[13] Mu R and Jong M D 2012 Establishing the conditions for effective transit-oriented development in China: the case of Dalian *Journal of Transport Geography* **24** 234-249
[14] Paladugula A L and Rathi S 2013 Strategies to Rduce Energy Use for Commuting by Employees *Procedia - Social and Behavioral Sciences* **104** 952-961
[15] Papa E Bertolini L 2015 Accessibility and Transit-Oriented Development in European metropolitan areas *Journal of Transport Geography* **47** 70-83
[16] Yudono A 2013 Isu, Prinsip, dan Ide Penataan Kota Makassar. *Badan Penerbit UNM Makassar*
[17] Feudo F L 2014 How to build an alternative to sprawl and auto-centric development model through a TOD scenario for the North-Pas-de-Calais region? Lesson from an integrated transportation-land use modelling. *Elsevier Science Direct Transportation Research Procedia* **4** 154-177
[18] Canepa B 2015 Determining the Transit Oriented Development's Walkable Limits 1992 *Journal of Cleaner Productions*
[19] Nasri A and Zhang L 2014 The analysis of transit-oriented development (TOD) in Washington, D.C. and Baltimore metropolitan areas *Transport Policy* **32** 172-179
[20] Zoltis L. 2015 7 Prinsiples for Transit-Oriented Development. *The City Fix* 1-6
[21] Jin W J and Jiao J 2013 The Study in Diamond Interchange Traffic Organization. *Elsevier* 591-598.
[22] Pacheco P and Caccia L 2015 How Transit-Oriented Development Benefits Local Economies. *The City Fix* 1-5
[23] Renne J L 2008 Smart Growth and Transit Oriented Development at the State Level: Lesson from California, New Jersey, and Western Australia *Journal of Public Transportation* **11** 77-108.
[24] Cervero D R 2012 Transit-Oriented Development and Land Use *Ensiclopedia* 10923-10934
[25] Shirly W and Natalia V V 2011 Integrated Spatial Planning dan Transportation Systm to Reduce Mobility in Sub Urban Area *The 14th FSTPT International Symposium, Pekanbaru* - 1-11
[26] Wey W M 2015 Smart growth and transit-oriented development planning in site selection for a new metro transit station in Taipei, Taiwan *Habitat International* **47** 158-168
[27] Hasibuan H S Soemardi T P Koestoer R and SetyoMoersidik 2014 The Role of Transit Oriented Development in Constructing Urban Environment Sustainability, The Case of Jabodetabek, Indonesia *Procedia Environmental Sciences* **2** 620-631.

Expert System Development for Urban Fire Hazard Assessment. Study Case: Kendari City, Indonesia

S Taridala[1], A Yudono[2], M I Ramli[3], A Akil[4]
[1, 2, 4] Department of Architecture, Hasanuddin University, Makassar, Indonesia
[3] Department of Civil Engineering, Hasanuddin University, Makassar, Indonesia

E-Mail: taridala@gmail.com

Abstract. Kendari City is a coastal urban region with the smallest area as well as the largest population in Southeast Sulawesi. Fires in Kendari City had rather frequently occurred and caused numerous material losses. This study aims to develop a model of urban fire risk and fire station site assessment. The model is developed using Expert Systems with the Geographic Information System (GIS). The high risk of fire area is the area which of high building density with combustible material, not crossed by arterial nor collector road. The fire station site should be appropriately close by high risk of fire area, located on arterial road and near with potential water resource.

1. Introduction

Event Development and growth of urban areas in Indonesia has increased in each year. This can be seen by the increasing area of urban, the new growth centres and the increasing number of population living and doing activity in urban area. Currently, there is 53 percent of Indonesian population being the urban community and living in urban area [1]. Generally, the driving factors for urban area development are the increasing population naturally and/or by the migration process. These factors impact on the urban itself, namely the increasing demand on the public service facility and infrastructure, service and trade service facility, transportation, clean water and other infrastructures. The urban population growth also impacts on the change of land use, population density, transportation demand, entertainment, increasing security and other aspects.

The general reality of urban development in Indonesia currently is the physical growth of urban space not supported by environment carrying capacity, strict devices and regulation. These lead to the uncontrolled urban growth and development. The appropriate urban spatial will create harmony between natural environment and artificial environment as well as create the protection on spatial function and prevention on negative impacts on the environment caused by urban spatial utilization. The urban spatial will also increase the area ability or decrease its vulnerability on various possible negative risks, both naturally or non-naturally. One of the important aspects and often neglected in urban and area spatial is the planning. Indonesia is a country with quite high disaster potency because geographically, the area is located in the equator area having morphology with various lands until high mountains, the movement activity of active tectonic plate around the Indonesian seas, so that it creates earthquake track, series of active volcanoes and geological fault as earthquake and landslide-prone areas. The disaster potency includes the main hazard potency, such as earthquake, landslide, volcano eruptions, tsunami, flood, and collateral hazard potency such as fire, epidemic and social conflict. The highest potency of collateral hazard is in urban area having high population density, land use complexity, urban population activity centre, building materials and urban slump areas.

The high risk of urban area in Indonesia on the disaster can be seen from the area vulnerability, including the physical area vulnerability, social and economic population. The physical vulnerability describes the estimation on damage level on the environment facility and infrastructure if there are any certain hazardous factors, in social and economic population aspects related to the community capacity and condition in facing the disaster. The fire is one of the disaster types often found in urban area, mainly in area in dense population or area with high activity, such as commercial area. The area with dense population contributes on the increasing vulnerability on urban fire. The danger risk of urban fire can be decreased by good urban planning concept, such as efficient land use and the road system supporting on the accessibility of fire engine [2].

Kendari City is one of the cities with the populous number of population in Southeast Sulawesi Province, in 2014, it reaches the number of 335.889 people, with the mean of population growth flow per year is 3,51 percent. Kendari City has the smallest regional area, namely 295,89 Km^2, or 0,78 percent of the area of Southeast Sulawesi Province region [3].

The large number of population by relatively small regional area has bigger potency on urban environmental problems, such as building density, disorderly of land use, bad sanitation, improper road facility, and the arising temporary semi-permanent or urgent buildings. Mantra wrote that the vulnerability on the fire disaster including the environment condition (width of entrance, availability of community field/parking lot), building materials, building structures and inter-building distances [4].

The fire disaster in Kendari City often and it causes quite large material losses. There was one of the biggest fire events in 2015, it caused losses until billion rupiah. The fire was in company building, namely PT. Daka Samudera. The firefighter arrived at the location in 15 minutes after the warning on the fire [5]. The high urban city intensity and risk level in Kendari city is mostly caused by the very limited number of firefighter station, namely one station is for the service of all urban area, as well as the slow firefighter response time service, namely the mean service time is ≥15 minutes since the fire is started.

The previous research on comparison of fire service response time on the number of firefighter station in Dubai shows that the 5 (five) minute-response time requires 13 stations, 4 (four) minute-response time requires 20 stations and 3 (three) minute-response time requires 25 stations [2][7], described most urban area in California had service standard, the firefighter arrived on the incident location with 9 minute-response time since the fire. This is addressed that the protection on urban fire disaster must be done by quick response. Standard of fire response time according to NFPA 1710 (2001) is four minutes (240 seconds) or less for the arrival of the first arriving engine company at a fire suppression incident and/or eight minutes (480 seconds) or less for the deployment of a full first alarm assignment at a fire suppression incident [8].

This research tries to conduct the assessment model on the fire disaster risk level and site of fire station, as the mitigation efforts of fire disaster in urban area based on some physical aspects and availability of facility and infrastructure in the affecting area.

Table 1. History of fires on 2012-2016 in Kendari City, Indonesia (Source: Fire Department of Kendari City, 2016)

Years	Location (Subdistrict)	Caused	Fire types	Constraint of firefighting	Material looses (IDR)	Fataliti es
2012	Abeli, Baruga, Kadia, Kambu, Kendari, West Kendari, Mando-nga, Poasia, Pu-watu, Wua-wua	Candle lamp, electrical short circuit, fireworks, bursting stove, mosquito coils, garbage burning, fireplace, undetected	Bladygrass, storehouse, kiosk, cooperation, stall, market, coconut trees, residence, shophouse	High elevation, narrow roads, throng, water exhaust-machine not optimal	22.493.500.000	None
2013	Abeli, Kadia, Kambu, Kendari,	Electrical short circuit, bursting stove, mosquito	Bladygrass, machine shop, hotel, office, kiosk, car, market	Fire engine not optimal, electric current hazard,	24.227.000.000	None

Years	Location (Subdistrict)	Caused	Fire types	Constraint of firefighting	Material looses (IDR)	Fataliti es
	Mandonga, Poasia, Puwatu, Tobuuha and Wua-wua	coils, land combustion, garbage burning, stub, fuel oil spilled, fireplace, burnt by person, undetected	residence, shophouse, place of business	road damaged, narrow roads, unavailable water for firefighting		
2014	Abeli, Baruga, Kadia, Kambu, Kendari, West Kendari, Mandonga, Poasia, Puwatu and Wua-wua	Candle lamp, electrical short circuit, bursting stove, oil overflowed, land combustion, garbage burning, stub, gasometer, undetected	Bladygrass, bulk asphalt, tire, warehouse, compa-ny building, worker base camp, machine shop, powerhouse, generator, storehouse, office, homestead garden, kiosk, peatland, sago trees, car, motorcycle, crackers factory, market, residence, shophouse, garbage, bunk, woodpile	Fire engine not optimal, smoke, electric current hazard, traffic jam, road damaged, narrow roads, the rising road, far travelled distance, traffic density, not accessed by roads, woodpile	21.679.500.000	2 person (pass away)
2015	Abeli, Baruga, Kadia, Kambu, Kendari, West Kendari, Mandonga. Poasia. Puwatu and Wua-wua	Electrical short circuit, fuel pipe leakage, regulator hose leakage, fireworks, mosquito coils, land combustion, garbage burning, stub, piece of electrical weld, fireplace, undetected	Bladygrass, warehouse, company building, worker base camp, machine shop, powerhouse, storehouse, college, office, home-stead garden, kiosk, peatland, sago trees, car, market, hospital, school, residence, shophouse, garbage, gasometer, weld tube, woodpile, sawmill, shrubland	Smoke, road damaged, narrow roads, electric current hazard, high elevation, slippery road, traffic jam, roadwork, not accessed by roads	141.724.500.00 0	None
2016 (inclu-sive to August)	Abeli, Baruga, Kadia, Kambu, Kendari, Mandonga, Poasia, Puwatu, Wua-wua	Electrical short circuit, gas leakage fireworks, bursting stove, gas explosion, land combustion, garbage burning, stub, undetected	Bladygrass, powerhouse, motor boat, kiosk, peatland, school, residence, shophouse, barbershop, gasometer	Electric current hazard, traffic jam, road damaged, narrow roads, fire on the sea, not accessed by roads, un- available fire-escape	10.225.000.000	1 person (injur ed)

2. Methods

According The high urban population activity leads to bigger opportunity of fire in urban area. The general conditions including physical, social economic and political factors are one of the potencies for a group of community to be more vulnerable to get a disaster. Toki stated that most fire in dense areas in urban area relates to human habits or behavior, but the fire disaster risk can be reduced or eliminated overall by the availability of appropriate environmental infrastructures, such as the availability of hydrants in dense residence area [9]. Murray (2013), wrote that all of urban areas must be protected from fire disaster hazard, the firefighter must give rapid responses and services. The fire station services have long been a perquisite in the availability of urban facilities [7]. The use of service response time standard by the fire station tries to reduce the risk for human and property losses.

Expert systems are computer programs that apply artificial intelligence to narrow and clearly defined problems. Expert systems typically combine rules with facts to draw conclusions; the process relies heavily on theories of logical deduction. Both heuristic methods and conventional computer programs (e.g., FORTRAN programs) are often used in expert systems. The expert system for site selection helps a user develop a set of site attributes and determine their weighted relative importance. The site selection knowledge base of expert system consists of four parts: the knowledge acquisition, induction, design, and decision analysis units [10].

IOP Conf. Series: Earth and Environmental Science **79** (2017) 012035 doi:10.1088/1755-1315/79/1/012035

The major research area for artificial intelligence includes natural language processing, symbol processing, rule-based system and logic systems. The development of functional expert systems is always centered on the organization of a knowledge base. Knowledge engineers collects and organizes knowledge gathered from domain expert then convert the expert knowledge into a form which computer expert system understands and save those converted knowledge into the knowledge base. Users enter the collected facts into the system via the user interface and save the data into the fact base. Finally, users get the results, recommendations and explanations from the system [11].

Study about urban fire hazard which covered fire risk assessment and fire station site was done by expert system approach method. This expert system came from knowledge-based expert system, i.e. a system using human knowledge which then was inserted into computer to solve problem that generally needs an expert's expertise. This expert system procedure works by imitating knowledge and thinking process of an expert in solving complicated problem. The system designing uses hardware, i.e. personal computer (PC) or notebook and software, i.e Geographic Information System (GIS) application (ArcGIS, ArcView, etc.), Fortran and Quick Basic. This system was operated in Microsoft Windows Operation System.

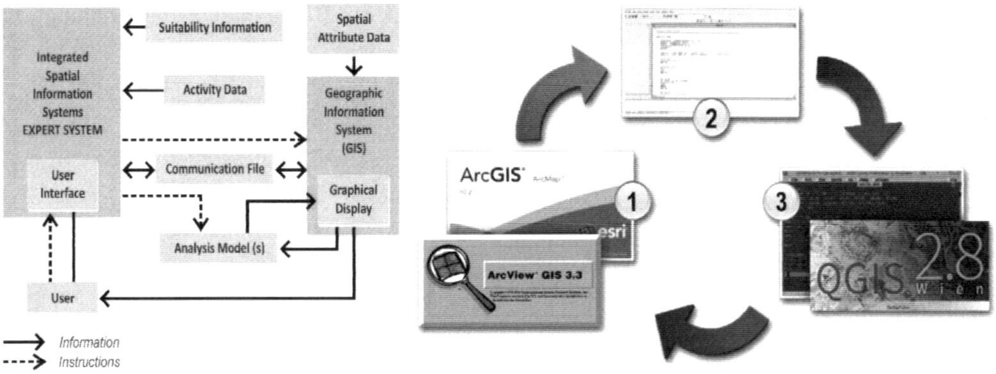

Figure 1. The expert system model procedure for urban fire. Source: Kim *et al*, 1990, Author, 2016

The development of GIS has provided a powerful tool for managing and solving emergency management problems. GIS is a professional computer system for collecting, storing, managing, retrieving, transforming, analyzing, and displaying of spatial data. It can be used for many kinds of purpose in both macro and micro scales. GIS were designed to support geographical inquiry and, ultimately, spatial decision making. Especially in the natural disaster area, GIS has been applied in the simulation and early warning system, emergency management system, and disaster damage assessment, etc. [12]

GIS approach can be assess of the fire risk, based on historical and current data and translated under cartographic shape, can be a remarkable contribution to the forest managers and a tool for a better preventive decision, based on logical bases. Indeed, these cartographic documents of the degree of risk reveal sectors of high sensibility at the fire risk [13]

Substances used in this urban fire research, were obtained from several sources which covered primary and secondary data. Primary data was obtained from field survey activity, included data about land use, building material types, and potential water source location. Secondary data included satellite imagery, administration boundaries map, forest area status map, road hierarchy map, etc. Apart from those two, we also use our own analytic data i.e. GIS grid based map (amount 4529 grids, each 250 m x 250 m or 62.500 m^2) and map of high risk fire area based on grid.

3. Result and Discussions

3.1 Urban fire risk

According to The urban fire risk was analyzed by determining factors which influence urban fire rate, i.e. combustible or non-combustible building material, building density, whether the area was crossed by arterial or collector road, and the distance from potential water source. Every influencing factor was then given weight based on its influence on fire risk, in the form of Certainty Factor (CF). Those factors were then made into a knowledge represented method i.e. risk rule in the form of the pair of condition-action or if-then. Risk rule is stated as implication of two part i.e. "if" part and "then" part. If "if" part is fulfilled, then "then" part will be valued as right. If "if" part is not fulfilled, then "then" part will be shifted to the following "then" part.

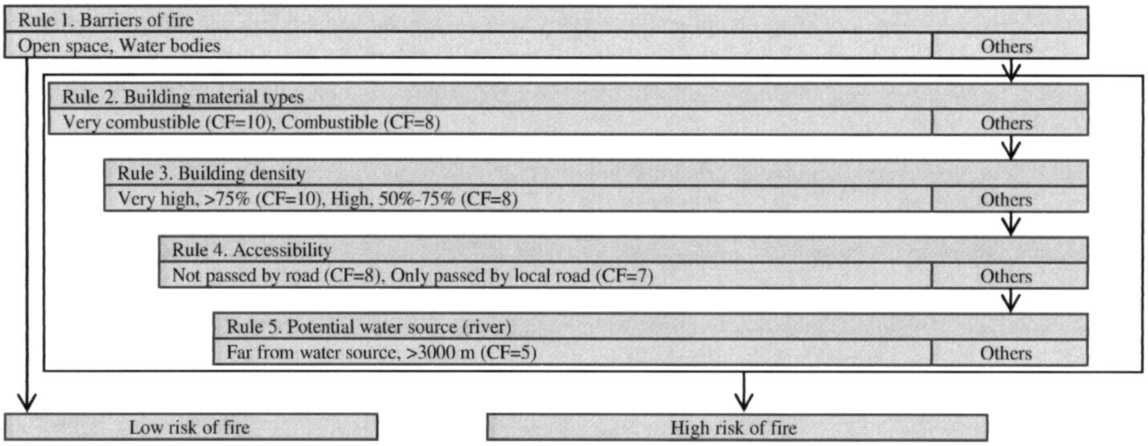

Figure 2. System chart for urban fire risk (Source: Authors, 2016)

The fire risk in Kendari City, was rated by expert system approach by using GIS Grid shows the risk degree which was then classified into four classes, i.e. (1) Very high risk, amounting to 586 grids, (2) High risk, amounting to 441 grids, (3) Low risk, amounting to 1861 grids, (4) Very low risk, 1641 grids. The influencing factors of fire risk rate are: (1) area infrastructure, included road dimension and fire engine accesibilty, and water source for fire fighting, included hydrant, (2) building material type, included combustible and non-combustible building, (3) response time service of firefighting.

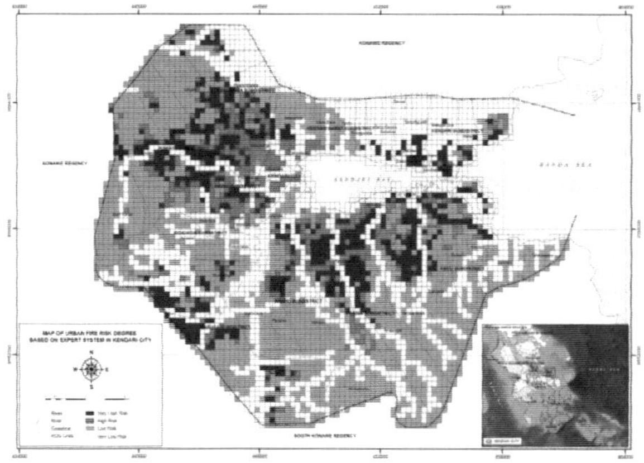

Figure 3. Map of urban fire risk degree based on grid.
Source: Authors, 2016

IOP Conf. Series: Earth and Environmental Science **79** (2017) 012035 doi:10.1088/1755-1315/79/1/012035

3.2 Fire station site

Fire station location determination was based on several supporting factors i.e. it's near with high risk of fire area, being crossed by arterial and collector road, the distance from potential water source (<3000 m), and categorized as unused area. In addition to supporting factors, the analysis was also conducted to several inhibiting factors in fire station site, i.e. natural disaster area, conservation area, slope degree >15%, restricted area, school zone, and place of worship area. Supporting factors are scored, ranged from lowest score 4 (CF=4) to highest score i.e. 10 (CF=10). While inhibiting factors are scored 1 if it exists and 2 if it doesn't.

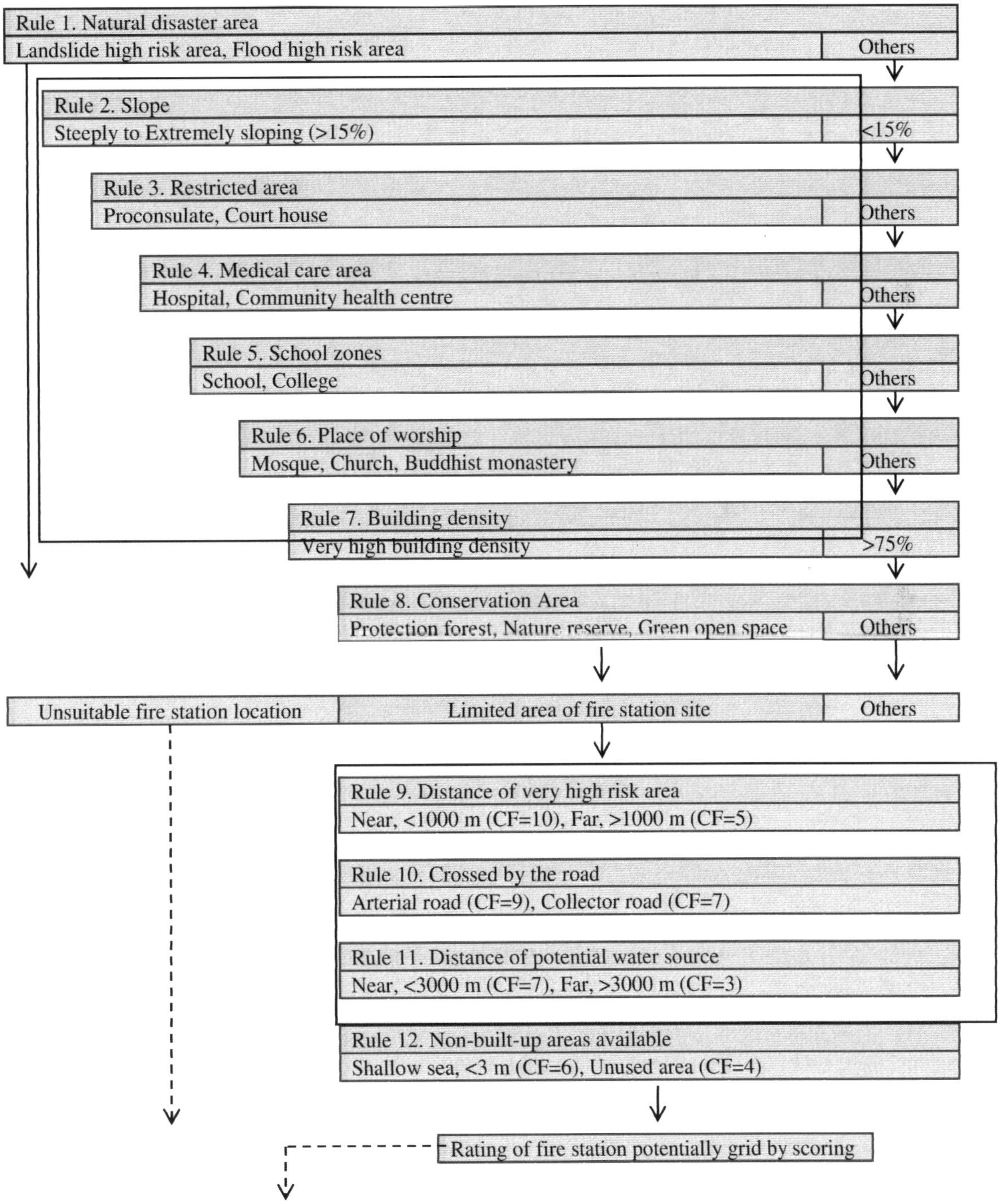

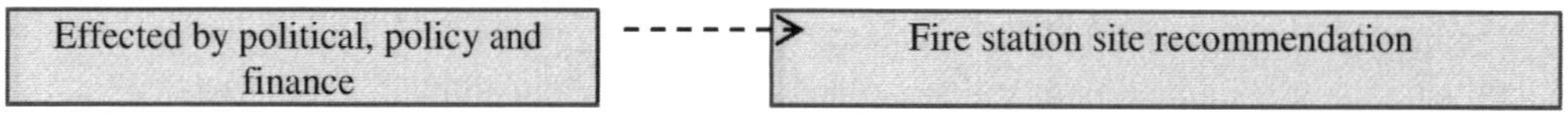

Figure 4. System chart for fire station site (Source: Authors, 2016)

The analysis of fire station location based on expert system approach by using GIS Grid resulted fourty nine chosen locations, i.e. (1) Suitable I, consists of grid no. 1268 and 1337, (2) Suitable II, amounting to 33 grids, (3) Suitable III, amounting to 14 grids. Out of these 49 grids, the others are unsuitable areas to set the location of the fire station.

Table 2. Chosen grid for fire station site suitability in Kendari City

No.	Suitability	Grid number	Value
1.	Suitable I	1268, 1337	37-40
2.	Suitable II	1033,1034,1037,1042,1043,1044,1110,1115,1118,1123,1183,1198,1267,1269,1335,1396,1496,1501,1552,1553,1609,1807,1808,1812,1967,1968,2040,2041,2202,2277,2460,2559,3507	33-36
3.	Suitable III	1167,1178,1201,1237,1343,1345,1450,1451,1752,1834,1985,1986,3328,3496	29-32
4.	Unsuitable	Out of these 49 grids	<29

Source: Authors, 2016

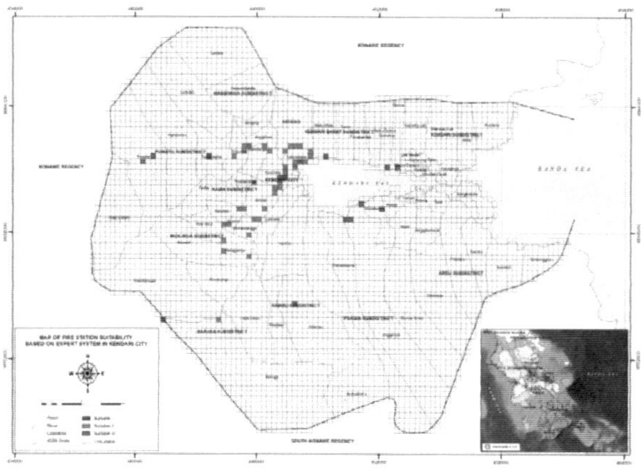

Figure 5. Map of fire station site suitability based on grid. Source: Authors, 2016

4. Conclusion

The research result studied in Kendari City through Expert System Approach, based on GIS Grid shows that:

1. The high risk of fire area is the area which of high building density with combustible material (wood and mixed material) also not crossed by arterial nor collector road (only crossed by local road). The low risk area is generally the non-built-up areas which consist of water body, open space and vegetation.
2. The fire station site should be appropriately close by high risk of fire area, located on arterial road and near with potential water resource to extinguish the fire. The response time service of fire extinguishing (from chosen fire station) toward high risk of fire area should be 4-5 minutes since fire alert.
3. The Expert System approach as a model for urban fire hazard assessment can be shows of the fire risk, including the degree of fire risk, based on existing data and selecting one or more of the fire

station site available alternatives).

References

[1] Ministry of Public Work and People Housing 2015 The Sixth Asia-Pacific Urban Forum Conference, Jakarta

[2] Chainey S and Ratcliffe J 2005 *GIS and crime mapping* (West Sussex: John Wiley & Sons)

[3] Central Agency on Statistics, Southeast Sulawesi Province 2015 *Kendari City in Figures, 2015* (Kendari: BPS Catalog)

[4] Mantra I B G W 2005 Study of Fire Preventive on Housing. A Base Study in Sarijadi Residence Bandung *Journal of Natah Settlement* **3**

[5] Badri *et al.* 1998 *Theory and Methodology. A Multi-Objective Model for Locating Fire Stations* European Journal of Operational Research **110** 243-260

[6] Fire Department Kendari City 2015 History of Fires on 2012-2016,

[7] Murray T A 2013 Optimising the spatial location of urban fire stations *Fire Safety Journal* **62** 64–71

[8] NFPA 1710 2001 Standard for the Organization and Deployment of Fire Suppression Operations, Emergency Medical Operations, and Special Operations to the Public by Career Fire Departments. National Fire Protection Association, USA

[9] Toki *et al* 2011 Protection of Cultural Heritage from Post-Earthquake *Fire Journal of Disaster Research* **6** 4-10

[10] Kim T J *et al* 1990 Expert Systems: Applications to Urban Planning. Springer-Verlag. Department of Urban and Regional Planning (Champaign: University of Illinois)

[11] Ikram A and Qamar U 2014 Developing an expert system based on association rules and predicate logic for earthquake prediction. *Knowledge-Based Systems Journal* **75** 87–103

[12] Gai *et al* 2011 *GIS-based Forest Fire Risk Assessment and Mapping. Fourth International Joint Conference on Computational Sciences and Optimization* Institute of Public Safety Research, Tsinghua University Beijing, China

[13] Guettouche MS *et al* 2011 A Fire Risk Modelling and Spatialization by GIS: Application on the Forest of Bouzareah Clump, Algiers (Algeria) *Journal of Geographic Information System*, 2011, 3, 254-265 Sciences and Technology University of Houari Boumediene, Algiers, Algeria

Acknowledgements

Author would like to say thank you to everybody contributing in this research. Especially Mr. Ananto Yudono, Professor of Architecture Department of Hasanuddin University, Makassar, who has given wonderful guidance and motivation to complete the research and paperwork. Also to Mr. Arifuddin Akil who has enriched author's knowledge.

IOP Conf. Series: Earth and Environmental Science **79** (2017) 012036 doi:10.1088/1755-1315/79/1/012036

Project delay analysis of HRSG

Silvianita[1], A S Novega[1], D M Rosyid[1], Suntoyo[1]
[1] Department of Ocean Engineering, Faculty of Marine Technology, Institut Teknologi Sepuluh Nopember, Indonesia

E-Mail: silvianita@oe.its.ac.id

Abstract. Completion of HRSG (Heat Recovery Steam Generator) fabrication project sometimes is not sufficient with the targeted time written on the contract. The delay on fabrication process can cause some disadvantages for fabricator, including forfeit payment, delay on HRSG construction process up until HRSG trials delay. In this paper, the author is using semi quantitative on HRSG pressure part fabrication delay with configuration plant 1 GT (Gas Turbine) + 1 HRSG + 1 STG (Steam Turbine Generator) using bow-tie analysis method. Bow-tie analysis method is a combination from FTA (Fault tree analysis) and ETA (Event tree analysis) to develop the risk matrix of HRSG. The result from FTA analysis is use as a threat for preventive measure. The result from ETA analysis is use as impact from fabrication delay.

1. Introduction

In Indonesia, most electrical procurements are still using diesel power plants – which are old and inefficient. But in recent development, both government and IPP (Independent Power Producer) are starting to use gas and steam power plants / HRSG (Heat Recovery Steam Generator). Gas and steam power plant is a combination of steam power plant and gas power plant, where the heats from gas power plant excess gases are used to produce steam that will generate the steam power plant.

In HRSG manufacturing process, there is always a contract between the owner and the fabricator. The contract contains of total manufacturing cost, time needed to complete the project, terms of payment and so on. According to Rosdianto[1] it is important to state the amount of fine that the fabricator has to pay to the owner if they cannot fulfil the contract.

But in reality, the existence of obstacles that happens usually caused a project to be late. In this paper, the author is using FTA (Fault Tree Analysis) and ETA (Event Tree Analysis) method to do an analysis on project delay.

The *FTA* (Fault Tree Analysis) method was used in analysing the failure or delay on a project, while the ETA (Event Tree Analysis) method was analysing the result from a project delay. From both analyses the bow-tie diagram will be produce and used as tool to analyse what are the impact that can be reduced by making a project mitigation so manufacture project scheduling can go as smooth as initial planning.

The Problems that will be accounted in this paper are:

1. What are the causes of HRSG pressure part fabrication project delay – using Fault Tree Analysis method?
2. What are the impacts that will be posed from HRSG pressure part fabrication project delay - using Event Tree Analysis method.

2. Fault Tree Analysis

According to Rosyid [2], Fault Tree Analysis is a method to identify all probable causes (component failure or other failure events that happened on its own or simultaneously) that caused a system failure and gave a footing for failure possibility calculation. Based on Ericson [3] there are eight steps on FTA, these are the steps that need to be followed:

1. Understanding the system design and operation. Gaining the current design data (drawing, scheme, procedures, diagram, etc.)
2. Descriptively defining the problems and determining components that are absolutely doesn't want to be analyse
3. Specified analysis' base rules and scope boundaries, also write down all of the base rules.
4. Follow the construction process, rules, and logics to build the system fault tree model.
5. Producing cut set and probability, then identify the weak-link and problems in design security.
6. Let the respondent check the fault tree model to determine its actuality, generality, and accuracy.
7. Modify the fault tree model following the facts acquired during validation or due to design system changes

3. Event Tree Analysis

According to Mahandeka [4], Event tree analysis is a method to analyse various impacts caused by an event. This method is used to estimate and judge the probability from each consequence that can emerge from a single event. Therefore, this method is used as a reference in anticipating various consequences.

First step in analysis process using event tree analysis method is by drawing parts of system that are correlating with main event as detail as possible. This is done to obtain estimation of events that probably happen after the main event. This process is very dependent to the part of system drawn; the more detailed the picture the more events that can be estimated. As a result, the consequences or scenario that's being predicted is more valid.

The second step is to draw a tree diagram event in accordance with all the events that have been estimated. Every event in each diagram, form a question that can be answered with "yes" or "no". Every answer initiates other related events and continues until the end of the known consequences of any branch of the estimated incidence. The third step is the time to find the value of the probability for every answer from each event that's written on the diagram. The total value of probability for each event then multiplied with the value of each answer's probability that correspondence with certain consequence sequence wanted, and by that the value of each consequence's probability are obtained.

Kurniawan [5] explains Bow-tie analysis is a method to form a diagram that showed and analyse the sequence of a risk from its cause to its impact. This method is often considered as a combination from FTA (fault tree analysis) method that analyse the cause of certain event using the ETA (event tree analysis) that analyse the impact. According to Gifford et.al [6], this method is called bow-tie because of the diagram produced resembling the shape of a bow-tie with cause and impact both forming the left and right wing – flanking the event's risk in the middle. Bow-tie is combining the components from Fault Tree Analysis, and Event Tree Analysis. Bow tie analysis has been used in many areas such as for identifying risk assessment of mobile mooring [7]. This paper will elaborate the useful of FTA and ETA to determining the delay of HRSG fabrication.

4. Elaboration of FTA and ETA

The elaboration of FTA and ETA can be explained as follows:

1. Problems identification and formulation
   Identification is a process to investigate a problem in HRSG and formulate the problems in order to find the solutions of project delay.
2. Data Collecting
   The required data used in this paper are:
   a. Initial schedule of HRSG pressure part fabrication data.
   b. Daily working hour data.

c. Data containing the number of workers involved.
d. Data containing production facilities (machinery and tools) owned.

3. Data Analysis and Discussion
From the data obtained, some analysis and discussion are done, amongst are:
a. Analyzing interview result to find items that experience some delay and what are the factors that affect those items.
b. Making FTA construction drawings from HRSG pressure part fabrication project.
c. Making ETA construction drawings from HRSG pressure part fabrication project.
d. Making bow-tie analysis from FTA and ETA output

5. The Result of FTA and ETA

FTA starts by formulating the root causes of Delay Fabrication of HRSG. In order to formulate the root causes of delay fabrication of HRSG, the experts in one of the fabrication company are involved. As can be seen in Figure 1, the fault diagram derived from the initial cause of HRSG pressure parts fabrication delay and broken down until the basic events are found.

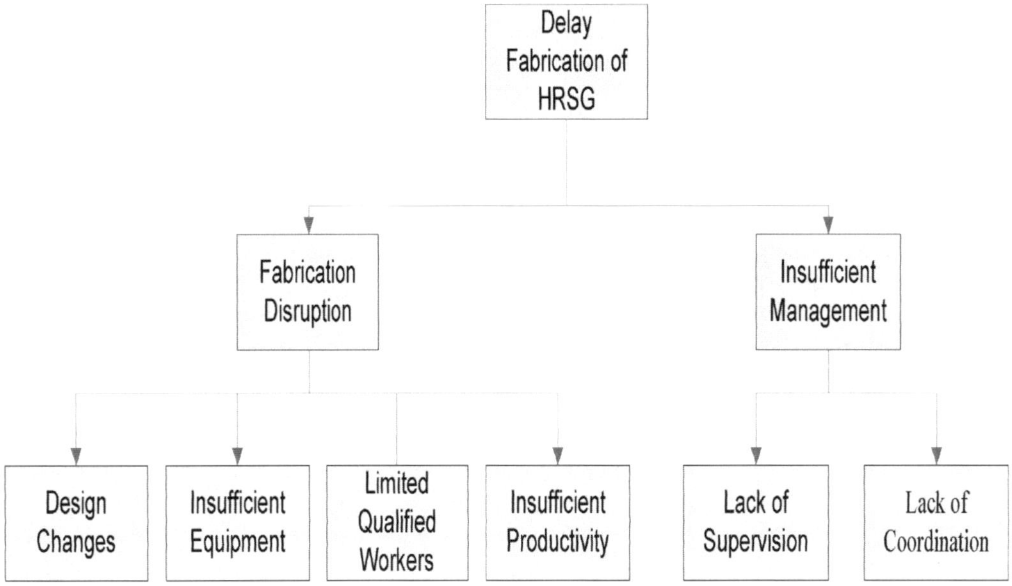

Figure 1. Fault Tree Main Branch

Once the fault tree diagram has been accomplished the next step is to determine the probability of each basic event. The probabilities of each of basic event on HRSG fabrication process are obtaining from questionnaires and interview with respondents. The frequency of the FTA basic events is referring to probability as follows:

Table 1. Probability

FI	Rating	Qualitative	Quantitave
5	Frequent	Likely to occur every new production HRSG	10^{-1}
4	Reasonably Probable	Likely to occur per 5 times new production HRSG	10^{-2}
3	Remote	Likely to occur per 25 times new production HRSG	10^{-3}
2	Extremely Remote	Likely to occur per 75 times new production HRSG	10^{-4}
1	Extremely Improbable	Likely to occur per 100 times new production HRSG	10^{-5}

After obtaining the probability for each basic event the next step is to find the minimum cut set from the two main branches which are "Fabrication Disruption" and "Insufficient Management". Based on the FTA calculation the probability of top event of "Delay Fabrication of HRSG" is 0.0635. The next step is to formulate the event tree of Delay Fabrication of HRSG as seen in Figure 2:

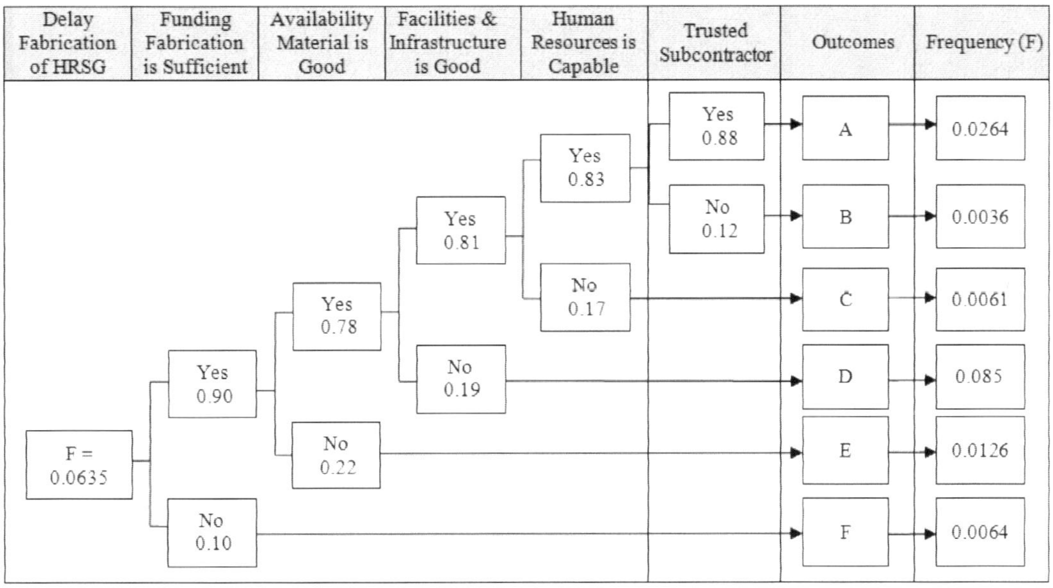

Figure 2. Event Tree Analysis Result

Each ETA result has consequences. The result for each part will be shown below:

1. Output A : the HRSG is completed but experienced some delay between 1 day to 2 week. Output A happens with probability of:
   0,0635 x 0,9 x 0,78 x 0,81 x 0,83 x 0,88 = 0,0264
2. Output B : the HRSG is completed but experience some delay between 2 week to 1 month because of an inexperience subcontractor. Output B happens with probability of:
   0,0635 x 0,9 x 0,78 x 0,81 x 0,83 x 0,12 = 0,0036
3. Output C : the HRSG is completed but experience some delay between 1 months to 3 month

because of lack of manpower. Output C happens with probability of:
0,0635 x 0,9 x 0,78 x 0,81 x 0,17 = 0,0061

4. Output D : the HRSG is completed but experience some delay between 3 month to 6 month year because of facilities and infrastructure that's not supportive. Output D happens with probability of:
   0,0635 x 0,9 x 0,78 x 0,19 = 0,085
5. Output E : the HRSG is completed but experienced some delay between 6 month to 9 month because of material unpreparedness on the field.
   Output E happens with probability of:
   0,0635 x 0,9 x 0,22 = 0,0126
6. Output F : the HRSG is not completed in fabrication process because insufficient fund from the owner. Output F happens with probability of:
   0,0635 x 0,1 = 0,0064

The probability from ETA's result will be used in determining the consequences category in risk matrix. The first step is to determine Probability (FI) and *Severity Index* (SI) from the output of *Event Tree Analysis* (ETA), followed by calculating the *Risk Index* (RI) to be categorized into the *risk matrix*. The table above shows the categorization of the quantitative and qualitative data from the ETA to FI where the rating of complications explains the time ranges of those complications during the production of new ships. The making of these data also requires the agreement of the ETA respondent by interview.

Table 2. Severity Index

SI	Rating	Qualitative	Quantitave
1	Minor	New project HRSG get max penalty 1M & delay between 1 day - 1 month	< 0,01
2	Moderate	New project HRSG get max penalty 15M & delay between 1 month - 6 month	0,01 - 0,1
3	Serious	New project HRSG get max penalty 30M & delay between 6 month - 1 year	0,1 - 1
4	Catastrophic	New project HRSG failed	1 - 10

Table 2 explains the categorization of the quantitative and qualitative data from the ETA to SI where the rating of complications explains the categorization for the impacts of those complications on the production of new ships. The results of frequency and severity index are derived from the respondents as seen in the table 3.

Table 3. Frequency of Index and Severity Index for Risk Matrix

No	Output	Frequency Index (FI)		Severity Index (SI)		Risk Index (RI)	
1	Output A	4	Reasonably Probable	1	Minor	5	Moderate
2	Output B	3	Remote	2	Moderate	5	Moderate
3	Output C	3	Remote	2	Moderate	5	Moderate
4	Output D	4	Reasonably Probable	2	Moderate	6	Moderate
5	Output E	4	Reasonably Probable	3	Serious	7	Moderate
6	Output F	3	Remote	4	Catastropic	7	Moderate

Once the frequency and severity index have been obtained, the next step is tabulated into the risk matrix as can be seen in Table 4. All the output of the consequences are in the Medium Level.

Table 4. Risk Matrix Results from ETA

FI	Rating	Severity Index (SI)			
		1	2	3	4
		Minor	Moderate	Serious	Catastrophic
5	Frequent	6	7	8	9
4	Rasonably Probable	(5)	(6)	(7)	8
3	Remote	4	(5)	6	(7)
2	Extremely Remote	3	4	5	6
1	Extremely Improbable	2	3	4	5
		A	BC, D	E	F

6. Conclusion

Based on the results of this paper, the following conclusions are made:

1. The main factors for the delay of the fabrication process of HRSG using the Fault Tree Analysis consist of fabrication disruption and insufficient management.
2. The delay of HRSG fabrication processes are obtained from each path of critical event associated with all their frequencies of pivotal events4).

References

[1] Rosdianto and M Afif 2014 Analisa Percepatan Durasi Pengerjaan Proyek Pembangunan Jacket Platform di PT. Meindo Elang Indah (Surabaya: ITS)
[2] Rosyid D M 2007 Pengantar Rekayasa Keandalan (Surabaya: Airlangga University Press)
[3] Ericson A C 2005 Hazard Analysis Techniques for System Safety (Hoboken: John Wiley & Sons, Inc)
[4] Mahandeka. D S 2015 Analisis Perencanaan Proyek Berbasis Resiko; Wooden Sailing Boat Project – Maritime Challange ITS (Surabaya: ITS)
[5] Kurniawan R 2015 Studi Keterlambatan Proyek Pembangunan Kapal Kargo Dengan Motode Bow Tie Analysis (Surabaya: ITS)
[6] Gifford M, Giltert S and Bernes I 2003 Bow-Tie Analysis Equipment Safety Assurance Symposium (ESAS).
[7] Silviania, Khamidi M F, Kurian V J 2013 *Decision Making for Safety Assessment of Mobile Mooring System* Jurnal Teknologi **61** 41-52

A Conceptual Model for Water Sensitive City in Surabaya

A Pamungkas, K P Tucunan, A Navastara, H Idajati, N A Pratomoatmojo
Urban and Regional Planning Department
Institut Teknologi Sepuluh Nopember
Jalan Raya ITS, Keputih, Sukolilo, Surabaya, Indonesia

E-mail: adjie@urplan.its.ac.id; adjie.difi@gmail.com

Abstract. Frequent inundated areas, low quality of water supply, highly dependent water sources from external are some key problems in Surabaya water balance. Many aspects of urban development have stimulated those problems. To uncover the complexity of water balance in Surabaya, a conceptual model for water sensitive city is constructed to find the optimum solution. A system dynamic modeling is utilized to assist and enrich the idea of conceptual model. A secondary analysis to a wide range data directs the process in making a conceptual model. FGD involving many experts from multidiscipline are also used to finalize the conceptual model. Based on those methods, the model has four main sub models that are; flooding, land use change, water demand and water supply. The model consists of 35 key variables illustrating challenges in Surabaya urban water.

1. Introduction

Surabaya is a capital city of East Java Province and also the second largest city in Indonesia. The fast growing economic in Surabaya can be seen from percentage of economic growth around 6-7% annually [1]. The impacts of its economic performance are varies in terms of development. To some extent, the good economic performance leads to the increasing investments particularly for trading and housing investments. In other words, Surabaya becomes one of the interesting cities for investments. From 2010-2015, a steady increase of high-rise building indicates that the development process in Surabaya is at high rate. Consequently, the high demand will need a high supply of infrastructures including managing water.

In terms of water management, water sensitive city (WSC) has considered 6 stages of city form in terms of water management (Brown et al., 2009) as on Figure 1 [2].

Water Supply City

The Water Supply City represents the first modern urban water city-state in Australia, reflecting the colonization of Australia by Europeans in the early 1800s. The normative underpinning at the time was the effective provision of safe and secure water supplies for a growing urban population, and centralized provision, particularly for the elite, where the social movement of cleanliness was strongly linked with social status.

Sewered City

In concert with some of the newly developing American cities, from the late 1800s Australian cities invested in separate sewerage systems to avoid pathogen infection from wastes and industrial effluents in combined sewerage. Many cities also invested in on-site septic systems due to the perceived prohibitive cost of providing this infrastructure.

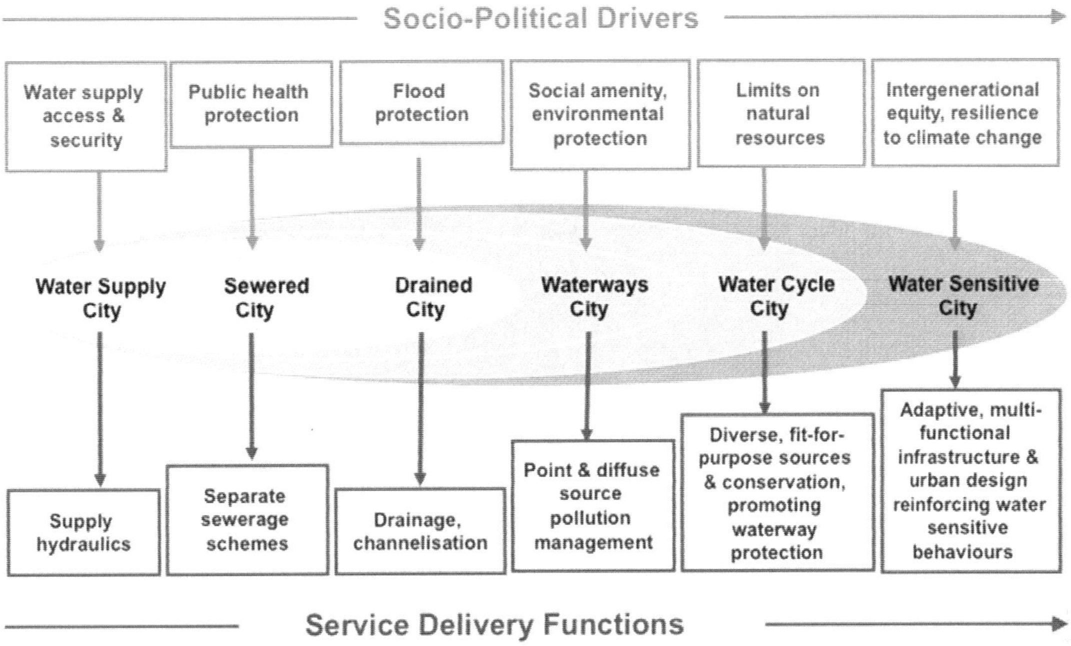

Figure 1 Socio-Political Derivers and Service Delivery Functions
Source: Brown et al., 2009

Drained City

Australia was a strong innovator in the professional community – focused on developing techniques and models that enabled the rapid and efficient conveyance of storm water out of cities, to receiving waterway environments. This substantially impacted the development patterns of Australian cities, with numerous waterways piped and located underground, and river systems channelized to allow for more urban development in floodplain areas. Many houses at this time were constructed facing away from waterways, which were often perceived as waste dumping grounds, and were therefore not a socially valued part of the urban landscape.

Waterways City

In this phase, water began to be integrated into planning functions as important visual and recreational features for communities, and measures were taken to reduce pollutant inputs into waterways, which involved regulating environmental discharges from wastewater treatment plants and industrial processes as well as replacing septic tanks with centralized sewerage systems. Science revealed the impact on waterways from diffuse-source storm water pollution, prompting researchers and practitioners to develop new technologies such as wetlands and bio-filtration systems to protect receiving waterways.

Water Cycle City

The Water Cycle City is a response to the recognition of the current 'limits' to traditional water sources for supplying ever growing populations and urban development, as well as the limits to waterways being able to assimilate pollution. It also reflects the growing normative acceptance of the need for social, economic and environmental sustainability.

Water Sensitive City

Water Sensitive City concept is a form of integration among the normative values of environmental repair and protection, supply security, flood control, public health, amenity, livability and economic sustainability, amongst others. Communities would be driven by the normative values of protecting intergenerational equity with regards to natural resources and ecological integrity, as well as by concern that communities and environments are resilient to climate change.

Surabaya has experienced in between of a water supply and a drainage water city level. It indicates that many program of infrastructure occurred in providing water supply and draining the water to avoid floods. Surabaya government has "Master Meter" program that provided clean water from PDAM to household [3]. Other programs that indicate a more sensitive water management are still in pilot project or in a small-scale program.

On the other hand, the city problem related to water is increase. Inundation is still key issues in Surabaya City. Although the average of height, duration and areas of inundation are decreases as in Table 1, number of inundation is still increase [4]. The case of inundation in Surabaya is still problematic.

Table 1 Inundation recapitulation in Surabaya from 2010-2015

Year	Total Inundation	Total Inundate Area (Ha)	Inundate Average Period (Minutes)	Inundate Average Depth (cm)
2010	148	2183.07	59.27	20.36
2011	172	2213.00	54.25	18.64
2012	195	1613.78	58.00	19.66
2013	195	1371.71	57.63	17.70
2014	195	1303.13	55.15	16.82
2015	195	1270.55	53.38	16.40

Source: BAPPEKO Surabaya, 2015

One of the challenges is the impact of climate change to the variability of rainfall. Pamungkas, 2013 [5] indicates that the rainfall pattern will have similar total amount of water for rainfall in a year with less numbers of rainfall days compare the previous years. Consequently, the number of extreme rainfall in a day will have higher probability than before. This phenomenon will challenge the Surabaya government in managing its run off to avoid inundation.

Furthermore, the challenge of water management is also arising in water supply. Kalimas River as the main source of water supply has a limited volume to support a fast-growing need of water in Surabaya. Even though, the coverage if water supply in Surabaya has 95.51 % of the population in 2016 based on PDAM Surabaya City Profile, 2016 [6], the fast-growing city may cause Surabaya has limitation in supporting the water supply in the future. A new promising water supply from Umbulan is also still in a questionable stage as the construction of the spring water is still in discussion among related municipalities, even Umbulan have maximum production 4000 liter/second that can be used by Surabaya 1000 liter/second, but nowadays Surabaya use 500 liter/second [7]. Supply from outside of Surabaya is indicated a non-sensitive city since it causes a dependency to outside of Surabaya. The worst situation for water supply is on the quality of water provided by Surabaya. In 2014, Department of Health Surabaya had a water quality inspection in 3 ways are microbiology, chemistry and physics. Based on the quality inspection in microbiology are 86.85% in good quality, in chemistry are 82.19% in good quality and in physics are 100% good quality [8]. The ideal water quality as drinking water is still far behind resulting the community has to buy other good quality water for drinking and cooking. It results a highly cost of living for Surabaya residents or poor quality of health related to water use.

Those complex problem based water in Surabaya indicates that Surabaya needs to evaluate their business process in managing the water. A modeling process is one of decision support systems (DSS) that can evaluate a system in an objective manner. It can also make an optimization in finding the most effective combination for a complex system such as water. Therefore, modeling in water to make Surabaya as water sensitive cities is important. This paper will explain the modeling process particularly in conceptual model building.

2. Methods

To have an urban water model that values the goals of water sensitive cities in Surabaya, we build a hypothesis model based on rational basis. This hypothesis is then assessed and formulated to reflect the real situation of urban water management in Surabaya as well as framing the water sensitive cities concept via serial group discussions. Serial group discussions consider many aspects of urban waters, that are; water resources, water quality, water quantity, weather, infrastructures, demography, social cultural, economic, engineering aspect, land use change, urban policy and technology. To increase validity, government and non-government publication in supporting the conceptual model accompany the rational basis judgments in serial group discussion. At this stage, the formulation on conceptual model of urban water in Surabaya is the key output of the research.

3. Result and Discussion

In building a conceptual model in Surabaya, we value catchment areas boundaries as the main significant division since water will follow the physical characteristics. Three main segments of the areas can be divided that are; from upper catchment to Jagir floodgate, local system which is around from Jagir floodgate to the estuaries, and around coastal areas. Those segments have difference characteristics and are very important to be considered for making a water sensitive city.

3.1 Upper Catchment Area up to Jagir Floodgate

This segment highlights that a vast catchment area called by Brantas Catchment Areas affect the Kalimas River. The main concern on this is the quantity of run off. Run off from Brantas River is divided into two main rivers before arrive in Surabaya, that is: Wonokromo River and Porong River in Sidoarjo. If the run off in Brantas River is in high volume, it could inundate the catchment in Porong and Wonokromo. Therefore, limited run off is needed to maintain the balance between water debit and river capacity in both rivers. To limit the debit of run off, the model will consider several key variables, those are;

1. Land use in non-local areas. Most of the land use in upper catchment is open space. An issue of deforestation due to the need of land for development has been escalated in all the municipalities in the upper catchment. Since there is a large of land conversion, the debit of run off in the river is increase. In terms of the land uses, Brantas catchment area has 280.258 ha critical areas. The critical areas indicate an alarming condition for run off. Among the critical areas, the middle critical dominated the areas [9]. Therefore, considering the land use in non-local areas is one of the key variables in building the model.

Table 2 Critical Land Areas in Brantas Catchment Areas

No	Area	Critical Forest Area (Ha)				Total
		Very critical	Critical	Mildly critical	Potentially critical	
1	Protected forest	4.706	700	-	-	5.406
2	Conservation forest		7.067	3.930	-	10.997
3	Production forest	16.761	26.339	4.238	-	47.338
4	Protected area outside the forest	456	4.300	9.046	674	14.476
5	Agricultural cultivation	3.231	47.216	70.707	37.794	158.948
6	River border	-	3.284	18.530	93	21.907
7	Green open space at the urban settlements near the beach	-	1.014	1.815	-	2.829
8	Flood area	-	-	158	-	158
	Total	26.267	93.469	120.953	39.569	280.258

Source: Balai Besar Wilayah Sungai (BBWS) Brantas, 2011

2. Rainfall. In terms of meso-level, rainfall data will have a different pattern than before. Some of the situation highlight that the rainfall will have a high degree of variability. To capture rainfall variability, we use rainfall data daily then grouped into monthly data. For average rainfall, Brantas catchment area has 2.000 mm/year. In addition, average rainy day is 61-150 day/year and mostly occurs in rainy season [9].
3. Land Characteristics. As the fertile land in upstream, most of land in the upper catchment has high ratio of infiltration. The infiltration rate is a converse version of run off. Therefore, the original type of land in upper catchment can reduce the run off. Most of the type of land in Brantas catchment area is alluvial, which has erodibility between 0,12-0,17 and occupies the area until 320.285 Ha [9].
4. Water usage in upper catchment can affect the debit of water in the river. To date, the water usage in upper catchment areas can be in the form of irrigation and water supply. Water usage in Brantas catchment area is such as irrigation, domestic, non-domestic, industry and fishpond [9].

The interaction among those variables in debit of run off in upper catchment can be seen on Figure 2.

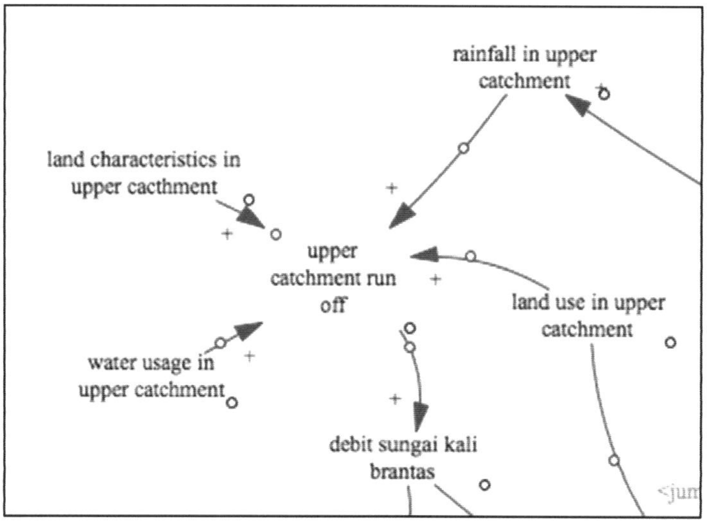

Figure 2 Conceptual Model for Debit Run off in Upstream

In terms of water quality, we value the quality of water in the Brantas River since the same water is used for water resources in Kalimas River. We consider several variables related to water quality, are:

1. Erosion. Erosion in upstream will make sedimentation in the river. The sediment in river will cause poor quality of water. The erosion will be depended on land use types and land characteristics. Since most of the land in Brantas catchment areas is alluvial, the possibility of erosion and sedimentation is high. Based on Table 3, we can see that the sedimentation level is high in upper catchment areas. This sedimentation is a result of erosion in land areas of upper catchment areas. The high sedimentation will directly affect the water quality [9].
2. Dumping waste along the river. Dumping waste could be from two main sources that are residential areas and industrial areas. Current situation indicates that the point sources pollution from both land uses is high. In residential areas, education, culture, urban system, and number of population of residents along the rivers influence the number of wastes dumping in the river. Numbers of industries, types of industries, law enforcements are some variables influencing the number industrial waste dumping on river. Those two types of point

source pollution will influence the quality of water [9]. For industry, Brantas catchment area has 483 industries that can be dumping waste on river. Based on Surabaya *River Pollution Control Action Plan Study* in 1999, BOD *net* from industry is 125 ton BOD/day and for domestic, BOD *net* is 205 ton BOD/day. These number of BOD based on Ministry of environment Regulation no 1 2010 about water pollution control, Allocations above allowable pollutant load.

Table 3 sedimentation caused by erosion in Upper Brantas Catchment Area

Basin Block	Sedimentation			
	A=RKLSCP t/ha/year	$M^3/km^2/year$	Mm/year	$M^3/year$
Upper Brantas	108,2	6.009,2	6,0	1.093.679
Bango-Sari	60,1	333.760	3,3	874.454
Amprong	172,5	9.595,6	9,6	3.335.779

Source: Balai Besar Wilayah Sungai (BBWS) Brantas, 2011

The interaction among those variables in water quality before Surabaya segment can be seen on Figure 3.

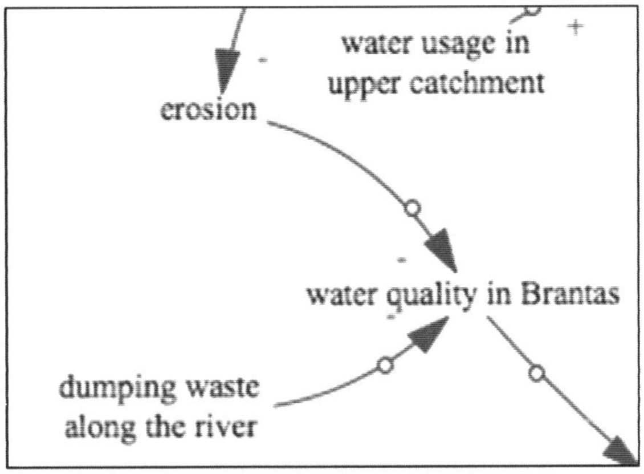

Figure 3 Conceptual Model for Water Quality in Upstream

Since Umbulan spring water is going to be operated, the upstream system particularly in Umbulan will also influence urban water in Surabaya. The supplies of Umbulan for Surabaya will depend on;

1. Spring water debit of Umbulan. Debit of Umbulan in the beginning is around 6,000 L/sec, but now, it is around 5,000 L/sec. among them, only 10% is used for water supply and 90% is discharge to the river (PDAM Surabaya, 2016). The debit of spring water will be a main consideration for allocating water for Surabaya. These sources will become significant additional sources for Surabaya.
2. Commitment among municipalities. Since Umbulan will support some municipalities including Surabaya, the commitment among parties will define the water allocation for Surabaya. Project that support umbulan is National Showcase Project which use method of Public Private Partnerships/PPP [10].
3. Infrastructure capacity. The dimension in every infrastructure built in connecting Umbulan to Surabaya will also influence the debit water for Surabaya. It also will influence the timing

aspect on providing additional water sources for Surabaya. Umbulan has transmission system and off take system. Those systems distribute water to Surabaya area.

The interaction among those variables in Umbulan water source can be seen on Figure 4.

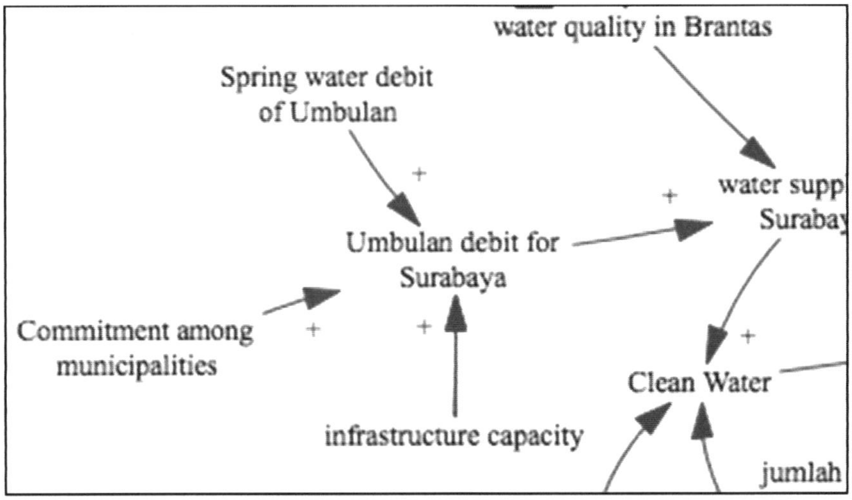

Figure 4 Conceptual Model for Umbulan Water Source

3.2 In local system

In the local system, we highlight some kinds of matters for urban water in Surabaya. Inundation is still one of the most problematic for Surabaya. Water supply and water use are other things for consideration in managing urban water in Surabaya.

For inundation, the related variables for the model are;

1. Drainage system. Most of the problematic condition related to inundation is drainage system. Within the system, there are primary, secondary and tertiary drainage. In Surabaya based on SDMP, there are 5 catchments for drainage service areas, that are: Rayon Wiyung, Rayon Tandes, Rayon Genteng, Rayon Gubeng and Rayon Jambangan. Rayon Wiyung have 38 sub-catchments, Rayon Tandes have 39 sub-catchments, Rayon Genteng have 12 sub-catchments, Rayon Gubeng have 15 sub-catchments, Rayon Jambangan have 11 sub-catchments. Figure 5 illustrates the drainage sub catchments in Surabaya [11].
2. Local rainfall. The variability of rainfall also occurs in local rainfall. Therefore, to consider local rainfall, we will highlight the variability of daily rainfall and sum up for monthly rainfall [12].
3. Local land characteristics. Inundation in Surabaya is also caused by local land characteristics. Mostly, in Surabaya, the land is formed from clay, which has limited capability to absorb the run off [13]. In addition, the low groundwater causes low level of infiltration in the area. The flat land in majority of Surabaya area makes Surabaya easily to be inundated.
4. Debit from Kali Brantas River. Since Kalimas River is connected with Kali Brantas River, some of the kalimas river capacity is used by water from kali brantas. The fluctuated pattern of brantas water discharge in Brantas will affect to the idle capacity of Kalimas River. Therefore, the idle capacity for local run off will be very much depended on the water from kali brantas.

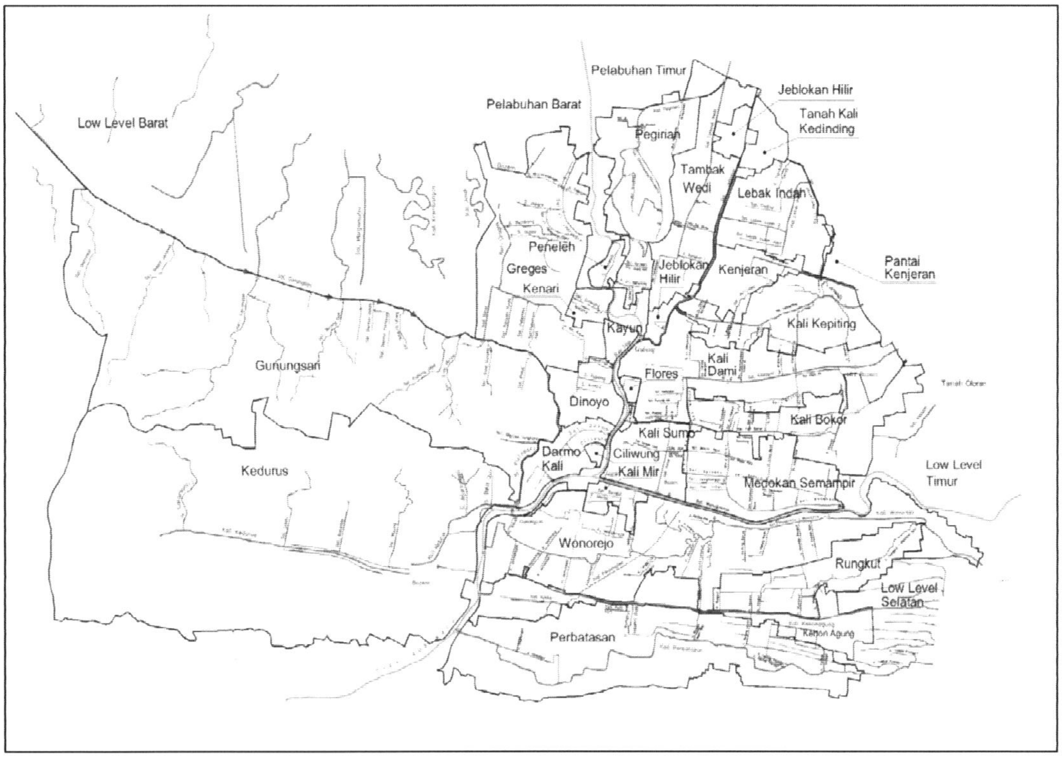

Figure 5 Drainage Sub Catchments Delineation in Surabaya
Source: Surabaya Drainage Master Plan, 2005

Table 4 Total Rainy days and Rainfall in Surabaya 2012-2014

No	Month	Rainy days (days)			Rain fall (mm)		
		2012	2013	2014	2012	2013	2014
1	January	28	26	29	499	648.7	302
2	February	22	20	23	273.8	197.3	266
3	March	22	20	19	315	197.3	251
4	April	12	12	19	60.3	141	253
5	May	9	16	13	70.6	179	51
6	June	4	17	13	3.1	235	68
7	July	1	9	9	0	91	6
8	August	0	0	0	0	0	0
9	September	0	0	0	0	0	0
10	October	5	5	1	57.3	6	8
11	November	13	13	13	49	80	70
12	December	26	26	24	215.4	371	337
	Average	12	14	14	129	179	161

Source: Surabaya dalam Angka, 2015

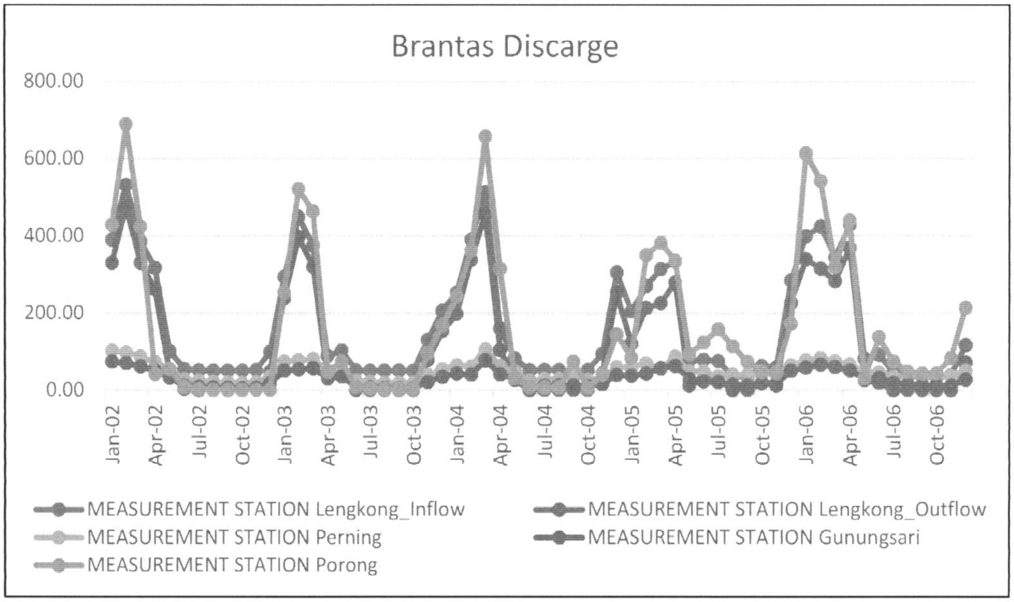

Figure 6 Monthly Brantas Water Discharge

5. Local land cover. Most of the inundation problems relate to the run off in Surabaya relates to local land cover. The land cover and soil types based on the Darcy Law define the run off itself. In Surabaya, since the rapid growing of build-up areas make the run off increase. Describe a fast conversion level in Surabaya. Residential has a very significant growing number resulting high run off. For the soil types, it has been discussed in point no. 3.
6. Outtake for water use. In Surabaya, outtake water from river is used for local water supply. This water supply covers all type of areas in Surabaya including residential, commercial, and industrial areas. Only limited number for irrigation since there is not many land for agriculture in Surabaya. The main agriculture is aquaculture in which the need of water is supplied by sea tide.

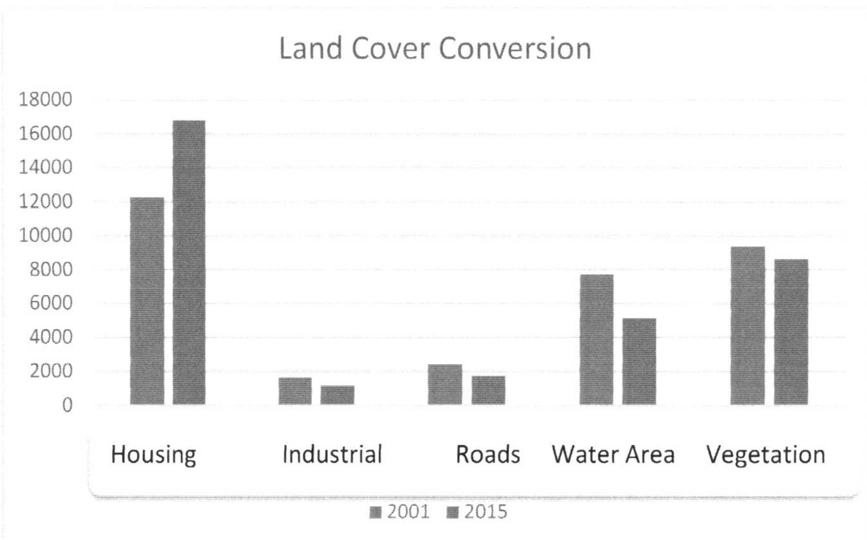

Figure 7 Land Cover Conversions in Surabaya 2001-2015
Source: Extract Citra Landsat

The interaction among those variables in inundation can be seen on Figure 8.

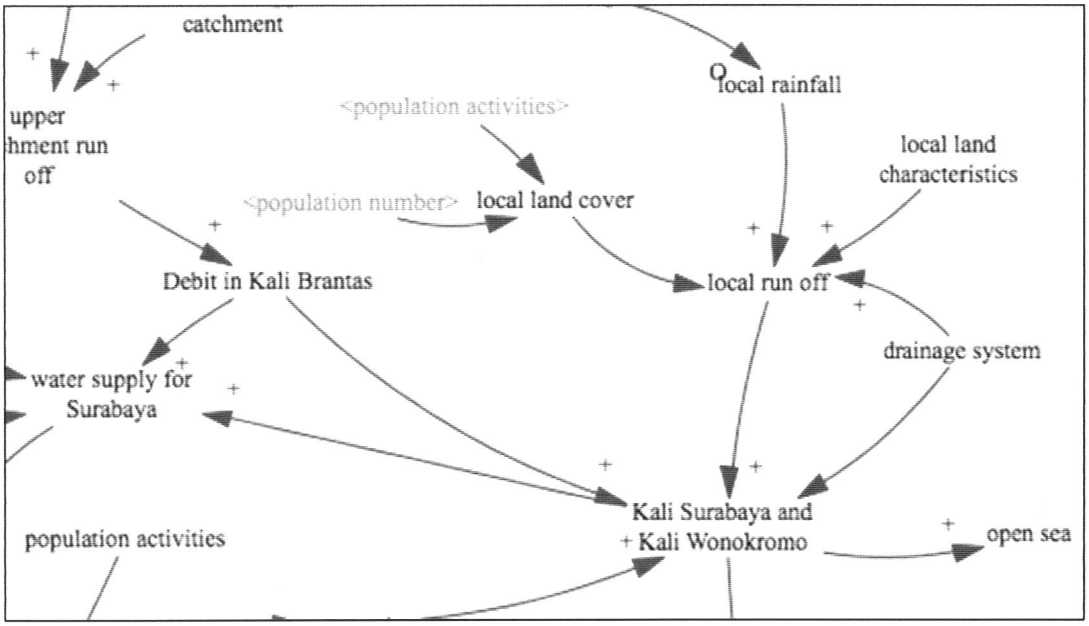

Figure 8 Conceptual Models for Inundation

For water supply, some key variables are discussed as follows;

1. River debit. Kalimas debit will be depended on water from Kali Brantas, intake water from local catchment areas and also the drainage capacity connecting to the main river. PDAM fulfill their need of water supply from two rivers, those are Surabaya River and Kali Wonokromo River. Kali Wonokromo has more supply with 0,425 m3/sec than Surabaya River with 0.119 m^3/s.

Table 5 Water Debit of Kali Surabaya and Kali Wonokromo for Water Supply

River	Length (m)	Debit (m^3/sec)
Kali Surabaya	17,400	0.119
Kali Wonokromo	12,100	0.425

Source: PDAM Surabaya

2. Water production. Debit in Kalimas River is purified to produce clean water. The production of clean water is varying from month to month indicating fluctuated production relies on river debit [14].
   Within the production on Figure 9, Surabaya water supply can cover the boundaries of adequate water supply in general. But, two months in 2015, the water supply is not adequate supplied with current system. In general, the set boundary for water supply is minimum 210 L/person/day. The boundary is from national guideline on construction and building (National Planning Department, Deputy of Water and Irrigation, 2006) [15].
3. Purification technology. Surabaya Local Water Enterprise (PDAM) uses Kali Surabaya and Kali Wonokromo as the main sources of water. The efficiency of water purification will be depended on the current water quality in both rivers and advance technology used in PDAM. In terms of water quality, since the intake of PDAM is just at the border of local system and Kali Brantas, the water quality will be considered following the discussion of water quality on

Figure 10. For the advance technology, purification technology will affect the amount of water can be used for water supply.

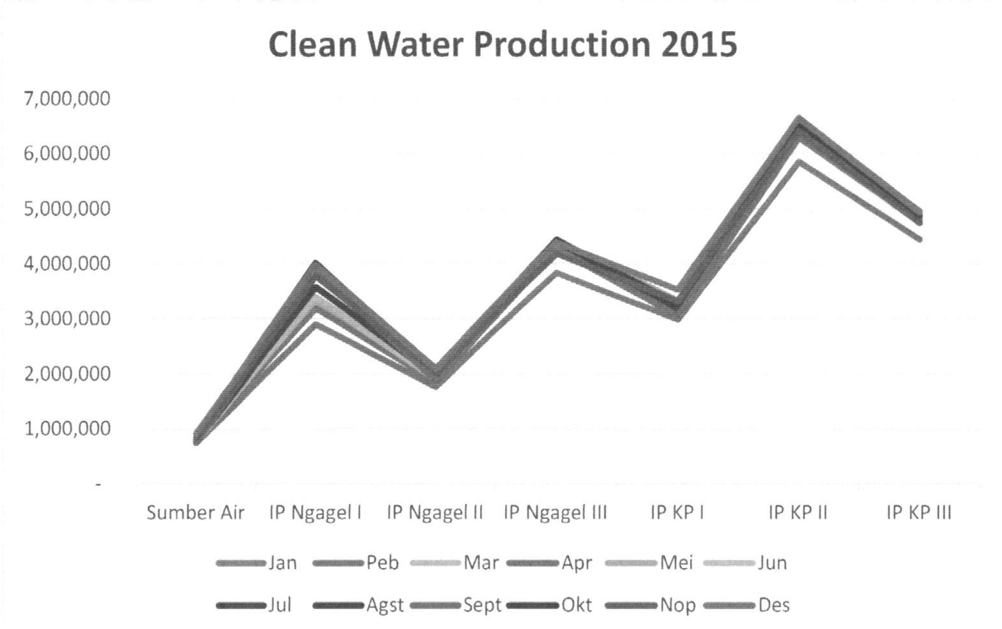

Figure 9 Clean Water Production in 2015
Source: PDAM Surabaya, 2015

Table 6 Adequacy of Water Supply in Surabaya

Month	Total production	Monthly water supply l/person/day	Adequacy (>210 l)
Jan	25,028,242	214.5	**adequate**
Peb	22,539,675	193.2	**inadequate**
Mar	25,051,053	214.7	**adequate**
Apr	24,454,682	209.6	**inadequate**
Mei	25,317,308	217.0	**adequate**
Jun	24,522,337	210.1	**adequate**
Jul	25,330,335	217.1	**adequate**
Agst	25,676,845	220.0	**adequate**
Sept	24,937,971	213.7	**adequate**
Okt	25,965,589	222.5	**adequate**
Nop	25,290,201	216.7	**adequate**
Des	26,508,732	227.2	**adequate**
Total	**300,622,970**	**2576.1**	

Source: Analysis, 2016

4. Pipeline system. The leakage is always one of the main problems for water supply in Indonesia. The usage, water quality, construction issues are the main causes for leakage. In

Surabaya, the percentage of leakage is still considered as at the level of reasonable. Water leakage consists of some types of leakages, those are: physical leak caused by the seepage or pipe leak and operational leak caused by waste water excess in operation process, and wasting water [16]. Based on PDAM, the percentage of leakages for whole Surabaya clean water system are 34%.

The interaction among those variables in water supply can be seen on Figure 10.

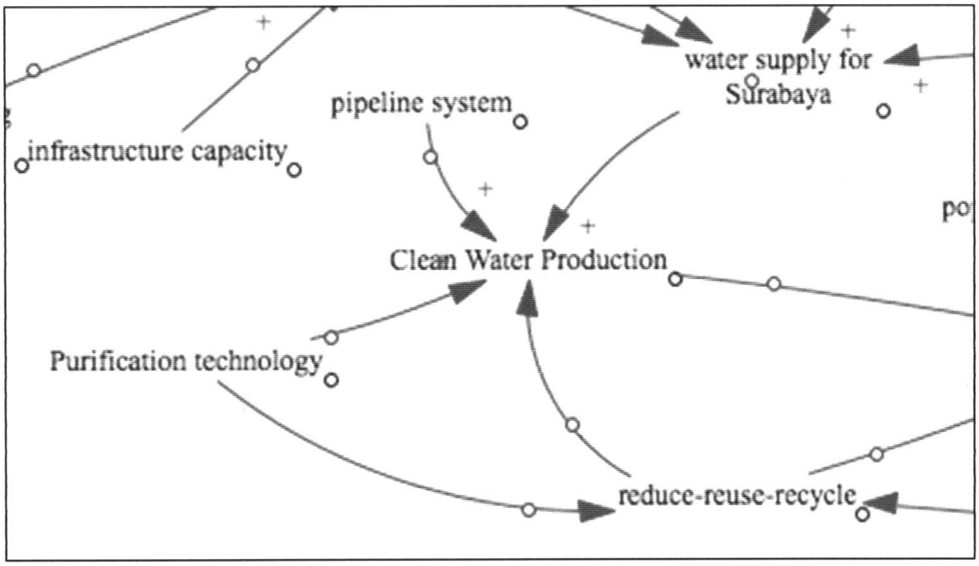

Figure 10 Conceptual Models for Water Supply

For water use, some key variables for urban water model in Surabaya are;

1. Population. Surabaya has varies increases in population growth. Most of the population has positive growth in the periphery of Surabaya while some decline indicates in the central of Surabaya. Based on Kota Surabaya dalam Angka BPS, In 10 years, Surabaya population has increased from 2.522.028 to 2.722.876 by growth 1% per year. The number of births and immigration causes the growth percentage [17].
   It also has wide difference in terms of daytime population and nighttime population. Daytime population indicates actual residents with additional people moving in to Surabaya for working while nighttime population indicates Surabaya residents only. Those two characteristics of population will impact to the water demand in Surabaya.

Table 7 Surabaya GDP from 2010 to 2014

No	Sector PDRB	2010	2011	2012	2013	2014
1	Agriculture	431.542,7	440.989,2	467.099,5	504.369,8	522.264,9
2	Mining and Quarrying	16.714,4	17.134,	17.440,3	18.095,3	18.674,9
3	Manufacturing	45.351.158,5	47.601.826	51.100.743,7	54.450.456,3	53.358.226,5
4	Electricity and Gas	1.948.227,1	1.828.007,4	1.643.231,5	1.610.562,4	1.59.076,3
5	Water supply, Sewerage, Waste Management and Remediation Activities	416.581,9	451.699,8	460.964,6	470.737	474.422,9
6	Construction	23.729940,1	25.457.717	27.182.986,4	29.357.611,8	31.368.882,7

7	Wholesale and Retail Trade	66.954.860,1	72.316.558,1	77.408.102,5	82.675.259,2	86.711.515,1
8	Transportation and Storage	10.966.478,7	11.742.106,7	12.667.180,2	13.686.933,8	14.843.763,2
9	Accommodation and Food Service Activities	31.651.758,6	34.358762,8	36.334.025,3	39.334.025,3	42.503.082,6
10	Information and Communication	13.366.551,4	14.710.691,8	16.403.598,2	18.494.072,8	19.701.365,9
11	Financial and Insurance Activities	9.947.341,9	10.687.470,1	11.781.104	13.778.309	14.842.875,7
12	Real Estate Activities	6.105.900	6.631.618,1	7.314.011,6	7.747.001,1	8.264.384,5
13	Business Activities	5.697.320,8	6.006.001,6	6.282.187,6	6.586.339,7	7.011.356,2
14	Public Administration and Defense, Compulsory Social Security	3.736.741	3.818.575,6	3.931.910,9	3.981.240,6	4.004.411,5
15	Education	5.444.974,1	5.685.743,6	6.276.776,6	6.758.833,1	7.144.694,9
16	Human Health and Social Work Activities	1.533.114,2	1.809.687,8	2.018.033,5	2.758.940,1	2.348.390,8
17	Other Service Activities	3.905.535,6	4.122.058,5	4.222.224,6	4.419.444,7	4.620.617,8
Total		**231.204.741**	**247.686.648,1**	**265.892.080,7**	**286.057.232**	**305.308.006,2**

Source: PDRB Kota Surabaya, 2015

2. Economic growth. The need of water will also be influenced by the economic growth. Point 1 indicates a need of population on water from quantity aspect while economic growth indicates a need of population on water from quality aspect. Growing economy can boast the need on water since it will make impact to the land uses character. In terms of economic growth, Surabaya economic growth in 2014 is around 6,73%. It shows that the growth is getting low compared to 2013 (7,58%). However, the growth is still remaining around 6-7% indicating a stable and manageable development growth. Based on Table 7, we can conclude that the economics of Surabaya is dominated by wholesale and trade with more than 28% of the total Surabaya GDP. Meanwhile, industrial sector takes the second place with around 17% [18].
3. Water usage habitual. The water use will also be very much influenced by water usage habitual from residents. Efficiency will be the key of habitual. The habit can be generalized into several groups such as high-income earners families, low-income earners families, industrial activities, commercial activities and residential activities.

Table 8 Land use change from 2001 and 2015

Land Use	Year	
	2001	**2015**
Housing	42.01%	42.00 %
Urban Farming	16.28%	16.24%
Fishpond	15.27%	15.20%
Commercial Use	1.76%	10.76%
Industrial	2.37%	7.30%
Vacant Land	1.78%	5.50%

Source: RTRW Surabaya, 2015

4. Land uses. Since water usage in general can be calculated based on land uses, the changes on land uses can be one of the key variables in predicting water use in Surabaya. The land use characteristics in Surabaya are dramatically changes especially for the growth of commercial and residential areas. Based on the data, the highest land use change occurs in retail sector which is increases from 1,7% until 10,7% up to the total land use in Surabaya. While land use for housing and moor are still constant. Land use for industry and non-productive land increase around 5% in the last 15 years [19]. Those changes lead to the escalated needs of water.
5. 3R culture. The water usage in a city is also influenced by 3R culture of society [20]. Reduce-reuse-recycle is part of cultural aspect in making efficiency in water usage. The 3R culture can be divided into several categories that are local 3R culture, industrial 3R culture, and commercial 3R culture. In local 3R culture, income and understanding are the key variables for efficiency in water usage. For commercial and industrial, a regulation and government incentives can be part of key variables for successes of water usage efficiency.
6. Government regulation. The efficiency of water usage can be established by implementing the supportive government regulation. The regulation of no ground water usage in Surabaya has increased the use of runoff water for daily use. Therefore, the government regulation plays important role in regulating water usage. Since 2012, Surabaya government does not allow the usage of ground water and the regulation is still discussed.
7. Water pricing. Water usage can also be influenced by the price of water. The low price of water can encourage community to have more water than the high price. The prices of water are varying from the type of usage.
8. Pipeline is one of the key variables that have been discussed in earlier section (water supply in local system).

Table 9 Water Pricing

No	Classification	Pricing code	Usage	Water pricing	Usage/month
1	Classification I	1	Non- progressive	600	10
2	Classification II	2A	0-10	350	10
			11-20	600	
			21-30	900	
			>30	1.800	
3	Classification III	2B	0-10	500	10
			11-20	1.000	
			>20	2.250	
4	Classification IV	3A	0-10	500	10
			11-20	1.200	
			>20	1.900	
5	Classification V	3B	0-10	1.500	10
			11-20	3.500	
			>20	6.000	
6	Classification VI	3C	0-10	2.300	10
			11-20	4.000	
			>20	5.500	
7	Classification VII	4A	0-10	1.000	10
			11-20	1.500	
			>20	2.500	
8	Classification VIII	4B	0-10	1.500	10
			11-20	2.200	

			>20	3.500	
9	Classification IX	4C	0-10	4.000	10
			11-20	6.000	
			>20	7.500	
10	Classification X	4D	0-10	6.000	10
			11-20	8.000	
			>20	9.500	
11	Classification XI	5	Non-Progressive	10.000	10

Source: PDAM Surabaya, 2008

The interaction among those variables in water usage can be seen on Figure 11.

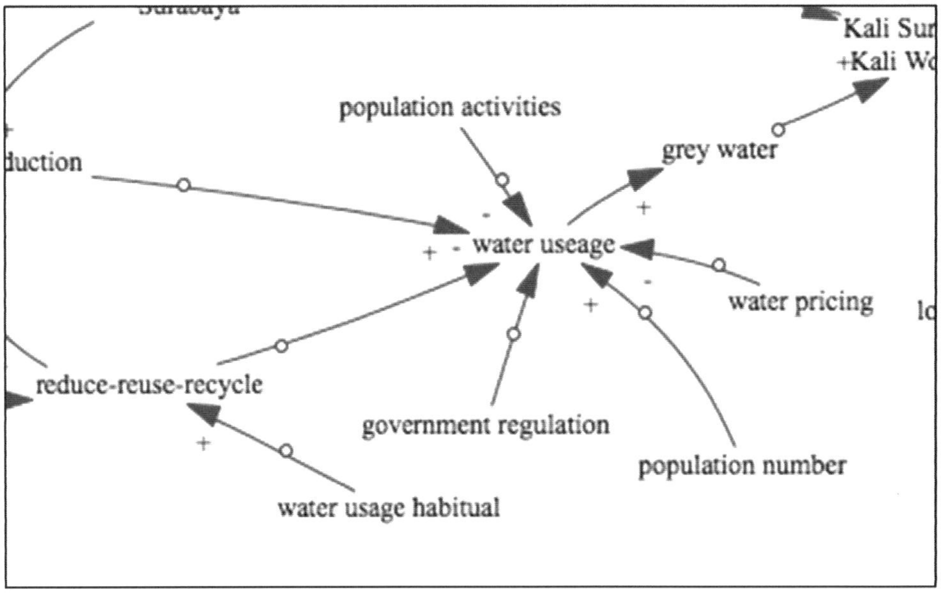

Figure 11 Conceptual Models for Water Usage

3.3 Around coastal areas

In around coastal areas, key variables in affecting urban water model are;

1. Sea tide. In Surabaya coastal areas, the tide gap only 2-meter difference between high tide and low tide (Figure 12). Since the Surabaya lies in the low-lying areas, this tide can still affect the water discharge system from river to the sea. Based on Meteorological, Climatological and Geophysical Agency in Surabaya, we can get the average value for the sea tide each day. The highest tide occurs at 10 p.m., meanwhile the lowest occurs at 6 a.m. every day [21]. Therefore, considering the tide in Surabaya coastal areas for urban water management is important.

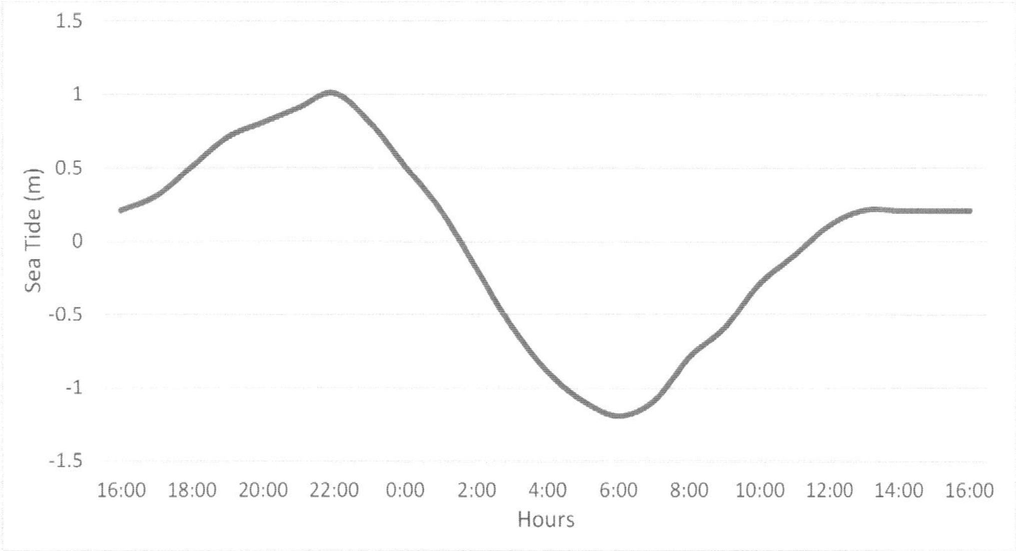

Figure 12 Sea Tide in Surabaya Coastal Areas
Source: BMKG Surabaya, 2005

2. Debit in Kali Surabaya and Kali Wonokromo. The debit in both river will depend on the level of sea tide. Currently, some local floods occur caused by large amount of local run off and worst by high sea tide for some areas near the coastline.

The interaction among those variables in around coastal areas can be seen on Figure 13.

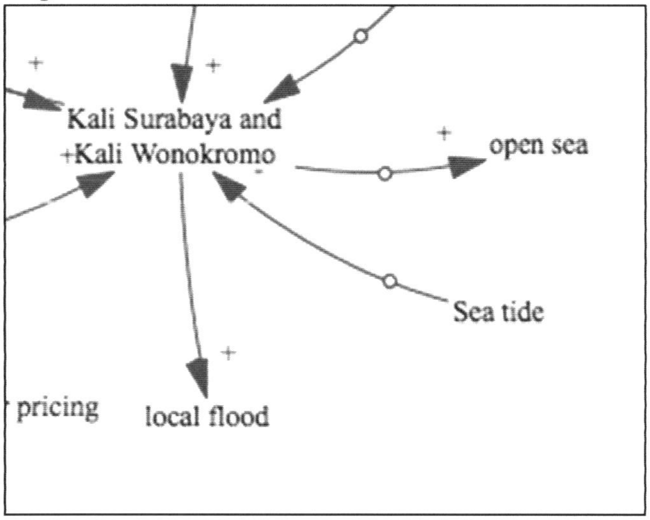

Figure 13 Conceptual Model for Around Coastal Areas

From those 7 figures above on the conceptual model, we can outline the conceptual model for the whole of urban water management in Surabaya as on Figure 14. The conceptual model has 35 key variables for all the three segments of urban cycle for Surabaya. It also covers the key challenges of urban water in Surabaya that are; inundation, water supply, 3R in water usage, technology, high demand due to urban development, tidal flood and run off.

IOP Conf. Series: Earth and Environmental Science **79** (2017) 012037 doi:10.1088/1755-1315/79/1/012037

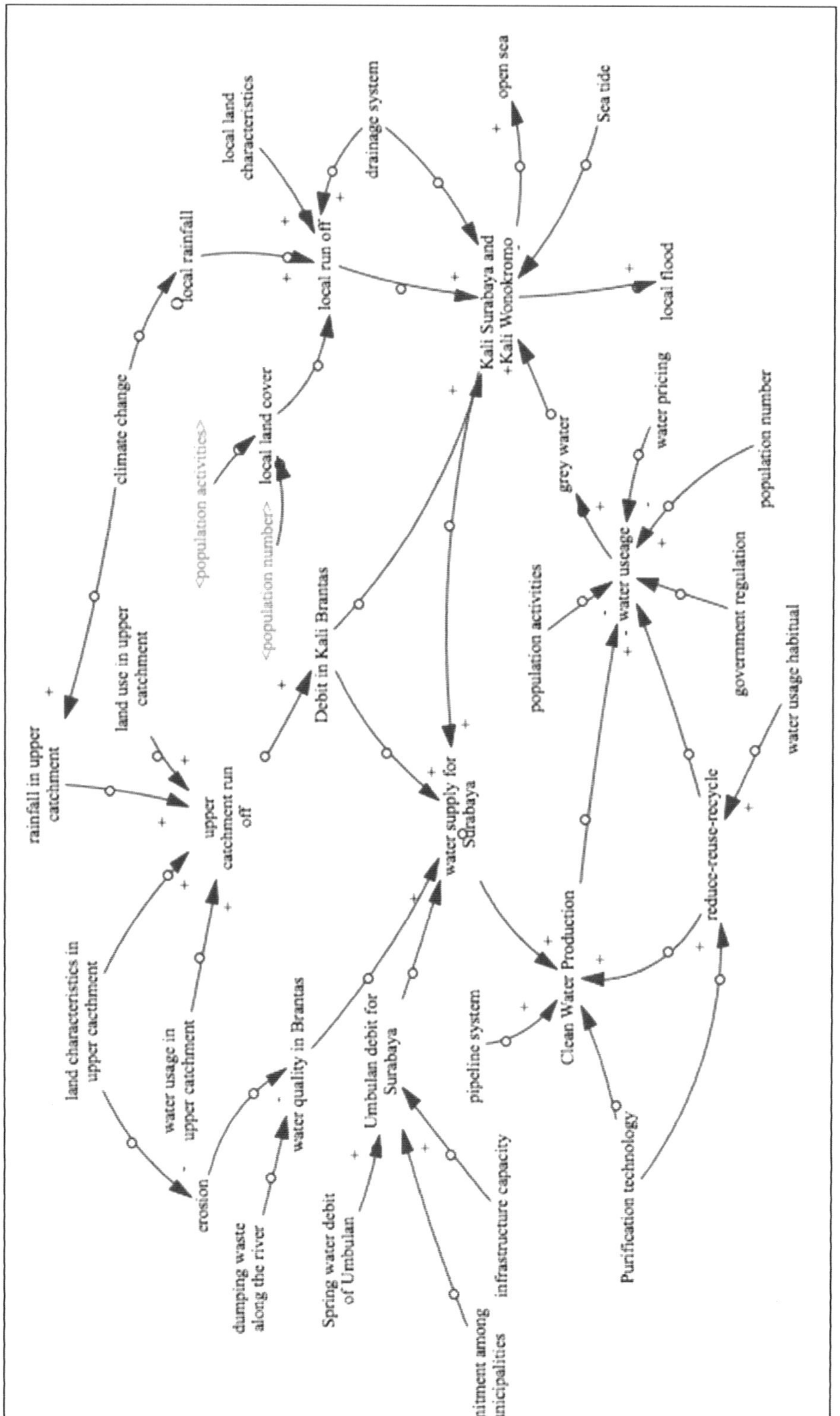

Figure 14 Conceptual Model on Water Sensitive City in Surabaya

4. Conclusions

The conceptual model has been built by understanding the characteristic of urban water in Surabaya. Some supporting data are used to ensure the validity of conceptual model. In addition, we also include some experts in constructing the conceptual model. A variety of experts also validate the conceptual model.

Within the model, 35 key variables are used to uncover the system in three main water segments that are: upper catchment area up to Jagir floodgate, local water system and around the coastal areas. By highlighting the three main key segments, the conceptual model also response to local challenges that affect Surabaya urban water.

Reference

[1] Badan Pusat Statistik. 2015. *Produk Domestik Regional Bruto Kota Surabaya.*

[2] Brown, R., Keath, N and Wong, T. 2009. *Urban water management in cities: historical, current and future regimes.* Water Science and Technology, 2009.

[3] PDAM Surabaya City. 2012. *Walikota Surabaya Resmikan Master Meter PDAM Surabaya,* retrieved from http://www.pdam-sby.go.id/page.php?get=tampil_berita&id=617001&bhs=0#.WLVevH89uu0 at 3 February 28, 2017

[4] Bappeko Surabaya. 2014. *Rencana Tata Ruang Wilayah Kota Surabaya, 2014-2034*

[5] Pamungkas, A., (2013), *Vulnerability assessment for disaster risk management,* PhD thesis at The RMIT University

[6] [6] PDAM Surabaya City, (2016), *Annual Customer Services,* retrieved from http://www.pdam-sby.go.id/page.php?get=jumlah_pelanggan_tahunan&bhs=1 at November 20th 2016

[7] Pemerintah Provinsi Jawa Timur. 2016. *Profil Proyek Kerjasama Pemerintah Swasta Sistem Penyediaan Air Minum Umbulan Provinsi Jatim.*

[8] Dinas Kesehatan Kota Surabaya. 2014. Pembinaan dan Pengawasan Kualitas Air Minum PDAM retrieved from http://dinkes.surabaya.go.id/portal/berita/pembinaan-dan-pengawasan-kualitas-air-minum-pdam/ at 6 February 2017

[9] Balai Besar Wilayah Sungai (BBWS) Brantas, 2011

[10] Pemerintah Provinsi Jawa Timur. 2016. *Profil Proyek Kerjasama Pemerintah Swasta Sistem Penyediaan Air Minum Umbulan Provinsi Jatim*

[11] *Strategi Pengembangan Permukiman dan Infrastruktur Perkotaan Kota Surabaya.* 2010. Dinas Cipta Karya dan Tata Ruang Kota Surabaya.

[12] *Surabaya Dalam Angka.* 2015. Badan Pusat Statistik

[13] *Surabaya Dalam Angka.* 2015. Badan Pusat Statistik

[14] *Produksi Air Bersih Kota Surabaya 2015.* PDAM Kota Surabaya

[15] Direktorat Pengairan dan Irigasi Bappenas, 2006

[16] Antara Jatim. 2012. *Tingkat Kehilangan Air di Surabaya 34 Persen* retrieved from http://www.antarajatim.com/lihat/berita/86575/tingkat-kehilangan-air-di-surabaya-34-persen at 14 February 2017

[17] Bappeko Surabaya. 2014. *Rencana Tata Ruang Wilayah Kota Surabaya,* 2014-2034

[18] Badan Pusat Statistik. 2015.*Produk Domestik Regional Bruto Kota Surabaya.*

[19] Bappeko Surabaya. 2014. *Rencana Tata Ruang Wilayah Kota Surabaya,* 2014-2034

[20] *Surabaya Kota Percontohan Pengolahan Sampah Terbaik Indonesia.* retrieved from http://www.mongabay.co.id/2014/02/27/surabaya-kota-percontohan-pengolahan-sampah-terbaik-indonesia/ at 14 February 2017

[21] BMKG Surabaya. 2005